HOOVER DAM 75TH ANNIVERSARY HISTORY SYMPOSIUM

PROCEEDINGS OF THE HOOVER DAM 75TH ANNIVERSARY HISTORY SYMPOSIUM

October 21–22, 2010
Las Vegas, Nevada

SPONSORED BY
The History and Heritage Committee
of the American Society of Civil Engineers

EDITED BY
Richard L. Wiltshire, P.E., F.ASCE
David R. Gilbert, P.E., F.ASCE
Jerry R. Rogers, Ph.D., P.E., D.WRE, Dist. M.ASCE

1801 ALEXANDER BELL DRIVE
RESTON, VIRGINIA 20191–4400

Library of Congress Cataloging-in-Publication Data

Hoover Dam 75th Anniversary History Symposium (2010 : Las Vegas, Nev.)
 Hoover Dam 75th Anniversary History Symposium : proceedings of the Hoover Dam 75th Anniversary History Symposium, October 21-22, 2010, Las Vegas, Nevada / sponsored by the History and Heritage Committee of the American Society of Civil Engineers ; edited by Richard L. Wiltshire, David R. Gilbert, Jerry R. Rogers.
   p. cm.
 Includes bibliographical references and index.
 ISBN 978-0-7844-1141-4
 1. Hoover Dam (Ariz. and Nev.) -- Congresses. I. Wiltshire, Richard L. II. Gilbert, David R. III. Rogers, Jerry R. IV. American Society of Civil Engineers. Committee on History and Heritage of American Civil Engineering. V. Title. VI. Title: Hoover Dam Seventy-five Anniversary History Symposium.

 TC557.5.H6H684 2010
 627'.820979313--dc22   2010035841

American Society of Civil Engineers
1801 Alexander Bell Drive
Reston, Virginia, 20191-4400

www.pubs.asce.org

Any statements expressed in these materials are those of the individual authors and do not necessarily represent the views of ASCE, which takes no responsibility for any statement made herein. No reference made in this publication to any specific method, product, process, or service constitutes or implies an endorsement, recommendation, or warranty thereof by ASCE. The materials are for general information only and do not represent a standard of ASCE, nor are they intended as a reference in purchase specifications, contracts, regulations, statutes, or any other legal document. ASCE makes no representation or warranty of any kind, whether express or implied, concerning the accuracy, completeness, suitability, or utility of any information, apparatus, product, or process discussed in this publication, and assumes no liability therefore. This information should not be used without first securing competent advice with respect to its suitability for any general or specific application. Anyone utilizing this information assumes all liability arising from such use, including but not limited to infringement of any patent or patents.

ASCE and American Society of Civil Engineers—Registered in U.S. Patent and Trademark Office.

Photocopies and reprints.
You can obtain instant permission to photocopy ASCE publications by using ASCE's online permission service (http://pubs.asce.org/permissions/requests/). Requests for 100 copies or more should be submitted to the Reprints Department, Publications Division, ASCE, (address above); email: permissions@asce.org. A reprint order form can be found at http://pubs.asce.org/support/reprints/.

Copyright © 2011 by the American Society of Civil Engineers. All Rights Reserved.
ISBN 978-0-7844-1141-4
Manufactured in the United States of America.

Preface

The dedication of Hoover Dam 75 years ago on September 30, 1935, culminated what many regard as the world's greatest dam design and construction project of the 20th Century. Hoover Dam has remained highly regarded within the national and international civil engineering profession, which lead ASCE to designate Hoover Dam as its Monument of the Millennium dam in 2000 after a vote by ASCE's members. Also, Hoover Dam has been designated an ASCE National Historic Civil Engineering Landmark in 1984, a National Historic Landmark in 1985, and was listed on the National Register of Historic Places in 1981. Hoover Dam is still the highest concrete dam in the Western Hemisphere, was the world's largest hydropower installation from 1939 to 1949, and annually produces over 4 billion kilowatt-hours of electricity, which still ranks it as one of the largest hydropower installations in the USA.

The sixteen papers included in these proceedings are a wide-ranging collection that gives the reader an excellent understanding of the outstanding engineers and the magnificent engineering and architecture that created Hoover Dam. There is probably no other dam in the world that has received as much public interest and press coverage as Hoover Dam. The hardships overcome by the engineers and the construction workers at the Black Canyon site are hard to imagine today, but President Roosevelt's dedication speech on September 30, 1935, gives a good picture:

> "Ten years ago the place where we are gathered was an unpeopled, forbidding desert. In the bottom of a gloomy canyon, whose precipitous wall rose to a height of more than 1,000 feet, flowed a turbulent, dangerous river. The mountains on either side were difficult of access, with neither road nor trail, and their rocks were protected by neither trees nor grass from the blazing heat of the sun."

Since being established in 1964, ASCE's History and Heritage Committee has sponsored and organized a number of symposiums intended to enhance and preserve the knowledge and appreciation of our civil engineering history and heritage. While many publications and books have been written about Hoover Dam, we hope this collection of symposium papers focused on the engineering associated with the design and construction of Hoover Dam, its performance over the last 75 years, and the lessons learned will be of historic value to civil engineers in the USA and internationally.

The Steering Committee for the Hoover Dam 75th Anniversary History Symposium included Richard L. Wiltshire (Symposium Chair), David R. Gilbert, and Dr. Jerry R. Rogers. The Steering Committee members reviewed and provided the authors with comments on the papers included in these proceedings.

Richard L. Wiltshire, David R. Gilbert, and Dr. Jerry R. Rogers

Acknowledgements

The Steering Committee for the Hoover Dam 75th Anniversary History Symposium would like to take this opportunity to thank all of the authors of the symposium papers in our proceedings publication. The authors have spent many hours in preparing and revising their papers, most of which will be presented at ASCE's 140th Annual Civil Engineering Conference during its Hoover Dam 75th Anniversary History Symposium on October 21-22, 2010. These papers have been reviewed by the members of the Steering Committee who put in their valuable time and helped make these papers even better – thank you. The Steering Committee would also like to acknowledge the assistance of ASCE's Carol Reese, the History and Heritage Committee's Liaison, and Donna Dickert of ASCE's Book Production Department.

Contents

Keynote Paper

Politics and Dam Safety: The St. Francis Dam Disaster and the Boulder Canyon Project Act 1
    Donald C. Jackson

Politics, Economics, Technology, and History

Building Blocks of Hoover Dam: Technology, Politics, Economics 25
    Brit Allan Storey

The New Town of Boulder City: City Planning and Infrastructure Engineering for Hoover Dam Workers 40
    Jerry R. Rogers

Reclamation, the Army, and Hoover Dam during World War II 48
    Jim Bailey

Concrete Technology

Advances in Mass Concrete Technology—The Hoover Dam Studies 58
    Timothy P. Dolen

Long-Term Properties of Hoover Dam Mass Concrete 74
    Katie Bartojay and Westin Joy

Engineering, Hydraulics, and Structural

Hoover Dam: Evolution of the Dam's Design 85
    J. David Rogers

Hoover Dam: First Joint Venture and Construction Milestones in Excavation, Geology, Materials Handling, and Aggregates 124
    J. David Rogers

Hoover Dam: Construction Milestones in Concrete Delivery and Placement, Steel Fabrication, and Job Site Safety 163
    J. David Rogers

Hoover Dam: Operational Milestones, Lessons Learned, and Strategic Import 189
    J. David Rogers

Hoover Dam: Scientific Studies, Name Controversy, Tourist Attraction, and Contributions to Engineering 216
    J. David Rogers

75 Years of Hydraulic Investigations—Hoover Dam 249
    Philip H. Burgi

Performance of Spillway Structures Using Hoover Dam Spillways As a Benchmark 267
William R. Fiedler

Seismic Evaluation of Hoover Dam Powerplant 288
Adam Toothman, David Gold, Tim Brown, and Mary Beth Schuetz

Civil Engineers, Architecture, and Construction

Frank Crowe: General Superintendent of the Six Companies, Inc. Hoover Dam Project 307
Philip Dunn, Jr.

Engineering and the Sculptural Program of Hoover Dam 318
Alfred Willis

Megaproject Success: Hoover Dam Construction and Pre-Construction Management Ingenuity 329
John Walewski and Hessam Sadatsafavi

Construction Management of a Mega Project 340
Charles R. Parrish

The Construction of Hoover Dam: A Case Study from a Builder's Perspective 346
Tamiko Powell-Melhado, Michael Hein, and Linda Cain Ruth

Building Hoover Dam (Men, Machines, and Methods) 360
Raymond Paul Giroux

Indexes

Author Index 411
Subject Index 413

Politics and Dam Safety:
The St. Francis Dam Disaster and the Boulder Canyon Project Act

Donald C. Jackson[1], Ph.D.

[1]Professor of History, Lafayette College, Easton, PA, 18042, jacksond@lafayette.edu

ABSTRACT: In March 1928, the U.S. Congress was poised to approve the Boulder Canyon Project Act (BCPA) and authorize federal support for building what would become Hoover Dam. California Congressman Phil Swing and Senator Hiram Johnson had championed the BCPA since the early 1920s and now – despite continued opposition from private power interests and Arizona politicians – their politicking appeared ready to bear fruit. But minutes before midnight on March 12th the St. Francis Dam gave way, releasing a deadly torrent down the Santa Clara Valley in Southern California and killing more than 400 people. This paper examines the relationship of Los Angeles (and St. Francis Dam designer William Mulholland) to the BCPA, and considers how a key investigation of the disaster was tied to the proposed Hoover/Boulder Dam. While assessing the technology of concrete curved gravity dam design in the 1920s, it also considers how the disaster prompted authorization of a special Colorado River Board and delayed congressional passage of the BCPA until December 1928. It concludes with consideration of how the disaster spurred revision of California's dam safety regulations and professional concerns this initiative engendered among members of the ASCE.

INTRODUCTION

Shortly before midnight on March 12, 1928, the 205-foot high St. Francis Dam collapsed and during the early morning hours of March 13th, 38,000 acre-feet of water surged from the reservoir. Wreaking havoc through San Francisquito Canyon and Southern California's Santa Clara Valley, by the time the flood washed into the Pacific Ocean (some fifty-five miles downriver), over 400 people lay dead. Considered one of the greatest civil-engineering disasters in United States history, the St. Francis Dam tragedy generated intense interest not only because of the deaths and destruction it caused, but also because it involved the failure of a concrete curved gravity dam, the design type then planned for the massive Hoover/Boulder Dam on the Colorado River.[1] The disaster emboldened critics of the Boulder Canyon Project, in no small part because William Mulholland, longtime head of Los Angeles's Bureau of Water Works and Supply (BWWS) and the official in charge of the failed dam's design and construction, had been prominently involved in advocating the Boulder

Canyon Project Act (BCPA, popularly known as the Swing/Johnson Act in recognition of its primary supporters, California Congressman Phil Swing and California Senator Hiram Johnson). This paper is not intended to tell the full story of either the St. Francis tragedy or the complex political interactions that underlay federal approval of the BCPA, but it does bring to the forefront an appreciation of how the St. Francis Dam collapse – and the subsequent engineering investigation authorized by California Governor C. C. Young – were tied to the BCPA. And it serves as a reminder that politics and safety, particularly as they involve large public works projects, are not always easily separated.[2]

ORIGINS OF HOOVER/BOULDER DAM

Hoover/Boulder Dam is rooted in a privately-financed project to irrigate Southern California's Imperial Valley. Starting in 1900 the Colorado Development Company diverted water from the Colorado River to nourish a huge tract of desert land just north of the California/Mexico border.[3] Soon thousands of acres of in the Imperial Valley were "under the ditch" and the company's prospects appeared bright. However, the company's canal connecting to the Colorado River kept clogging with silt.[4] Because silt accumulation was worst in the section of the canal closest to the river – and because the company sought to move the headgates beyond U.S. jurisdiction in order to more readily irrigate land in Mexican territory – in 1904 the company excavated a more direct opening at a site a few miles south of the California/Mexico border.[5]

To the company's misfortune, in June 1905 heavy storms washed away the new headgates and soon the entire flow of the river was pouring into the Imperial Valley and the Salton Sink. Eventually, the flooding was brought under control by the Southern Pacific Railway Company but it took almost two years to close the breach. In the meantime, thousands of acres of low lying land were inundated under what is now known as the Salton Sea.[6] By 1909, the California Development Company had transferred most of its assets to the Southern Pacific and, in 1916, the recently formed Imperial Irrigation District purchased the water supply system from the railroad.[7] Once the flooding had stopped, agricultural production resumed. Nonetheless, fear that a devastating uncontrolled "break" might recur was never far from the minds of valley residents and investors alike. Soon the district began clamoring for federally-supported flood protection.

Although the federal government had refrained from fighting the floods of 1905-07, the lower Colorado River had not been ignored by the Reclamation Service. As early as 1902, Arthur Powell Davis (at that time Assistant Chief Engineer of the Service) had considered development of the basin.[8] For many years the issue of a major storage dam across the lower Colorado River was overshadowed by other Reclamation Service projects, but by the end of World War I the Service was seeking new venues for its dam-building skills. In 1915 Davis became Director of the Reclamation Service and he appreciated that controlling the lower Colorado could involve construction of one of the largest dams in the world. At the urging of the Imperial Irrigation District, in May 1920 Congress authorized the Reclamation Service to develop preliminary plans for a Colorado River storage dam.

In 1922, these plans spawned the "Fall/Davis Report" a major study prepared under the auspices of Secretary of the Interior Albert Fall and Reclamation Service Director Davis. In the report, Davis advocated hydroelectric power development because only power revenues could repay construction costs with any degree of certainty. From a practical point-of-view, the development of hydroelectricity made sense as a dam over 500 feet high and impounding more than 20 million acre-feet of water could generate many millions of kilowatt hours per year. But from a political perspective, the use of power revenues to finance the dam invited controversy. Privately-financed companies controlled America's electric power grid in the 1920s and they viewed askance any legislation that would authorize a huge federal dam financed by hydroelectric power.[9] As historian Paul Kleinsorge later noted:

> "The controversy over the power aspects of the [Boulder Canyon] project… [involved] a clamorous argument that took on the aspects of a nation-wide debate, chiefly because it involved the whole question of whether or not the federal government should enter large-scale power production activities . . ."[10]

In the face of opposition from private power interests, the Fall/Davis Report nonetheless advocated construction of a high dam and hydroelectric powerplant in the vicinity of Boulder Canyon. [Note: both Boulder Canyon and Black Canyon (lying about twenty miles downstream) feature dramatic, narrow gorges with steep granite walls extending upwards for several hundred feet. While initial investigations focused on Boulder Canyon, by at least 1924 Service engineers recognized that the Black Canyon dam site was preferable and – despite retention of the name Boulder or Boulder Canyon Dam – this became the site where the dam was built.[11]]

LOS ANGELES AND THE METROPOLITAN WATER DISTRICT OF SOUHERN CALIFORNIA (MWD)

Between the initiation of the project in the early 1920s and congressional authorization in December 1928, the City of Los Angeles and other southern California communities assumed an essential role in promoting the Boulder Canyon Project. As early as July 1921 Los Angeles had expressed interest in helping build Boulder Canyon Dam in return for control over the hydroelectric power plant.[12] And by 1924 this interest had expanded into a formal water claim filed on the city's behalf for 1,"500 cubic feet per second of Colorado River flow.[13] With this claim, the City of Los Angeles served as the catalyst for the Colorado River Aqueduct and for what evolved into the Metropolitan Water District of Southern California (MWD). Most importantly in terms of Congressional approval for the Boulder Canyon Project, the MWD would comprise the most important customer for hydropower. As noted in the MWD's first annual report:

> "It was early recognized that to secure favorable [congressional] consideration [the BCP] must be self-supporting and that the power to be generated from any development… must find a market which would eventually return all costs of the entire project to the Government. As

additional engineering work for a Colorado River Aqueduct was done it became evident that any practicable diversion of the river must… involve pumping. Such pumping was practicable only if a large amount of power could be obtained at low price. This created, at once, a potential market for a substantial part of the power from any major Colorado River development. When these facts… were laid before Congress support for the Swing-Johnson measure became easier to obtain."[14]

In other words, the need to use huge amounts of electric power to pump water through the Colorado Aqueduct would prove essential in convincing Congress – already subject to intense anti-dam lobbying by private power advocates – that Hoover/Boulder Dam would not generate huge quantities of unmarketable power. The MWD could sign contracts guaranteeing power sales, and, in turn, the federal government could be assured that such contracts would be honored because the MWD possessed the right to directly tax land within its service area. The Boulder Canyon Project was born out of the flooding of the Imperial Valley, but its long-term viability depended upon the ability of greater Los Angeles to absorb its enormous cost.[15]

WILLIAM MULHOLLAND

Although he was not in charge of the City of Los Angeles' municipally owned electric power system (that job was handled by electrical engineer E. C. Scattergood), by the 1920s William Mulholland reigned supreme over the operation – and growth – of the city's water supply system. As Chief Engineer of the Bureau of Water Works and Supply (BWWS) he was in charge of dam design and construction as well as the development of new water supply sources. Thus, in the early 1920s he was responsible for designing and building the St. Francis Dam in northwestern Los Angeles County and for planning the aqueduct that would be used to tap into the Colorado River and carry water more than 200 miles across the Mojave desert.

Mulholland assumed an important public role in advocating authorization of the Boulder Canyon Project because, without it, operation of the Colorado Aqueduct would prove impossible. As early as 1921 he journeyed with Scattergood to explore power dam sites along the Colorado and the next year spoke at the city's Sunset Club on the importance of the Boulder Canyon Project.[16] In November 1923 he took a well publicized five-day journey down the Colorado, proclaiming that a dam should be built at Boulder Canyon and that transporting Colorado River water to Los Angeles was viable. This excursion was quickly followed by a trip to Washington D.C. in January/February 1924 where he consulted with the Department of the Interior on planning for the Boulder Canyon Project and, even more importantly, publicly testified before Congress in support of the Swing/Johnson Act.[17]

During 1925 Mulholland lobbied for creation of the Metropolitan Water District of Southern California and at the end of the year he again testified at a hearing related to the Boulder Canyon Project. In November he returned for another public inspection trip of the Boulder/Black Canyon dam site.[18] And when, in January 1928, it appeared that the Swing/Johnson Act was close to being approved by Congress, Mulholland made yet another trip to Washington D.C. to testify in support of the bill. As it turned out, the proposed hearings were delayed and Mulholland returned to Los Angeles without testifying, but he left Congress with a report arguing Los Angeles' need for the Boulder Canyon Project in order to maintain the city's long-term economic viability.[19] When he arrived back on the Pacific Coast, the collapse of the St. Francis Dam lay but a few weeks in the future.

To insure clarity, it should be emphasized that Mulholland was *never directly involved in the design of Hoover/Boulder Dam*. But his public advocacy of, and association with, the Boulder Canyon Project throughout much of the 1920s was nonetheless quite significant.

In the midst of his advocacy of the BCP and his planning for the Colorado Aqueduct, Mulholland undertook the design and construction of St. Francis Dam. Significantly, no outside contractor was engaged for the project and construction was carried out by the BWWS between the summer of 1924 and the spring of 1926. As head of the Bureau of Water Works and Supply, Mulholland had charge over all aspects of the project. What is perhaps most surprising to the modern eye is that in exercising this responsibility Mulholland sought no outside review by consulting engineers to review plans for the St. Francis Dam. By the 1920s it was common practice for dam engineering projects to involve design review by consulting engineers and, since its origin, the Reclamation Service/Bureau had relied upon engineering boards to review (and approve) plans before construction could proceed.[20] Immediately after the St. Francis collapse, the Bureau rushed to remind the public and fellow engineers of this policy: "The recent unfortunate failure of the St. Francis Dam in California," announced the Bureau in *Engineering News-Record*, justifies "special mention of the extensive geological and engineering investigations that preceded the approval of the site and designs for the Owyhee Dam" that included three geologists and three engineers not on the Bureau staff.[21]

In California, dam projects in the 1920s were also subject to state supervision under legislation enacted in 1917; this legislation required the California State Engineer to review and approve all dams over 10 feet high unless: A) they were to be built by a corporation under the jurisdiction of the State Railroad Commission; or B) they were to be built by a municipality with a department of engineering. It was this latter "municipal exemption" from the state dam safety law that allowed Mulholland to design and build the St. Francis Dam without any review by engineers who worked outside of the BWWS.[22]

By the 1920s Mulholland had achieved enormous prestige because of his success building the Los Angeles Aqueduct that brought the water from the Owens River Valley into the city limits, a distance of more than 230 miles. Work on the aqueduct had been carried out between 1907 and 1913 and involved some extremely difficult construction, most notably the completion of the 5-mile-long Elizabeth Tunnel. When water from the Sierra Nevada first entered San Fernando Valley in November 1913, it ushered in the beginning of modern day Los Angeles. No longer would the city be economically constrained by the water resources of the Los Angeles River. Now the incredible commercial potential of Southern California could be unlocked without fear of impending drought. The person most responsible for opening up these possibilities was William Mulholland.

True, there had been criticism of the aqueduct's construction (in particular, complaints about the quality of the concrete) but this was relatively modest given the overall scope of the project. Nonetheless, Mulholland bristled at such sniping and it seems to have encouraged him to keep the work of the BWWS within the department and beyond the scrutiny of outsiders who might not agree with his plans or ideas. In light of the incredible benefits that the aqueduct brought to the political economy of the Southland, elected officials and politicians in Los Angeles were more than happy to let Mulholland carry on his work however he saw fit. Mulholland's engineering success had brought with it the privilege for him to do whatever he wanted without outside review.

Unlike many of his contemporaries and peers, such as Arthur Powell Davis, John R. Freeman, and Carl E. Grunsky, Mulholland possessed no university training and was essentially self-taught, deriving the core of his hydraulic-engineering knowledge from on-the-job experience. He had a quick mind, a remarkable memory, and, apparently for much of his early career, an appetite to supplement his extensive practical work with knowledge gleaned from technical books and articles.[23] Early on he apparently devoted much attention to self-education as he sought to expand his horizons beyond what he could learn from practical experience. But as he became older (he was born in 1855) his interest in self-education apparently waned. Or at least his interest in state-of-the-art concrete gravity dam design and construction was limited when he set out to construct the St. Francis Dam in 1924. Because if he had being paying attention to what other engineer's concrete gravity dam designs of the 1920s were like, it is difficult – if not impossible - to imagine that he would have designed and built St. Francis as he did.

THE ST. FRANCIS DAM DESIGN IN CONTEXT

From its intake in the Owens Valley, the Los Angeles Aqueduct descends over 2,500 feet; the biggest drop comes when it passes from the western Mojave Desert (or Antelope Valley) into coastal southern California. Here, the five-mile-long Elizabeth Tunnel feeds into San Francisquito Powerhouse No. 1. Then water flows south through a six-mile-long tunnel in the east canyon wall before dropping down to San Francisquito Powerhouse No. 2 along the banks of San Francisquito Creek. A broad, open area bounded by a narrow gorge lies but a short distance upstream from Powerhouse No. 2. This became the site of the St. Francis Dam and Reservoir.

In terms of topography the St. Francis site was suitable for storing a large quantity of water but in terms of geology the location was less than ideal.[24] For example, the city's 1911 annual report on the aqueduct's construction described the rock along the eastern side of San Francisquito Canyon as "exceedingly rough, and the dip and strike of the slate [schist] such as to threaten slips."[25] Joseph B. Lippincott, Mulholland's Chief Assistant Engineer during aqueduct construction, recalled the difficult geological character of the east canyon ridge and, acknowledging he had been "intimately connected with the driving of a series of tunnels for our aqueduct through the range of mountains on which the left or east abutment of the dam rested," later declared: "The rock that we encountered was a broken schist and a good deal of it expanded when it came in contact with the air and was what the tunnel men called 'heavy ground.' We had great difficulty in holding this ground [for the aqueduct tunnel] before it was lined with concrete."[26]

The west (right) abutment of the dam site was comprised of a red sandstone conglomerate and a fault was formed where the eastside schist formation and the westside conglomerate met. Although geologically inactive, the presence of this fault (that extended through the dam site generally parallel to the creek bed) did not offer an ideal foundation for a concrete gravity dam. While the fault and the differing geological characteristics of the two sides of dam site are readily apparent, there is no evidence that Mulholland undertook any significant analysis of the site geology prior to the dam's construction.

In designing the St. Francis Dam, Mulholland did not prove technologically adventuresome in selecting a concrete gravity structure. Instead, he opted for a simple design that incorporated but few features oriented toward coping with the effect of uplift.

The phenomenon of "uplift" (so-called because it tends to lift the dam upward) destabilizes gravity dams by reducing the structure's "effective weight," thereby lessening its ability to resist horizontal water pressure. Uplift can act through bedrock foundations that, in the abstract, are strong enough to bear the weight of the dam, but are fractured or fissured and thus susceptible to seepage and water saturation. The deleterious effect of uplift upon a concrete gravity dam can be countered in various ways: 1) excavating foundation "cut-off" trenches; 2) grouting the foundation; 3) draining the foundation through use of relief wells; 4) draining the interior of the dam through use of porous pipes and tunnels; and -) increasing the dam's thickness or cross-sectional profile to counter the destabilizing effect of water pressure pushing upward.[27]

For a distance of about 120 feet in the center of the St. Francis dam site, Mulholland did place ten drainage wells. But for the remainder of the dam's 600-foot long main section he did not grout the foundation, excavate a cut-off trench, or install a drainage system up the sides of the canyon walls. In addition, during construction the height of the dam was raised at least ten feet, but no apparent attempt was made to increase the thickness of the dam to accommodate the increased hydrostatic forces created by the deeper reservoir.[28]

Mulholland's consideration of the effect of uplift (or lack thereof) did not accord with other major concrete gravity dam designs built contemporaneously with St. Francis. Apprehension about uplift intensified among American engineers following the collapse of a concrete gravity dam in Austin, Pennsylvania on September 30, 1911.[29] John R. Freeman, the prominent New England-based engineer and future President of the ASCE, soon visited the site of the Austin tragedy. In a lengthy letter published in *Engineering News* describing his visit, he implored engineers to understand that "uplift pressures may possibly occur under or within any masonry dam and should always be accounted for."[30]

Also taking the Austin failure very seriously was Arthur Powell Davis, chief engineer (later director) of the U.S. Reclamation Service. After visiting the disaster site, Davis expressed concern about the possible effect of uplift on the Service's Elephant Butte Dam, a concrete gravity structure over 200 feet high to be built across the Rio Grande in southern New Mexico.[31] The agency soon approved a design for Elephant Butte Dam that included extensive foundation grouting, placement of a drainage system along the length of the dam, and a deep cut-off trench.[32] For the Service's 354-foot-high concrete gravity Arrowrock Dam built in 1913-1915 near Boise, Idaho, Davis reported: "In order to prevent leakage in the foundation of the dam, a line of holes was drilled into the foundation just below the upstream face of the dam to depths of 30 to 40 feet. They were grouted under pressure... another line of holes was drilled to serve as drainage holes to relieve any leakage under the dam. These were continued upward into the masonry and emerged into a large tunnel running the entire length of the dam."[33]

After the Service became the Bureau of Reclamation in 1923, concern about uplift continued. For example, Black Canyon Dam in southern Idaho, a 184-foot-high concrete-gravity structure completed in 1924, featured two rows of grout holes "drilled into the bedrock along the upstream edge of the dam along its entire length... [A] row of drainage holes was drilled 8 feet downstream from the second row of grout holes. . . . The water from them is carried to a tile drain embedded in the concrete parallel with the axis of the dam." In case anyone missed the point, *Engineering News-Record* declared: "The purpose of this drainage system is to collect and lead off any water that might accumulate and to prevent an upward pressure under the dam."[34]

The Black Canyon Dam was constructed contemporaneously with development of the Bureau's Hoover/Boulder Dam design and the latter project paid close attention to the possible effect of uplift. This included in-depth foundation excavation, extensive pressurized grouting of the entire foundation, and elaborate drainage systems for both the foundation and the interior of the dam.[35] Although the designs for St. Francis and Hoover/Boulder were both for concrete curved gravity structures, they were light years apart in terms of how they addressed the possible effect of uplift. The Bureau's work at Hoover/Boulder built upon the foundation established at Elephant Butte, Arrowrock, and Black Canyon Dams to create a highly sophisticated – although not necessarily inexpensive – design. In contrast, Mulholland's St. Francis design barely took note of gravity dam development in the wake of the Austin dam disaster and, as constructed, bore scant resemblance to what was proposed for the Boulder Canyon Project.[36,37]

UPLIFT AND THE CAUSE OF THE ST. FRANCIS DAM COLLAPSE

Clearing of the St. Francis reservoir site commenced in late 1923, but the first concrete was not cast until August 1924. Construction then proceeded for close to two years until the dam topped out at about 205 feet in May 1926.[38] After completion, the reservoir was not immediately filled, although it did come to within three feet of the spillway in May 1927. Nine months later, in February, 1928, the water level came to within a foot of the spillway and, on March 7, 1928, the reservoir reached three inches below the spillway crest. It stayed at that elevation – which equated to over 12 billion gallons of stored water – until late in the evening of March 12.[39] Then disaster struck. This paper is not the place to recount the horror of the flood that swept down San Francisquito Canyon and then into the Santa Clara River Valley during the early hours of March 13th, but suffice it to say that the damage wrought by the flood was horrendous, with more than 400 people left dead.

Bluntly stated, the cause of the collapse can be traced to the effect of seepage into and saturation of the broken mica schist forming the east (left) canyon wall. This created significant uplift pressure acting against the eastern section of the dam and, in the absence of any measures taken to mitigate or counter the effect of uplift, resulted in the instability and collapse of the structure. In April 1928 this failure mechanism was identified by civil engineers Carl E. Grunsky and his son Eugene L. Grunsky and Stanford University geologist Bailey Willis who investigated the collapse at the behest of the Santa Clara River Conservation District.[40] Their work culminated in two reports published in *Western Construction News* (the Grunskys' in May 1928 and Willis' a month later). On June 25th, Charles Lee, a San Francisco-based engineer, published his own analysis of the failure in *Western Construction News* that followed directly from the Grunkys' report and emphasized that "a logical sequence can be predicated upon a failure commencing at the east end of the dam." Lee also observed that initiation of the failure on the west side "does not satisfactorily explain many of the known facts."[41] More recently in the 1990s, geotechnical engineer J. David Rogers affirmed that the cause of the St. Francis Dam collapse closely conforms to the conclusions reached by the Grunskys, Willis and Lee.[42]

To synopsize the Grunskys' and Willis' analysis (affirmed by Lee), the foundations underlying both sides of the dam were deemed to be to be unsuitable, "but the critical situation developed more rapidly in the east abutment" where "the schist is... traversed by innumerable minute fissures, into which water would intrude under pressure and by capillary action." Also, the "east abutment was located on...the end of an old landslide... [and] when it had become soaked by the water standing in the reservoir against its lower portion, it became active and moved." That movement resulted from "a great hydrostatic force under its [the dam's] foundation surface from end to end," which triggered the collapse of the east abutment. And finally, "the old slide against which the dam rested at the east . . . offered only insecure support to the dam, and this was rendered more precarious by [Mulholland's] adoption of a design which did not include adequate foundation drainage."[43] In focusing attention on the "old landslide" that formed the east abutment, the Grunskys and Willis were acknowledging the effect of the tremendous "new" landslide that dislodged much of the east abutment and the concrete structure that rested upon it. This new landslide – which was activated by seepage emanating from the storage reservoir – brought an estimated 500,000 cubic yards of disintegrated mica schist into the flood surge and left behind a huge scar on the landscape. In his report, Charles Lee emphasized that "the far greater bedrock erosion on the east side, as compared with the west, indicates that a greater volume of water flowing for a longer period of time and under greater head, was pouring through this gap than on the west."[44]

The Grunskys also noted the overall instability of the St. Francis Dam's gravity design, calling attention to a crack near the base of the dam's surviving center section. As will be discussed below, other engineers would herald the surviving center section as affirming the strength of gravity dam technology. But the Grunskys pointed out – and included pictures as proof – that the center section had tipped backwards sufficiently to allow a wooden ladder to enter the crack; this ladder became trapped in the crack when the structure rocked back into position. Rather than standing as a beacon of strength, the center section apparently had survived by only a very thin margin.[45]

THE GOVERNOR'S COMMISSION

The collapse of St. Francis Dam prompted the creation of several panels of engineers and geologists, but only the Grunskys/Willis investigations were sponsored by interests tied to the Santa Clara River Valley. Others were sponsored by California Governor C.C. Young, the Los Angeles County district attorney, the Los Angeles County coroner, and the Los Angeles City Council.[46] These latter panels quickly concluded that the collapse initiated within in the red sandstone conglomerate beneath the western abutment. The conclusion of the Governor's Commission (which was the first to report its findings and was chaired by A. J. Wiley, an accomplished dam engineer connected to Hoover/Boulder Dam through his chairmanship of the Bureau of Reclamation's Boulder Dam Board) established the essential parameters of the "eastside/conglomerate" failure hypothesis synopsized below. The conclusions reached by the Governor's Commission provided a template closely followed by other investigations that quickly completed their work, including most notably the Los Angeles City Council committee chaired by Elwood Mead, Commissioner of the Bureau of Reclamation.[47] Before the public airing of the Grunskys/Willis' reports, *Engineering-News Record* proclaimed in an April 19th headline: "All St. Francis Dam Reports Agree as to Reason for Failure."[48]

While the Governor's Commission headed by Wiley believed that "the foundation under the entire dam left very much to be desired," the west abutment emerged as the culprit. "The west end," stated the Governor's Commission, "was founded upon a reddish conglomerate which, even when dry, was of decidedly inferior strength and which, when wet became so soft that most of it lost almost all rock characteristics." The softening of this "reddish conglomerate" undermined the west side. "The rush of water released by failure of the west end caused a heavy scour against the easterly canyon wall… and caused the failure of that part of the structure." There then "quickly followed… the collapse of large sections of the dam." In making this assessment the Commission did acknowledge that "as yet the manner and chronological order in which the failure of the various sections of the structure occurred are not yet certain [49] No small part of the uncertainty attending exactly how various parts of the dam failed involved the tremendous landslide that had carried away much of the eastern abutment. It was one thing to propose that flow originating from the west abutment somehow had crossed over the canyon and "caused a heavy scour against the eastern wall." But to ascribe the erosion of an estimated 500,000 cubic yards of schist to such scour was something that the Governor's Commission could only countenance as "not yet certain."[50]

The Governor's Commission reached their conclusions within a week after initiating study of the failure (and less than two weeks after the collapse). But such haste in completing their study and acknowledgement that "the manner and chronological order" of the collapse was uncertain nonetheless fostered little doubt. "With such a formation [the red conglomerate]," concluded the Governor's Commission:

"the ultimate failure of this dam was inevitable, unless water could have been kept from reaching the foundation. Inspection galleries, pressure grouting, drainage wells and deep cut-off walls are commonly used to prevent or remove percolation, but it is improbable that any or all of these devices would have been adequately effective, though they would have ameliorated the conditions and postponed the final failure."[51]

As far as the Governor's Commission was concerned, the poor quality of the foundation material on the west side of the canyon (and "defective foundations" generally) rendered all other design issues – including uplift – irrelevant.[52] Similarly, the Mead's committee's report (completed on March 31st) concluded that "the dam failed as a result of defective foundations."[53]

In their conclusion, the Governor's Commission did not contain itself to simply stating "The failure of St. Francis Dam was due to defective foundations." It also offered reassurance that "there is nothing… to indicate that the accepted theory of gravity dam design is in error…" and even proclaimed that "the middle section which remains standing even under such adverse conditions [offers] most convincing evidence of the stability of such structures when built upon firm and durable bedrock."[54] In championing the great strength of gravity dam technology it appears that the goal of the Governor's Commission was to do more than simply offer a dispassionate analysis of why the dam failed. And – in the face of clear physical evidence that the center section was at the brink of instability as water surged from the reservoir – it would also seem disingenuous that they would trumpet the surviving center section as a symbol of great strength.

At one level of analysis it is difficult to disagree with the Governor's Commission (and Mead's committee) in terms of "defective foundations" being the cause of the collapse. But the overarching simplicity of this conclusion and the erroneous emphasis placed on the "west side/red conglomerate" hypothesis prompts the questions: Why did the Commission seek to complete its work so quickly and why might they have been willingly to focus sole attention on "defective foundations" as the essential cause of the collapse? And why was it so important to offer affirmation of the validity of gravity dam technology? Answers to these questions can be tied to Hoover Dam and the impending Congressional enactment of Swing/Johnson's Boulder Canyon Project Act.

In this context, it should be noted that C.C. Young had been elected California's Governor in 1926 on a platform that strongly endorsed authorization of the Boulder Canyon Project Act and his support for the project was featured in his Jan. 4, 1927 inaugural address.[55] The Swing/Johnson Act was entering a crucial stage in Congress' deliberations and, in fact, the front page of the Thursday March 15, 1928 *Sacramento Bee* featured photos of the St. Francis carnage as well as the story "Boulder Dam Favorably Reported to Congress." This latter article noted that the House of Representatives' Irrigation Committee had "reported the [Swing/Johnson] bill [to the full house] with a favorable recommendation." Finally, after more than six years from the time it was first proposed, the Swing/Johnson Act appeared poised for passage. But the St. Francis disaster now introduced a cloud of uncertainty about the safety of large dams, and in particular the safety of large concrete curved gravity dams.

Although William Mulholland had not played any role in the Hoover/Boulder dam design, his association with – and public advocacy of – the Boulder Canyon Project made it imperative for supporters of the impending legislation that questions about the cause of the disaster be quickly resolved. In concert with this, public apprehension about the safety of large dams – which would include the proposed Hoover/Boulder Dam – also needed to be assuaged.

In charging his commission with the task of "learning just what caused the failure of the St. Francis Dam" Governor Young stressed to them that:

> "the prosperity of California is largely tied up with the storage of its flood waters. We must have reservoirs in which to store these waters if the state is to grow. [And] we cannot have reservoirs without dams."[56]

The task was thus twofold; 1) find the cause of the failure; and 2) discern "the lesson that it teaches us [that] must be incorporated into the construction of future dams."[57] In other words, faith in dam construction must be affirmed as the state and society moved into a future where disasters like St. Francis could never be allowed to recur.

Records documenting the deliberations of the Governor's Commission are essentially non-existent as it appears that any drafts, memos, or other written records relating to the committee's work were destroyed once the final report was completed. But a three-page memorandum outlining the basic work of the committee survives in the Division of Safety of Dams files and offers a modicum of insight into the Commission's work. The memo reveals that "Mr. A. J. Wiley, chairman, arrived at Sacramento the morning of [Sunday] March 18th, spent that day in consultation with Governor Young, Director of the Department of Public Works B. B. Meek, and State Engineer Edward Hyatt."[58] Nothing further is told of their discussion, but simply the fact that Wiley and the governor met directly (and before the committee as a whole convened) lends credence to interpretations of the Governor's Commission work that acknowledge its political dimensions.

After meeting with Governor Young, Wiley immediately headed for Los Angeles. The next morning (Monday March 19th) most of the committee (consisting of Wiley, F. E. Bonner, H. T. Cory, F. H. Fowler, G. D. Louderback, and F. L. Ransome) along with Meek and Hyatt developed "plans for procedure to some extent and [then] adjourned to meet Mr. Mulholland by appointment at his office." The next day the commission spent a long day at the St. Francis site. On Wednesday the commission convened in Los Angeles and was finally joined by member Ransome ("who had just arrived from Nevada"). The full committee returned to the dam site on Thursday so that Ransome could inspect it and also to facilitate an effort "to identify the various fragments of the dam that had been washed downstream."[59]

On Friday March 23rd the Commission witnessed "testing of the samples of concrete and conglomerate" and then "started work on their report." On Saturday the committee "continued work on report," and at 5pm on Sunday March 25th were "able to sign [the] completed report agreed upon by all members [although] photographs, maps, etc. were not yet complete." The following day the complete report was "assembled with maps, photographs etc. and three copies were taken north [i.e. to Governor Young in Sacramento] by Mr. Hyatt."[60]

Immediately upon completion of the Commission's report – and even before the Governor had seen it – Hyatt wired Congressman Swing in Washington, D.C. to apprise him of its conclusion. Specifically, he reassured Swing that:

> "Report of the Investigating Committee St. Francis Dam just completed but not yet in hands of Governor Young Stop Statement to you to the effect that there is absolutely no relation between the failure of the St. Francis Dam and the safety of the proposed Boulder Canyon Dam can be sent best advantage tomorrow morning after conference between Governor Young and A J Wiley Chairman of the investigating commission Stop Please wire advice if this is satisfactory or if statement absolutely necessary today."[61]

The next day Governor Young did indeed meet with Wiley, and soon thereafter he sent his own telegram to Congressman Swing. In this he avowed that what occurred at St. Francis bore no relation to anything that could be ascribed to the Hoover/Boulder site:

> "I have positive assurance from A. J. Wiley, Chairman of Commission and of Dr. F. L. Ransome Professor of Economic Geology at California Technical Institute (*sic*), who is also a member of the St. Francis Dam investigating commission, both of whom have examined the Boulder and Black Canyon dam sites that the bedrock there is so sound, hard and durable and so different from the very soft foundation of the St. Francis Dam, that the failure of St. Francis Dam need cause no apprehension whatever regarding the safety of the proposed Boulder Canyon Dam."[62]

Governor Young further counseled Swing that the failure was not related to gravity dam design – which might somehow get people worried about Hoover/Boulder – and, drawing upon the Commission's conclusion, averred that the surviving middle section of the St. Francis structure actually affirmed the strength of gravity dam technology if built on solid foundations:

> "The report of the investigating committee also states that there is nothing in the accepted theory of gravity dam design that is in error or that there is any question about the safety of concrete dams designed in accordance with that theory when built upon ordinarily sound bed rock but that on the contrary the action of the middle section of the St. Francis Dam that remained standing even under such adverse conditions is most convincing evidence of the stability of such structures when built upon such firm and durable bedrock as is present in Boulder Canyon."[63]

In response, Swing advised the Governor Young that "I think the report of the Engineering Commission will give me just the information I desire."

THE COLORADO RIVER BOARD

With Arizona adamantly opposed to the Boulder Canyon Project (on grounds that it served the interest of Southern California at the expense of Arizona) and other states in the Colorado River basin fearful that California might advantage of them, the congressional political coalition structured by Swing and Johnson was fragile.[64] And protestations by opponents of public power did not abate after the collapse of Mulholland's dam. For example, the *Wall Street Journal* advised readers that the failure of the St. Francis Dam, a structure built by a governmental entity and not private enterprise, should give pause to those supporting the Boulder Canyon Project because "the St. Francis dam break is an indictment of municipal ownership."[65]

It had taken over six years to reach the point where passage of the bill seemed possible. Then a wave of uncertainty uncoiled by St. Francis threw the Boulder Canyon Project Act into limbo. The speedy completion of the Governor's Commission's investigation helped Swing and Johnson avoid complete loss of support for their bill in the spring of 1928. But significant momentum had been lost. At the end of May adjournment loomed, and a compromise was reached whereby a Senate vote on the bill would be delayed until Congress reconvened in December. In the interim, a "Colorado River Board" of engineers and geologists (chaired by Major General William Sibert) was authorized to independently evaluate the proposed Boulder Canyon Project including the Hoover/Boulder dam design; Sibert's board was to submit a report to Congress within six months.[66] In the absence of the St. Francis collapse it appears unlikely that such an engineering board would have been created at such a late date in the legislative process. But it was politically difficult for Swing and Johnson to block such a move because, after all, why not take the time to be extra careful? By December, public memories of St. Francis presumably would have faded away – at least for a national audience far removed from the Santa Clara Valley – so long as nothing appeared to challenge the conclusions of the Governor's Commission report.

At the end of November, the Colorado River Board endorsed the dam's basic design, but recommended that the maximum allowable stresses in the structure be reduced from 40 to 30 ton/ft^2. To the layperson this might appear as a straightforward way to increase the strength of the design, but it presented problems because strict adherence to a 30 ton/ft^2 limit would significantly add to the (already massive) bulk of the dam and dramatically increase construction costs. The Bureau did not overtly resist this directive, but nonetheless made no meaningful alteration to the profile of the existing design. Instead, the Bureau claimed that more sophisticated mathematical analysis (using the "Trial-Load Method") indicated that their proposed design afforded a maximum allowable stress of about 34 ton/ft^2, sufficient to meet the 30 ton/ft^2 criteria. In Commissioner Elwood Mead's phrasing: "It is not believed that the maximum stress as finally calculated will appreciable exceed the 30-ton limit. It is believed that the general plan of the dam can be agreed upon without serious difficulties."[67] In the end Colorado River Board's recommendation had no substantive effect on the final structural proportions as the Bureau simply asserted that they could adhere to the 30 ton/ft^2 limit without substantive design changes.[68]

When Congress convened on December 5th the political landscape shifted since the previous spring, most notably in regard to the "Power Trust." Hearings by the Federal Trade Commission had brought to light nefarious lobbying activities carried out by opponents of public power and the American public was outraged. By the end of the year, arguments portraying the Boulder Canyon Project as an unwarranted expansion of government into the hydroelectric power business carried far less weight.[69] With adroit legislative maneuvering Swing and Johnson obtained both House and Senate approval by December 18th; three days later President Coolidge signed the Boulder Canyon Project Act into law.[70] Much still needed to be done before construction could begin (most notably, the negotiation of power contracts), but the major legislative hurdle had been passed. Less than a year after the disaster, the specter of St. Francis had lost its force as a means of opposing both Hoover/Boulder Dam and large dams in general.

MULHOLLAND'S FATE

As for William Mulholland, the collapse of St. Francis Dam brought his career to an ignoble end. Given his sole authority over the design and construction of the dam, it would have been impossible for him not to accept responsibility for the collapse. At the Los Angeles County Coroner's Inquest Mulholland famously testified "Don't blame anybody else, you just fasten it on me. If there is an error of human judgment, I was the human."[71] As a result, he was publicly hailed in the engineering press as a "Big Man" for his forthrightness.[72] But any engineer with an understanding of gravity dam technology knew that Mulholland's design fell far short of acceptable practice and no effort was made to defend him on technical grounds. Outside the Santa Clara Valley, public approbation was rare, but this did not mean that Mulholland and the fruits of his labor at St. Francis were held in high regard.

For example, shortly after the disaster John R. Freeman privately acknowledged to J. B. Lippincott that "I have been careful… to say nothing [to newspaper reporters] regarding the Los Angeles dam which could come back to hurt Mulholland." But Freeman then candidly criticized Mulholland for his habit of not consulting independent experts: "[he] does not appreciate the benefit of calling in men from outside to get their better prospective and their independent point of view."[73] To another colleague, Freeman reinforced the point: "This [St. Francis Dam] site plainly required many precautions that were ignored, and while I have the highest personal regard for my good old friend William Mulholland, I can but feel that he trusted too much to his own individual knowledge, particularly for a man who had no scientific education."[74]

Perhaps it was coincidence that William Mulholland's formal resignation as "Chief Engineer and General Manager" of the BWWS occurred on November 13, 1928, just a few weeks before the Boulder Canyon Project Act came back for final congressional approval. Or perhaps his resignation represented one of the final pieces of the puzzle that would separate Los Angeles and its future association with the Boulder/Hoover Dam and the Colorado Aqueduct from William Mulholland and *his* association with the St. Francis Dam disaster. Accolades of appreciation rained down on Mulholland upon the occasion of his retirement, with public mention of the St. Francis Dam disaster notable only by its absence.[75] But the despite the official silence, the political and professional significance of the disaster he was responsible for long endured.

POST-ST. FRANCIS DAM SAFETY POLITICS

In light of how important dam safety regulation became in the post-St. Francis era, it is somewhat surprising that there is no record of any one questioning Mulholland as to *why* the municipal exemption in the 1917 California dam safety law was in place when the St. Francis Dam was built. But it is clear that the exemption did not exist by chance or happenstance. Specifically, an October 1928 letter to State Engineer Edward Hyatt from San Francisco's City Engineer Michael M. O'Shaughnessy (who, like Mulholland, was also covered by the municipal exemption) explained why he sought to be free of state regulation:

> "I had our City Attorney present objections to the State legislative body in Sacramento in 1917, against allowing [then State Engineer] Mr. McClure to have anything to do with our dams at Hetch Hetchy, as I did not think, from his previous experience and knowledge, he had the requisite experience to pass on such a subject and I did not care to be subject to his capricious rulings... I did not think that Mr. McClure's previous clerical and engineering experience entitled him to be czar over the plans for our dam."[76]

In the aftermath of the St. Francis disaster, public clamor arose calling for a new dam safety law that would eliminate the municipal exemption and place all authority in one state office (presumably the State Engineer and not the State Railroad Commission). In accord with this, the final recommendation offered in the Governor's Commission report urged that all dams be "erected and maintained under the supervision and control of state authorities... with the police powers of the state... extended to cover all structures impounding any considerable quantities of water."[77]

Given the horrible destruction wrought by a dam built without state supervision, it was difficult for anyone to overtly oppose such a proposal. But, looming behind the prospect of new legislation lay fear that, if carried too far, state regulation of dams could impede economic growth. After all, what if vital development of California's water resources were to be blocked by adherence to an unrealistic standard of safety? As early as April 1928 State Engineer Hyatt acknowledged problems that might result from excessive zeal in regulating dam construction:

> "[T]he failure of the St. Francis Dam has greatly disturbed public confidence in the safety of all dams, and for a time at least, proposals for the construction of new structures are going to face unmerited opposition no matter how carefully supervised by public authority. Even among competent engineers there will be a tendency toward undue conservatism..."[78]

Hyatt further elaborated on the possible financial effects of new regulation:

> We [in the California Department. of Public Works] are thoroughly in sympathy with the feeling that public interest requires dams... to be made

absolutely safe against failure and provided with adequate spillway capacity. At the same time, we feel that we must exercise great care to avoid insisting upon safeguards beyond the actual needs since many meritorious projects might be thereby rendered financially infeasible."[79]

As the issue of legislating dam safety played out in the months following the St. Francis tragedy, concern about unchecked governmental authority stirred within the state's civil engineering community. With the carnage wrought on the Santa Clara Valley still fresh in the public mind, few engineers wished to openly question the wisdom of a strict dam safety regime. But a file entitled "Am Soc C E Committee on Proposed Legislation on Design and Construction of Dams" within the professional papers of Hiram N. Savage (who served for several years as City Engineer of San Diego) provides insight into the professional complexity of dam safety regulation. This committee (with Savage as one of four members) was authorized by the ASCE Sacramento Section in June 1928 to:

> "assist in shaping and passing at the next session of the State Legislature suitable legislation governing the design, construction and operation of dams in the State of California."[80]

The committee's directive was not simply that legislation be passed, but that the interests of the professional engineering community be protected/promoted and not made unduly subservient to state officials.

It was generally easy for the committee and ASCE members to support the idea of having a single state authority (i.e. the State Engineer) bear responsibility for enforcing the police powers of the state in regulating dams. But questions arose as to whether new legislation should codify – and perhaps require – the creation of consulting boards of engineers (presumably staffed by ASCE members) to advise the State Engineer when reviewing proposed dam projects.[81] During the committee's deliberations, committee member Walter L. Huber (who later served as ASCE President) forcefully advocated the use of consulting boards and provided a justification that, while not specifically naming Mulholland, made undeniable reference to the way that the St. Francis Dam came to be built:

> "I think that for any of the larger reservoirs the services of a Board of Consulting Engineers should be mandatory [emphasis in original]. Otherwise picture one of our powerful municipalities proposing a great reservoir to be built from plans by an engineer of that municipality who has much prestige but who alone considered the plans and the location. His reputation might far outweigh that of the State Engineer and might therefore make a critical review by the latter appear foolish. With political pressure reaching even to the Governor, would not a perfunctory review by the State Engineer, his early approval and a permit from the Department, be almost certain to follow without reference to consultants. Thus we would have failed to provide that protection to the public which we are seeking to accomplish."[82]

In the end, the final wording of the 1929 dam safety law did not require use of consultants by the State Engineer, but the act did not preclude it.[83] As the work of State Engineer Hyatt proceeded into the 1930s to implement the new law, consulting boards were commonly convened at his direction. Thus, the fears expressed by Huber (who, significantly, served on at least one such board) were generally quelled.[84] But the power of these consulting boards always flowed directly from the authority of the State Engineer (subsequently passed to the Division of Safety of Dams) to regulate dam-building within the state.

Consulting engineers (and engineers working for private companies) accepted the necessity for dam safety legislation, but this does not mean that everyone was thrilled with the idea of a state bureaucracy dictating what could or could not be constructed. At the extreme, comments offered in the context of an ASCE-sponsored symposium on the "Public Supervision of Dams" by M. M. O'Shaughnessy referred to California's new law as "most drastic" and opined that St. Francis catastrophe had created "an hysteria" whereby citizens and legislators "have practically lost their heads on the subject of dam design and construction..." He also called attention to the key question of who would assume the power to exercise that state's regulatory authority: "The great problem is to find that 'some one' who has gathered enough wisdom from his experience and who has adequate force, technical knowledge, and authority to fill the job."[85]

In a paper published as part of the symposium, A. W. Markwart, Vice President in Charge of Engineering for the Pacific Gas and Electric Company, was more circumspect than O'Shaughnessy, but he expressed concern that "it is not improbable that the tendency will be to require dams to be constructed stronger than actually necessary. Such excess strength can only be had from capital expenditures greater than have been required in the past..."[86] No mention was made of the recommendation by Sibert's Colorado River Board that the already ample dimensions of the Hoover/Boulder design be increased by simplistically decreeing that compressive stresses be reduced from 40 tons/ft^2 to 30 tons/ft^2, but clearly this reflects the type of "tendency" that Markwart feared might find traction in the wake of St. Francis.

In other comments published as part of the ASCE symposium, the European-trained engineer Fred Noetzli – a prominent advocate of thin arch and multiple arch dams who in a 1924 ASCE *Transactions* article had opined that "the gravity dam is an economic crime -- also expressed concern that that California's new law would foster adaptation of massive gravity dam technology at the expense of other, potentially less expensive alternatives: "there is no good reason why the most expensive type, namely the gravity dam, should receive first and sometimes sole consideration."[87] Here, Noetzli implicitly acknowledged the success of the Governor's Commission in championing the great strength of concrete gravity technology as evidenced by "the action of the center section [of the St. Francis Dam] which remain[ed] standing..."[88]

Since 1929 California has cultivated a reputation for sustaining one of the most demanding dam safety bureaucracies in the world. Aside from the Baldwin Hills Reservoir collapse in 1963 (that killed five people and destroyed more than 250 homes) the state has succeeded in preventing major dam disasters.[89] Certainly nothing has occurred that even remotely approaches the carnage resulting from the St. Francis Dam collapse. But, harkening back to fear of a possible tendency toward "undue conservatism," it may also be that efforts to insure dam safety have worked to suppress innovation in the development of new designs and construction techniques. In his 1976 article "The Evolution of the Arch Dam" the Swiss engineer and dam historian Nicholas Schnitter pointedly noted that in California "the number of [thin] arch dams as well as their proportion in relation to other types decreased sharply in the 1930s."[90]

In a society that holds dear the benefits that can accrue through competition and the expansion of free markets, state regulation is often portrayed as a deterrent to creative innovation. Of course, the freedom allowed to Mulholland by the "municipal exemption" proviso in the 1917 dam safety law did nothing to spur development of a creative, innovative design for St. Francis. But it should not be thought that freedom from government regulation must always lead to inadequate design.[91] Political reality in the post-St. Francis era dictated that dam technology would be much more regulated moving forward and a key question still resonates across the decades: what is the economic effect of stringent regulation and how might it constrain innovation?

CONCLUSIONS

In early June 1928, the Grunskys and Willis reports were synopsized in the nationally distributed *Engineering News-Record* under the headline "Sixth Report on St. Francis Dam Offers New Theories," and the next month the journal published a brief follow-up letter from the Grunskys regarding their investigation.[92] But after this, little public discussion or debate about the Grunskys' or Charles Lee's findings – and the way they offered a superior explanation of the cause of the collapse – appeared in the engineering press. In 1929 the ASCE *Proceedings* published a paper titled "Essential Facts Concerning the Failure of the St. Francis Dam: Report of the Committee of the Board of Direction," but this proved to be little more than an inconclusive review of the various investigations; it presented no judgment or analysis of its own, blandly observing: "some of the [investigating] reports express the belief that the break-up started at the west abutment, others at the east, and others reach no conclusion on this point"[93] The *Proceedings* paper offered no opinion on the merits of the various investigations and it was never published in the far more prominent and widely available ASCE *Transactions*.

It was a great embarrassment to the profession that an engineer as well-known as Mulholland had built such a deficient structure, so why call any more attention to the disaster than absolutely necessary? Mulholland had publicly accepted blame and departed the professional arena. Never again would someone like him build a dam in California without outside review. So what was the harm in allowing the views of the Governor's Commission and others that ascribed the failure to the western abutment's conglomerate to predominate. For both the engineering profession and any party interested in large-scale Western water development there was little to be gained in revisiting the St. Francis disaster. How else to say it? For many people it was best to simply let sleeping dogs lie and not worry any more about what caused the collapse of the St. Francis Dam.

In the context of engineering and the systems that engineers provide to society in the form of "Public Works," the notion that politics might somehow be involved often assumes a pejorative character. After all, engineers are presumed to work in an environment where non-technical and non-scientific factors are irrelevant to analysis and design. But public works are by their very nature political in that they are built to serve the social, economic, and – yes, political – interests of society. As part of this, the engineering profession itself possesses economic and political interests.

With structures as huge and expensive and important as Hoover/Boulder Dam, how could their creation possibly exist in a world beyond the realm of politics? Only if engineering projects were unimportant would it be possible for them to lack political significance. While this paper has sought to foster understanding of some ways that politics and dam safety became intertwined in the 1920s, it should not be thought that this issue represents an archaic relic of a bygone era.

ACKNOWLEDGMENTS

The author appreciates the support of Lafayette College for research underlying this paper. He also gratefully acknowledges his personal and professional involvement with Norris Hundley Jr. in researching the history of the St. Francis Dam.

REFERENCES

[1] For consistency, this paper will refer to "Hoover/Boulder Dam" as the structure that was commonly referenced as Boulder Canyon Dam or Boulder Dam during the 1920s. In 1930, the structure was designated Hoover Dam by Ray Lyman Wilbur (President Herbert Hoover's Secretary of the Interior) and construction contracts were issued under that name. In 1933, Harold Ickes (President Franklin Roosevelt's Secretary of the Interior and no great admirer of the prior president) decreed that the name Boulder Dam be used in place of Hoover Dam. In 1947, Congress enacted legislation, signed by President Harry Truman, formally designating the structure Hoover Dam. What further confuses this issue is that the dam was not built in Boulder Canyon, but in Black Canyon about 20 miles downstream.

[2] For more on the history of the St. Francis Dam disaster see Donald C. Jackson and Norris Hundley, Jr., "Privilege and Responsibility: William Mulholland and the St. Francis Dam ," *California History*, 82 (November 2004), pp. 8-47; and Charles Outland, *Man-Made Disaster: The Story of St. Francis Dam* (Glendale, Calif.: Arthur H. Clark Co., 1977).

[3] Mildred de Stanley, *The Salton Sea. Yesterday and Today* (Los Angeles, California: Triumph Press, c. 1966): 17-24; and Kevin Starr, *Material Dreams: Southern California through the 1920s* (New York: Oxford University Press, 1990): 20-44.

[4] Starr, *Material Dreams*, pp. 25-29; Beverly Moeller, *Phil Swing and Boulder Dam*, (Berkeley: University of California Press, 1971): 5.

[5] Starr, *Material Dreams*, pp. 34-6.

[6] Starr, *Material Dreams*, pp. 38-40; de Stanley, *The Salton Sea*, pp. 33-43.

[7] Moeller, *Phil Swing and Boulder Dam*, pp. 10-1.

[8] Norris Hundley, *The Great Thirst. Californians and Water, 1770-1990* (Berkeley: University of California Press, 1993): 204, 449-50.

[9] Moeller, *Phil Swing and Boulder Dam*, pp. 17-18; Paul Lincoln Kleinsorge, *Boulder Canyon Project: Historical and Economic Aspects* (Palo Alto: Stanford University Press, 1941): 76.

[10] Kleinsorge, *Boulder Canyon Project*, pp. 281-2.

[11] Bureau of Reclamation, "Report on the Problems of the Colorado River, Volume 5, Boulder Canyon: Investigations, Plans and Estimates," February 1924, p. 3; National Archives, Denver; Records of the Bureau of the Reclamation, RG 115; General Administration and Project Records, Project Files, 1919-1945; Entry 7; Box 477; Colorado River, Weymouth Report, 1924, part 5.

[12] Moeller, *Phil Swing and Boulder Dam*, p. 24.

[13] Charles A. Bissell, *The Metropolitan Water District of Southern California: History and First Annual Report* (Los Angeles: Metropolitan Water District of Southern California, 1939) p. 36.

[14] Bissell, *The Metropolitan Water District of Southern California: History and First Annual Report*, pp. 38-9.
[15] When 50-year leases governing use of Boulder Dam power were authorized in 1930, 64 percent of the dam's power was reserved for use in Southern California: 36 percent went to the Metropolitan Water District of Southern California, about 9 percent to the Southern California Edison Company and other private power companies and about 18 percent to the City of Los Angeles and other municipally-owned utilities. Arizona and Nevada were each allotted 18 percent of the dam's power. See Kleinsorge, *Boulder Canyon Project*, pp. 138-66.
[16] Catherine Mulholland, *William Mulholland and the Rise of Los Angeles*, (Berkeley: University of California Press, 2000): 171, 179.
[17] Mulholland, *William Mulholland*, pp. 280, 283-4.
[18] Mulholland, *William Mulholland*, pp. 301-5.
[19] Mulholland, *William Mulholland*, pp. 316-7.
[20] The use of engineering boards is described in U.S. Reclamation Service, *Third Annual Report of the Reclamation Service, 1903-04* (Washington D. C.: Government Printing Office, 1905), p. 41.
[21] J. L. Savage, "Design of the Owyhee Irrigation Dam," *Engineering-News Record* 100 (April 1928): 663-667. This article identifies the outside consultants as "Dr. F. L. Ransome, professor of economic geology at the California Institute of Technology, Dr. Warren D. Smith, professor of geology at the University of Oregon, and Kirk Bryan of the U.S. Geological Survey. Three consulting engineers, A .J. Wiley of Boise, D. C. Henny of Portland, and Charles D. Paul of Dayton, also passed on the site and reviewed the designs and estimates."
[22] As enacted in May 1917, the state's dam safety law is recorded in *California Statutes*, chap. 337, sec 2, (1917): 517-518.
[23] J. B. Lippincott, "William Mulholland—Engineer, Pioneer, Raconteur," *Civil Engineering* 2 (Feb.-March 1941): 106; Catherine Mulholland, *William Mulholland, p.* 28; Harvey A. Van Norman, "William Mulholland," *American Society of Civil Engineers Transactions* 101 (1936): 1605.
[24] Mulholland's original plan for the aqueduct system did not include a reservoir at the St. Francis site. Instead, large-scale water storage was contemplated for Long Valley north of Bishop. The price demanded for parts of the valley not owned by Los Angeles prompted Mulholland to look elsewhere. In the early 1920s he selected San Francisquito Canyon because of its proximity to the aqueduct right-of-way and because it was relatively near the city. No mention of a San Francisquito Canyon reservoir is made in Allan Kelly, *Historical Sketch of the Los Angeles Aqueduct* (Los Angeles: Times-Mirror Printing and Binding House, 1913) or in Los Angeles Department of Public Service, *Complete Report on the Construction of the Los Angeles Aqueduct* (Los Angeles: Department of Public Service of the City of Los Angeles, 1916).
[25] Kahrl, *Water and Power*, pp. 245-262, 311-312, 340-347; Hoffman, *Vision or Villainy*, pp. 172-173, 260-261; Los Angeles Department of Public Works, Bureau of the Los Angeles Aqueduct, *Sixth Annual Report of the Bureau of the Los Angeles Aqueduct to the Board of Public Works* (Los Angeles: Department of Public Works, 1911), 42.
[26] Joseph B. Lippincott to John R. Freeman, March 29, 1928, box 54, John R. Freeman Papers (MC51), Institute Archives and Special Collections, Massachusetts Institute of Technology, Cambridge, Mass.
[27] Early publications that discuss the effect of uplift include C. L. Harrison, "Provision for Uplift and Ice thrust in Masonry Dams," *Transactions of the American Society of Civil Engineers* 65 (1912): 142-225; and Charles E. Morrison and Orrin L. Brodie, *Masonry Dam Design Including High Masonry Dams* (New York: John Wiley, 1916), pp. 1-15.
[28] See Jackson and Hundley, *Privilege and Responsibility*, for more on the St. Francis Dam design and its development by Mulholland and the Bureau of Water Works and Supply
[29] "The Failure of a Concrete Dam at Austin, Pa. on Sept. 30, 1911," *Engineering News* 66 (Oct. 5, 1911): 419-422.
[30] John R. Freeman, "Some Thoughts Suggested by the Recent Austin Dam Failure Regarding Text Books on Hydraulic Engineering and Dam Design in General," *Engineering News* 66 (Oct. 19, 1911): 462-463.
[31] Arthur P. Davis to L. C. Hill, Oct. 11, 1912, entry 3, box 793, Rio Grande Project, General Administration and Projects 1902-1919, Record Group 115, Federal Records Center, Denver.

[32] The Service's close attention to the Elephant Butte foundation is documented in E. H. Baldwin, "Excavation for Foundation of Elephant Butte Dam," *Engineering News* 73 (Jan. 14, 1915): 49-55; Baldwin, "Grouting the Foundation of the Elephant Butte Dam," *Engineering News-Record* 78 (June 8, 1917): 625-628; Arthur. P. Davis, *Irrigation Works Constructed by the United States Government* (New York: John Wiley, 1917), pp. 243-245. In his book, Davis describes a "variety of precautions... adopted to prevent percolation under the [Elephant Butte] dam, and to relieve any upward pressure."

[33] Davis, *Irrigation Works*, pp. 117.

[34] Walter Ward, "Building Black Canyon Irrigation Dam in Western Idaho," *Engineering News-Record* 93 (Nov. 20, 1924): 818-823, 818.

[35] For detailed discussion of the Boulder/Hoover dam design see Bureau of Reclamation, *Boulder Dam: Part IV – Design and Construction*, (Denver: Bureau of Reclamation, 1938).

[36] Freeman, Davis and the Reclamation Service/Bureau were hardly alone in addressing the perils of uplift. In 1912, C. L. Harrison brought together the views of twenty engineers on the subject in "Provision for Uplift and Ice thrust in Masonry Dams," *Transactions of the American Society of Civil Engineers* (1912) pp. 142-225; Chester W. Smith, *The Construction of Masonry Dams* (New York: McGraw-Hill, 1915), pp. 100-109 included a ten-page section describing how cut-off trenches, foundation grouting, and drainage systems could ameliorate the effects of uplift; Charles E. Morrison and Orrin L. Brodie's *Masonry Dam Design Including High Masonry Dams* (New York: John Wiley, 1916): 1-16, also described several ways in which upward pressure may be countered by a foundation cut-off trench, adding a section to the profile, and providing drainage wells and inspection galleries. A year later, William Creager's *Masonry Dams* (New York: John Wiley, 1917) similarly emphasized the importance of countering the effect of uplift.

[37] Mulholland's placement of drainage wells only in the center section of St. Francis Dam did not reflect standard practice in California in the early 1920s for large concrete-gravity dams. For example, San Francisco began construction in 1919 on a major water supply-dam in the Sierra Nevada. The concrete curved gravity Hetch Hetchy Dam (later renamed O'Shaughnessy Dam) featured an extensive drainage system consisting of 1,600 porous concrete blocks and a cut-off trench running up both canyon walls. As detailed by *Engineering News-Record* in 1922, "the porous concrete blocks are placed in the bottom of the cut-off trench for its full length, and also in vertical tiers." See "Plant and Program on the Hetch Hetchy Dam," *Engineering News-Record* 89 (Sept. 21, 1922): 464-468, 467.

[38] In the 1960s, Charles Outland's examination of construction photographs revealed that the proportions/dimensions of the dam as indicated in drawings provided by the Department of Water and Power did not accord with the structure as-built. Outland also called attention to how the height of the dam had been raised at least ten feet after the start of construction and how this issue had never been addressed by investigators in the 1920s. Because this issue was not something addressed in the 1920s, it will not be discussed in this already very lengthy paper. But it need be noted that the raising of the dam (without any apparent widening of the base) only exacerbated the destabilizing effect of uplift. See Outland, *Man Made Disaster* (1977), 29-30, 230. Also see Jackson and Hundley, *Privilege and Responsibility*, pp. 22-23.

[39] Outland, *Man-Made Disaster* (1977), 30-36, 46, 49, and 51.

[40] C.E. and E.L. Grunsky, "St. Francis Dam Failure at Midnight, March 12-13, 1928," *Western Construction News* 3 (May 25, 1928): 314-320; Bailey Willis, "Report on the Geology of St. Francis Damsite, Los Angeles County, California," *Western Construction News* 3 (June 25, 1928): 409-413.

[41] Charles Lee, "Theories of the Cause and Sequence of Failure of the St. Francis Dam," *Western Construction News* 3 (June 25, 1928): 405-408. Although less prominent than the Grunskys' and Willis' reports, Halbert P. Gillette (the editor of *Water Works*) also criticized the west side failure hypothesis in testimony at the Coroner's Inquest and in "Three Unreliable Reports on the St. Francis Dam Failure," *Water Works* 67 (May 1928): 177-178. Also see Los Angeles County Coroner's Inquest, 750-753.

[42] J. David Rogers, "Reassessment of the St. Francis Dam Failure," in Bernard W. Pipkin and Richard J. Proctor, eds., *Engineering Geology Practice in Southern California* (Belmont, Calif.: Association of Engineering Geologists, Southern California Section, Special Publication No. 4, 1992): 639-666.

[43] C.E. and E. L. Grunsky, "St. Francis Dam Failure at Midnight, March 12-13, 1928," pp. 314-320; Willis, "Report on the Geology of St. Francis Damsite, Los Angeles County, California," pp. 409-413.

[44] Lee, "Theories of the Cause and Sequence of Failure of the St. Francis Dam," p. 408.

[45] The reservoir gauge ("automatic water-stage register") survived the collapse and appeared to indicate that, about 30 minutes before the collapse, a large flow of water began to drain out of the reservoir. As Lee pointed out, there was no evidence of this flow coming down the canyon in the half hour before the collapse; instead, the apparent drop in reservoir level was better explained as being "due to a slow rising of the dam" reflecting the effect of uplift. Lee, "Theories of the Cause and Sequence of Failure of the St. Francis Dam," p. 406.

[46] Other investigations were sponsored by the Los Angeles County Board of Supervisor and the Los Angeles Board of Water and Power Commissioners.

[47] The report of the City Council committee chaired by Elwood Mead was published in "The St. Francis Dam Failure," *New Reclamation Era*, (May 1928): 61-71. Others signing the committee's report include L. C. Hill, General Lansing Beach, D.C. Henny, and R.F. Walter (who was the Bureau of Reclamation's Chief Engineer).

[48] "All St. Francis Dam Reports Agree as to Reason for Failure." *Engineering News-Record* 100 (April 19, 1928): 639.

[49] *Report of the Commission Appointed by Governor C.C. Young*, p. 16.

[50] J. David Rogers estimates that the east side landslide encompassed 500,000 cubic yards of disintegrated schist. See J. David Roger, "'A Man, A Dam and a Disaster: Mulholland and the St Francis Dam, "*The St. Francis Dam Revisited* (Los Angeles and Ventura: Historical Society of Southern California and Ventura County Museum of History and Art, 1995: 53.

[51] *Report of the Commission Appointed by Governor C.C. Young*, p. 16.

[52] *Report of the Commission Appointed by Governor C.C. Young*, p. 16.

[53] "The St. Francis Dam Failure," *New Reclamation Era* (May 1928) p. 71.

[54] *Report of the Commission Appointed by Governor C.C. Young*, p. 18.

[55] See http://www.californiagovernors.ca.gov/h/documents/inaugural_26.html. At his inaugural Young commented that: "The prospects are very bright that the Congress at its present session will furnish the needed relief for the south by passing the bill for the dam at Boulder Canyon. California will certainly do all she can toward this end by making clear her attitude through representatives of this administration in Washington, as well as by legislative action which will fully safeguard the interests of the states comprising the basin of the upper Colorado River. I feel assured that this Legislature will also meet the acute need of the south for an adequate domestic water supply by authorizing the formation of a metropolitan water district such as may permanently solve her difficulties along this line."

[56] *Report of the Commission Appointed by Governor C.C. Young*, 16.

[57] *Report of the Commission Appointed by Governor C.C. Young*, 16.

[58] Wiley was invited by the Governor to join the Commission on Thursday March 15th and he immediately accepted, See telegram from C.C. Young to A.J. Wiley, March 15, 1928; and telegram from A.J. Wiley to C.C. Young, March 15, 1928; Edward Hyatt, "Memorandum St. Francis Dam," no date, circa March 26, 1928, St. Francis Dam file, Division of Safety of Dams, Sacramento, California.

[59] Quotes from Edward Hyatt, "Memorandum St, Francis Dam," undated, circa late March 1928, both in St. Francis Dam Disaster file, Division of Safety of Dams, Sacramento, California.

[60] Quotes from Edward Hyatt, "Memorandum St, Francis Dam," undated, circa late March 1928, both in St. Francis Dam Disaster file, Division of Safety of Dams, Sacramento, California.

[61] Telegram from Edward Hyatt to Phil D. Swing, March 26, 1928; St. Francis Dam Disaster file, Division of Safety of Dams, Sacramento, California.

[62] Telegram from C. C. Young to Phil D. Swing, March 27, 1928; St. Francis Dam Disaster file, Division of Safety of Dams, Sacramento, California.

[63] *Ibid.*

[64] The tenuous status of the Boulder Canyon Project's authorization in early 1928 is well described in Moeller, *Phil Swing*, pp. 107-112.

[65] *Wall Street Journal*, March 16, 1928, quoted in Moeller, *Phil Swing*, p. 111

[66] For more on the politics that lead to authorization of the Colorado River Board, see Moeller, *Phil Swing and Boulder Dam*, pp. 114-118.

[67] The manner in which the design could be considered as meeting the 30 ton per square foot criteria are described in Elwood Mead, "Memorandum to the Secretary re the Meeting of the Consulting Engineers to Approve Detail Plans of Boulder Dam," December 28, 1929; National Archives, Denver; Records of the Bureau of the Reclamation, RG 115; General Administration and Project Records, Project Files,

1919-1945; Entry 7; Box 490; Folder 301.1, Colorado River Project, Board & Engineering Reports on Construction Features, 1929.

[68] The acceptance by the Colorado River Board of the Bureau's rationale for not revising the design is acknowledged in U.S. Department of the Interior, *Boulder Canyon Project Final Reports, Part IV- Design and Construction* (Denver: Bureau of Reclamation, 1941), p. 25.

[69] For more on the "Power Trust" and its effect on passage of the Boulder Canyon Project Act see Donald J. Pisani, *Water and American Government: The Reclamation Bureau, National Water Policy, and the West, 1902-1935*, (Berkeley: University of California Press, 2002): 237-41, 261-63.

[70] Moeller, *Phil Swing*, pp. 119-122.

[71] "Los Angeles County Coroner's Inquest, p. 378.

[72] William Mulholland Still a Big Man," *Western Construction News* 3 (April 10, 1928): 223.

[73] John R. Freeman to Joseph B. Lippincott, April 4, 1928, Box 54, Freeman Papers, MIT

[74] John R. Freeman to Caleb Saville, March 29, 1928, Box 54, Freeman Papers, MIT

[75] For evidence of how he was lauded by city officials when he retired, see *The Intake* 5 (December 1928), pp. 1-8. This city-sponsored publication makes no mention of the St. Francis Dam disaster and its effect on Mulholland's career and reputation.

[76] Michael M. O'Shaughnessy to Edward Hyatt, State Engineer, October 3, 1928, "Supervision of Dams, 1928" folder, Public Utility Commission Records, California State Archives, Sacramento, California.

[77] *Report of the Commission Appointed by Governor C.C. Young*, 18.

[78] Edward Hyatt to M.R. McKall, April 7, 1928, "St. Francis Dam" file, Division of Safety of Dams., Sacramento, California.

[79] Edward Hyatt to M.R. McKall, April 7, 1928, "St. Francis Dam" file, Division of Safety of Dams., Sacramento, California.

[80] Norwood Silsbee [Secretary-Treasurer of the ASCE Sacramento Section] to C.B. Sadler[Secretary-Treasurer of the ASCE San Diego Section], June 19, 1928, Folder 25, Hiram N. Savage Papers, Water Resources Center Archives, University of California, Berkeley, California

[81] For example, see F.D. Howell to George Pollack, Hiram Savage and W.L. Huber, Sept 21, 1928, Folder 25, Savage Papers, WRCA, Berkeley

[82] W. L. Huber to George Pollack, Dec. 3, 1928; for more on use of consulting boards by the State Engineer see D. L. Bissell to H. N. Savage; both in Folder 25, Savage Papers, WRCA, Berkeley

[83] *Calif. Statutes*, chap. 766 (1929): 1505-1514.

[84] See Jackson, *Building the Ultimate Dam*, p. 241-242 for discussion of a 1931 state committee convened by Hyatt, and chaired by Huber, to study multiple arch dam technology.

[85] M. M. O'Shaughnessy, comments on "Public Supervision of Dams: A Symposium," *Transactions of the American Society of Civil Engineers* 98 (1933): 853.

[86] A. W. Markwart, "Recommendation for Legislation and Application of Law" in "Public Supervision of Dams: A Symposium," *Transactions of the American Society of Civil Engineers* 98 (1933): 828-635.

[87] Fred Noetzli, "An Improved Type of Multiple Arch Dam" *Transactions of the American Society of Civil Engineers* 87 (1924): 410; Comments by Fred Noetzli in "Public Supervision of Dams: A Symposium," p. 865.

[88] *Report of the Commission Appointed by Governor C.C. Young*, p. 18.

[89] R. B. Jansen, "Review of the Baldwin Hills Reservoir Failure," *Engineering Geology* 24 (December 1987): 7-81.

[90] N. J. Schnitter, The Evolution of the Arch Dam – Part Two," *Water Power and Dam Construction* 28 (November 1976: 19.

[91] Anyone interested in *this* subject is encouraged to read about the history of California's Littlerock Dam; see Jackson, *Building the Ultimate Dam*, pp. 198-205,

[92] "Sixth Report on St. Francis Dam Offers New Theories," *Engineering News-Record* 100 (June 7, 1928): 895. C. E. Grunsky and E. L. Grunsky, "The Grunsky Report on the Failure of the St. Francis Dam," *Engineering News-Record* 101 (July 26, 1928): 144.

[93] "Essential Facts Concerning the Failure of the St. Francis Dam: Report of the Committee of the Board of Direction," *Proceedings of the American Society of Civil Engineers* (October 1929): 2147-2163.

Building Blocks of Hoover Dam:
Technology, Politics, Economics

Brit Allan Storey[1], Ph.D

[1] Senior Historian, Mail Code: 84-53000, Bureau of Reclamation, P.O. Box 25007, Lakewood, Colorado 80225-0007; bstorey@usbr.gov

ABSTRACT: In 1902, when Congress passed the Reclamation Act, damming the Colorado River was a dream of a young engineer in the U. S. Geological Survey, Arthur Powell Davis, and water visionaries in California. Reclamation staff overcame unprecedented technological, political, and economic challenges to pave the way for Hoover Dam. This paper reviews some of the key issues, the accomplishments of Reclamation prior to and during construction of Hoover Dam, and the political and economic innovations associated with the dam.

INTRODUCTION

Founded in 1902, the U.S. Reclamation Service (Reclamation), renamed the Bureau of Reclamation (Reclamation) by the Secretary of the Interior in 1923, was the congressional response to increasing agitation in the West for Federal investment in irrigation projects. In addition, the U.S. Geological Survey (USGS) had identified a series of needs during its irrigation survey, 1888-1892, namely topographic mapping, planning, and hydrologic knowledge. According to John Wesley Powell's USGS annual report in 1889, the topographic study's function "is that of discovery, for it reveals the most important facts necessary for the planning of a system of irrigation works. . . ." The hydrologic knowledge sought was a long record of water supply variability, which would permit more accurate forecasts of future water supply. The planning knowledge sought resulted in USGS engineers exploring the West for possible agricultural water development projects—many of them later built by Reclamation.

Reclamation's first director, Charles Doolittle Walcott, served as the director of two organizations—the USGS and Reclamation. Frederick Haynes Newell, from the USGS's Division of Hydrography, served as the chief engineer of the new organization. Together Walcott and Newell acted quickly to hire a staff of some thirty 30 engineers as the core professional staff of Reclamation. These government engineers were to study and develop authorized water projects.

The result was a group of men who both expanded the edges of engineering knowledge and absorbed rapidly expanding civil engineering knowledge outside Reclamation as they tackled irrigation water development in the arid West. They began to examine dam design and construction techniques with an eye to some of the more challenging dam sites in their territory. The USGS irrigation survey had already identified many potential projects, and westerners, eager to see that water development occurred in their areas, had identified favored sites. By the time the Congress established Reclamation, westerners had identified virtually all possibly suitable major dam sites in the West.

Eight years after establishment, Reclamation topped out Shoshone Dam, later renamed Buffalo Bill Dam, near Cody, Wyoming, at 325 feet. For five years, this dam just outside the eastern boundary of Yellowstone National Park, reigned as the highest dam in the world. Then in 1915 Arrowrock Dam near Boise, Idaho, topped it by twenty-five feet and assumed the throne. In 1932, some seventeen years later, Arrowrock lost its crown to Owyhee Dam in eastern Oregon when that dam topped out at 417 feet, sixty-seven feet higher. Three years later Owyhee lost the title to Hoover Dam at 726 feet, and Hoover held the crown until deposed by Mauvoisin Dam in Switzerland in 1957—twenty-two years later. Each of these Reclamation dams in its turn caught the attention of the engineering world.

Each of these "highest" concrete dams as well as the other major dams Reclamation constructed in this period served as an evolutionary pathway for Reclamation engineering design and construction—a primer, if you will, for the construction of Hoover Dam. Pathfinder (1909), Theodore Roosevelt (1911), Lahontan (1915), Elephant Butte (1916), Tieton (1925), McKay (1927), Gibson (1929), and Deadwood (1931), each exceeded 150 feet in height. There were many other lesser dams in this period which joined the more recognized works. They also served as training grounds during which design of valves and gates, spillways, powerhouses, and the dams themselves pushed the edges of engineering knowledge and adjusted to the increasing demands of high water pressure; properties of concrete; differing foundations; spillway sizing and design; and myriad other details. For instance, Reclamation engineer O. H. Ensign, an electrical engineer, designed the Ensign Valve, several of which Reclamation installed on the upstream face of Theodore Roosevelt Dam. The Ensign Valve was the first step in Reclamation's design of the jet valves ultimately installed at Hoover and other Reclamation projects.

Parallel to development of new engineering skills and knowledge was the evolution of construction techniques and materials. For instance, in Reclamation's early days concrete was considered an expensive material, was generally mixed in small batches, was not well understood chemically, was somewhat unreliable in quality—though Reclamation sought to solve this problem through specifications and testing—and was made more expensive by railroad shipping costs combined with wagon freighting costs to reach Reclamation's often isolated construction sites. Reclamation's earliest large masonry structures included Pathfinder Dam and Theodore Roosevelt Dam—both of which Reclamation built using locally quarried stone for facing and as "plum" stones in the core of the dam to reduce the use of concrete. In fact, Reclamation found the raw materials needed for concrete in the vicinity of

FIG. 1: Plum stones in use at Buffalo Bill Dam (Shoshone Dam) to decrease the amount of concrete needed during construction. Date unknown. Bureau of Reclamation photograph.

Theodore Roosevelt and built a cement manufacturing plant on-site. Reclamation even shipped cement some thirty miles down its rugged construction road, the Apache Trail, on wagons behind mule teams to build various other facilities on the Salt River Project. At Buffalo Bill, construction of the cyclopean concrete structure introduced "plum" stones to reduce the quantity of concrete required.

FIG. 2: Derricks in use during construction of Theodore Roosevelt Dam on the Salt River Project in Arizona. December 1909. Bureau of Reclamation photograph.

These early construction projects relied heavily on human and animal muscle power combined with relatively simple cableway systems and derricks to move and place materials. On some projects, Reclamation could use steam-powered tractors, and steam-powered shovels. Cableways became more sophisticated over time, and Reclamation began to use construction railroads where feasible—as for instance, at Elephant Butte Dam, Belle Fourche Dam, Arrowrock Dam, and on the Yuma Project.

FIG. 3: A steam-powered tractor fording the Belle Fourche River to deliver a Dinky locomotive to the Belle Fourche damsite in March of 1906. Bureau of Reclamation photograph.

Hydroelectricity, from its beginnings on the Strawberry Valley Project in Utah, at Theodore Roosevelt Dam in Arizona, and at Minidoka Dam in Idaho was used to power machinery to ease construction burdens at Reclamation sites (Linenberger, 2002). Reclamation and contractors deemed electricity so useful that Reclamation even built petroleum and coal powered generating plants, as occurred at the Gunnison Tunnel on the Uncompahgre Project. Drills, pumps, small cement mixers, rock crushers, hoists, fans, and lights were among the useful tools used in tunnels, cableways, and dam construction.

FIG. 4: Interior of the powerhouse at the river portal of the Gunnison Tunnel in 1907. Bureau of Reclamation photograph.

Between 1902 and initiation of construction at Hoover some thirty years later, a virtually amazing transformation occurred in the technology of large-scale construction projects. Where gravel was obtained by hand on the Yakima Project in 1918, construction at Hoover relied on an automated gravel quarry and a railroad to deliver the materials to two automated concrete plants. The Hoover concrete plants used 4-cubic yard mixers which delivered concrete in 8-cubic yard buckets at the rate of 280 cubic

FIG. 5: Left--The upper concrete mixing plant at Hoover during construction. Date unknown. Right—A bucket used to deliver concrete at Hoover. Bureau of Reclamation photograph.

yards/hour. The buckets went to rail lines for delivery to five 20-ton temporary cableways serving the dam and the powerhouses. Fully operational, the concrete plants could deliver a total of 13,440 cubic yards of concrete each day.

FIG. 6: Obtaining sand and gravel on the Yakima Project in 1918. Bureau of Reclamation photograph.

Between 1902 and Hoover, materials transportation evolved from horses and mules hitched to wagons to dump trucks, and railroad use became more sophisticated. Gasoline engine powered equipment replaced the muscle powered equipment of such tools as Fresno Scrapers and horse drawn graders. Buses could carry workers to the jobsite more efficiently than the previous transportation.

FIG. 6: A "jitney bus" used to transport workers to the jobsite at Hoover Dam. Bureau of Reclamation photograph.

Other engineering problems tackled by Reclamation engineers and planners had to do with the concrete for Hoover. One issue was removing the heat of hydration from the masses of concrete necessary to build Hoover Dam. Left to itself, the heat of hydration in Hoover Dam would have expanded and contracted the body of the dam over a long period, and cracks would have developed. Instead of allowing that to occur, Reclamation engineers experimented in the labs and during the construction of Owyhee Dam (built 1928 to 1932) to develop a refrigeration system which would remove the heat of hydration thereby preventing untended expansion and contraction. In addition, Reclamation learned a great deal about the chemical interactions of cement, aggregates, and water in the years between 1902 and 1931, when construction on Hoover began. For instance, at the Belle Fourche Project in western South Dakota, built mostly 1905 to 1916, alkaline reactions resulted in replacement of much of the concrete work within ten to fifteen years of construction. So, by the time of Hoover's construction, Reclamation developed concrete specifications tailored to local aggregate and water supply conditions.

FIG. 7: Night picture of Hoover Dam under construction showing concrete forms and various catwalks, bridges, and ladders. October 18, 1934. Bureau of Reclamation photograph.

Reclamation learned a great deal from hydraulic modeling of the engineering issues at Hoover Dam. This modeling was accomplished at two primary locations—the laboratories at Colorado State University in Fort Collins where there were extensive tests and on the Uncompahgre Project in western Colorado where it was possible to develop high-head models of Hoover Dam engineering problems.

The state of hydrologic and hydraulic understanding was somewhat limited at the time of Hoover Dam's construction. Carl Hoffman, who worked on the design of the spillways, for instance, reminisced in February of 1995, that:

> Well, at the time that we worked up the scheme–let's say the scheme, not the design–on the spillway, and put it in the specifications, that was the time that we were making model testing. So, when we issued the specification, [Erdman B.] Debler told us we should have a capacity of the spillway of 250,000 cubic feet per second. I can't remember exactly. Later on he had a change of heart, and we didn't dispute him ever, as far as hydrology was concerned. . . . we designers didn't have a lot of information on foundation. Testing was done, always, but never in sufficient amounts. So we were really working on meager information. And so, . . . after specifications were written, as construction developed, why then of course there always were changes in concept, changes in actual conditions as they were encountered in the field, and so on. So you had to have changes. . . . you make modifications and changes based on conditions. So this is engineering, you see. . . .

Okay, so *this* is a [maximum] release that *could* happen at Hoover Dam. [But] at Hoover Dam, one way or another, Debler changed his mind. Finally decided that we have to have a spillway capacity of 400,000 cubic feet per second. So we designed it, and this is a potential release that can happen from Hoover Dam.

FIG. 8: Water flowing into the Arizona Spillway during high water in 1983. Bureau of Reclamation photograph.

However, all Reclamation projects depended on more than simply the engineering of the individual project. A wise former chief engineer of Reclamation once told me that all Reclamation projects are like a three-legged stool—that they rest on three legs:
- Technology—does the engineering technology exist;
- Politics—are the powers that be willing to support the project; and,
- Economics—are the powers that be willing to provide the money for the project.

The first leg of the stool, the engineering technology, did exist since Reclamation engineers built it up over the years from the beginning of the twentieth century. So,

Reclamation engineers resolved the technological aspects of the project through the expansion of their knowledge. The funding, however, did not exist because the Reclamation Act in 1902 provided that the Reclamation Fund would fund Reclamation projects. That fund, in turn, was created in the U.S. Treasury by the deposit of revenues from the sale of public lands. Construction expenditures on Reclamation's earliest projects depleted the Reclamation Fund, and by the time Hoover Dam became a realistic possibility there was no money available for such a massive project. This in spite of the fact that the Congress had added some mineral revenues to the sources of funds.

In addition, it took some twenty years to develop sufficient political support to build Hoover Dam. Arthur Powell Davis, Reclamation's chief engineer after Frederick Haynes Newell became director in 1907, had long supported development of the Colorado River, and there were numerous studies of the Colorado during the 1900s, 1910s, and 1920s.

Political will stumbled over many issues, but there were two primary ones besides money. Since the West adopted prior appropriation for divvying up western water, it was uncertain how the Colorado River Basin states were going to lay claim to the waters of that river. The other primary issue was the debate over public power.

Prior appropriation developed deep, thirsty roots in the West beginning in the California Gold Rush when the doctrine began to develop. The various versions of the doctrine of prior appropriation stated that the first person to use water was entitled to continue to use water in the same amount over time in perpetuity at the same location for beneficial use, and that water right took precedence over any water right established later. Variations in the details of prior appropriation doctrine developed around the West.

Interestingly, a very large percentage of the water in the Colorado, Laramie/North Platte, South Platte, Republican, Arkansas, Rio Grande, and San Juan Rivers originated in Colorado. Over time, Colorado developed the view that it, as the origin or source of the water, could claim as much of the water it originated as it could use—regardless of any uses for water that people in downstream states might claim. This doctrine evolved variations as some accepted having to protect downstream senior water rights and others mixed in the issue of "states rights." The Colorado River's water was a source of great concern to the states in its basin, and the other states began to fear that California interests might build a project or projects to establish claims appropriating the entire balance of that river's water.

The U.S. Supreme Court's 1922 decision in *Wyoming v. Colorado*, a case initiated in 1911, did not put the basin states' fears to rest. In this case, Wyoming argued it had the right to appropriate water in the Laramie River, a tributary of the North Platte that largely originated in a vast intermountain valley called North Park in Colorado. Colorado, on the other hand, argued that since the water originated in Colorado, it had the right to appropriate all it wanted. In fact, Colorado had its eye on transferring a big part of that water in a transmountain diversion to the agriculturally rich northern Colorado Front Range. The Supreme Court did not uphold Colorado's argument that it alone controlled water originating within the state's boundaries. Nor did it uphold Wyoming's argument that Colorado could not transfer water from one drainage basin to another. What the court did say was that the concept of prior appropriation applied across state borders to determine water rights–where states endorsed the principle of prior appropriation. Both Colorado and Wyoming used prior appropriation doctrine in their water law.

So, the case of *Wyoming v. Colorado* suddenly gave strength to the specter that California interests could legally claim the vast majority of the water of the Colorado River under the doctrine of prior appropriation—thus blocking development using Colorado River water in the other six Colorado River Basin states. In fact, a Colorado water attorney, Delphus E. Carpenter (Tyler, 2003), was involved not only in *Wyoming v. Colorado* beginning in 1911, but also in other interstate disputes between Colorado and Nebraska on the South Platte River and Colorado and Kansas on the Republican River. He early began to sense that Colorado might not win its arguments against the claims of the other states and began to look toward a way of dividing the flows of Colorado's rivers with downstream states. Ultimately, the solution was identified in a 1921 statute of Congress which authorized the Colorado River Basin states to "negotiate and enter into a compact or agreement not later than January 1, 1923, providing for an equitable division and apportionment among the said States of the water supply of the Colorado River and the streams tributary thereto . . ." (42 U. S. Statutes at Large) The idea of a negotiated compact was a fundamental departure from the West's established legal principles regarding prior appropriation.

Ultimately, Secretary of Commerce Herbert Hoover represented the United States and chaired the meetings, which led to the Colorado River Compact (Hundley, 1975; Terrell, 1965). Discussions were held at the Bishop's Lodge north of Santa Fe, and each state's representative and Herbert Hoover signed the resulting compact on November 24, 1922, in the historic Palace of the Governors on the square in Santa Fe. Because of a dispute between California and Arizona over the meaning of the compact relative to water in the Colorado's tributaries, ratification of the compact by the Congress occurred only on December 21, 1928, when the Congress determined that the compact could go into effect upon ratification by six of the seven Colorado River Basin states. The Congress in the Boulder Canyon Project Act, which also authorized construction of Hoover Dam and the All-American Canal, ratified the compact. The Supreme Court settled the disagreement between Arizona and California only in 1964 when the Supreme Court ruled in *Arizona v. California*, a case brought soon after Arizona finally ratified the compact in 1944. While the Colorado River Compact was the first interstate river compact signed in the West, it was not ratified in Congress until after the South Platte River Compact became the first western water compact ratified by the Congress in 1923.

Before authorization of Hoover Dam, however, it was necessary for the Congress to work its way through the issue of public power versus private power. At the time, this was highly controversial, and there was very strong opposition to the Congress approving a public power project that would compete with private power companies. Sarah Elkind has skillfully demonstrated the compromise approach used by the Congress to settle the issue (Elkind, 2008). Hoover Dam's powerhouse was to be built and owned by the U.S. Government, but the generating units would be leased for 50 years to private power companies, which exercised responsibility for operation and maintenance and owned the power. That arrangement ended in 1987 at the end of the fifty-year contracts with the power companies. Hoover is now operated by Reclamation under a thirty-year contract, and there is a working arrangement in which power customers fund operating expenses up-front and cooperate with Reclamation and the Western Area Power Administration, the marketer of the Federal power from Hoover, in long term planning for the operation and maintenance of the facility.

With funding as the big issue remaining, the Congress went to the U.S. Treasury to appropriate money for construction at Hoover Dam. The Congress had used this method in the past, but those appropriations were set up as loans to be repaid by the Reclamation Fund. In this case, the Congress specified that contracts for the sale of power had to repay construction costs at Hoover Dam. In effect, this was a tacit admission that the funding mechanism originally established for the reclamation program was insufficient to the needs and that irrigation projects, for the most part, did not do well at repayment. Congress here set the precedent of using electricity sales revenue, often called the cash register of Reclamation projects, to finance projects.

Hoover Dam met the needs of many diverse parties when built. The authorization recognized that Hoover Dam was to deliver water for irrigation, municipal, and industrial use. The primary purposes were outlined as: controlling floods; improving navigation; regulating the flow of the Colorado River; providing storage and water delivery for irrigation and other beneficial uses; and ". . . generation of electrical energy as a means of making the project herein authorized a self-supporting and financially solvent undertaking . . ." To the Imperial Valley and the Palo Verde Irrigation District, major California appropriators on the lower Colorado, the control of floods in the capricious lower Colorado River was very important in protecting their rich agricultural lands. In California, a steady supply of water to the Metropolitan Water District of Southern California pumping plant on Lake was very important. To the Lower Colorado River Basin states, the compact assured the capture and delivery of water to California and stored water for Nevada and Arizona when they could exercise their rights under the Colorado River Compact. To the other six basin states, the agreement to divide the waters of the river and assign quotas to each of the states in development of the basin's water supply was of great importance since it meant that California could not use the doctrine of prior appropriation to claim the lion's share of the waters in the river. To the Upper Colorado River Basin states the Colorado River Compact assured they could later develop Colorado River water resources using, among others, the Colorado-Big Thompson Project (1937), the Colorado River Storage Project Act (1956), the San Juan-Chama Project (1962), and the Colorado River Basin Projects Act (1968). In addition, Hoover Dam assured that the United States would be able to meet the 1944 treaty commitments to Mexico, to provide 1.5 million acre feet a year of water south of the international border.

CONCLUSIONS

Altogether, the technical, political, and economic mix necessary for a successful project in the Black Canyon of the Colorado River took about thirty years. During that time, Reclamation engineers developed the necessary technical skills, construction techniques evolved rapidly with the introduction of motorized transport and more sophisticated equipment, and politicians worked through the political and funding issues that stood in the way of the project.

The Great Depression loomed suddenly as construction planning proceeded, and Americans, and particularly American politicians, quickly understood that Hoover Dam was an examplar of a large job-creating project. That boosted Reclamation's program and resulted in many new projects funded from the U.S. Treasury, including three other of Reclamation's largest projects: the Colorado-Big Thompson Project, the Central Valley Project, and the Columbia Basin Project—those three projects plus the All-American Canal, out of over 180 Reclamation projects, irrigate over thirty percent of Reclamation acreage. Coming at the beginning of the Great Depression, Hoover was, for Reclamation, a fantastic public relations coincidence that boosted its program immeasurably—it was a grand and internationally lauded construction project that put thousands of Americans to work. And, it established many precedents that carried over into subsequent authorizations, appropriations, and designs of Reclamation projects.

REFERENCES

Billington, David P., with Donald C. Jackson. *Big Dams of the New Deal Era: A Confluence of Engineering and Politics.* Norman: University of Oklahoma Press, 2006.

Billington, David P., with Donald C. Jackson and Martin V. Melosi. *The History of Large Federal Dams: Planning, Design, and Construction.* Denver, Colorado: Bureau of Reclamation, 2005. Available from the Government Printing Office.

Elkind, Sarah S., 2008. "Private Power at Boulder Dam: Utilities, Government Power, and Political Realism," in Department of the Interior, Bureau of Reclamation, *The Bureau of Reclamation: History Essays from the Centennial Symposium,* Volume II, Bureau of Reclamation, Denver, CO: 447-65.

Hundley, Norris Jr, 1975. *Water and the West: The Colorado River Compact and the Politics of Water in the American West*, University of California Press, Berkeley and Los Angeles, CA.

Linenberger, Toni Rae, 2002. *Dams, Dynamos, and Development: The Bureau of Reclamation's Power Program and Electrification of the West*, GPO, Washington, D.C.

Terrell, John Upton, 1965. *War for the Colorado River*, 2 volumes, Arthur H. Clark Company, Glendale, CA.

Tyler, Daniel, 2003, *Silver Fox of the Rockies: Delphus E. Carpenter and Western Water Compacts*, University of Oklahoma Press, Norman OK.

42 U.S. Statutes at Large, Part 1, pp. 171-2. An act of August 19, 1921.

THE NEW TOWN OF BOULDER CITY: City Planning and Infrastructure Engineering for Hoover Dam Workers

Jerry R. Rogers[1], Ph. D., P.E., D. WRE, Dist. M. ASCE

[1] Department of Civil & Environmental Engineering, University of Houston, Houston, TX 77204; (rogers.jerry@att.net)

ABSTRACT: In 1930, the Bureau of Reclamation selected Saco Reink DeBoer as the city planner for the company town of Boulder City for the residences of workers on Hoover (Boulder) Dam. Over 1,500 buildings were constructed to house the 5,000 plus dam workers, making Boulder City the third largest city in Nevada. DeBoer gained experience with the Denver Parks and Parkway System and other towns throughout Colorado, New Mexico, and Wyoming. The Radburn, New Jersey superblocks around central park spaces were utilized in the Boulder City plan. In American city planning history, Boulder City was the first fully developed new town during the Community Planning Movement.

Named after the dam general superintendent in 1981, the Frank T. Crowe Memorial Park was a large neighborhood park. The civic center of public buildings was the center of a triangular-shaped city plan with the Bureau of Reclamation administration building at the apex. The engineering basic support infrastructure of water, sewer, electricity and roadways was constructed. In 1931, the water for Boulder City came six miles from the Colorado River in a twelve-inch water main pumped up 2,000 feet in elevation. A pre-sedimentation basin, filtration system, water softening, and chlorination were needed. Four pumping stations and five water storage tanks were required. The town was administered by city manager, Sims Ely, who reported only to the Construction Engineer for the dam: Walker R. Young.

The winning construction was by Six Companies Inc. Six Companies Inc. set up a separate subsidiary: the Boulder City Company to complete workers' lodging, boarding, recreation, transportation to and from the dam and other services. Six Companies Inc. agreed to finance the operation of the grade school and pay the twelve teachers' salaries with Congress paying $70,000 for the school building.

INTRODUCTION

The Bureau of Reclamation (Reclamation) studied areas around the dam site for soil and climatic conditions for living quarters for dam workers. Reclamation engineer Walker R. Young (Boulder City History, 2009) selected a town location on a high plateau seven miles southwest of the dam (that was 10 degrees cooler than the dam site). The city site was at an elevation of 2,500 feet above sea level and about 1,200 feet higher than the final crest of Hoover Dam. Reclamation hired General

Construction that sub-contracted work to R. G. LeTourneau Inc. on the asphalt-surfaced highway 22 feet wide (to withstand 50-ton trucks and 150-passenger transports) built from the dam site to Boulder City. The Union Pacific Railroad was extended by a spur of 22.7 miles to Boulder City with 10,000 spectators at the Silver Spike Ceremony September 17, 1930. Reclamation selected Lewis Construction Company to build the railroad spur of seven parallel tracks for steady materials shipments from Boulder City going 10.5 miles to the Black Canyon rim, extended by the contractor to the bottom of Black Canyon. About 300 railcars of materials were hauled daily to the dam site at the peak.

TOWN PLANNING AND INFRASTRUCTURE ENGINEERING

In 1930, the Bureau of Reclamation hired Saco Reink DeBoer as the city planner for the company town of Boulder City for the residences of Hoover Dam workers. DeBoer gained experience with the Denver Parks and Parkway System and other towns throughout Colorado, New Mexico, and Wyoming. Other than the initial Washington, D.C. layout, Boulder City was among the first federal efforts to plan and construct a new town in U.S. history.

Boulder City streets were laid out and paved in a triangular arrangement with the apex pointing north where the Bureau of Reclamation Administration Building (now Lower Colorado Regional Office) was built (see FIG. 1). The seven main streets were named after the seven states (Arizona, California, Colorado, the main street: Nevada also known as Boulder Highway, New Mexico, Utah, and Wyoming) drained by the Colorado River.

FIG. 1. Lower Colorado Regional Office (From Nevada Way - Park Street)(Formerly Bureau of Reclamation Administration Building)original photo

The Radburn, New Jersey superblocks were utilized in the Boulder City plan. Wilbur Weed, an Oregon landscape architect, was hired to select plant species from Phoenix, Palm Springs, and Hawthorne, California and to design and field supervise all landscaping. In 1932, parks and lawns were planted after eleven carloads of trees (9,000 from California and Nevada nurseries (Stevens, 1988)) and shrubs were delivered (Rocca, 2001, 2007). The Frank T. Crowe (Hoover Dam General Superintendent) Memorial Park was a large neighborhood park, and was dedicated in 1981 by Boulder City at its 50th Anniversary (see FIG. 2).

FIG. 2. Frank T. Crowe Memorial Park (Statutes with Hard Hat Used at the Dam and Bust of Frank T. Crowe)original photo

Two statutes and plaques, one of a hard hat (said to be first used in Hoover Dam construction) and one of the bust of Frank T. Crowe (Hon. M. ASCE 1943) are inside the park entrance. Crowe had gained valuable construction experience with the Bureau of Reclamation from 1905-1925 when he left to work on dams built by Morrison-Knudsen in Boise before his selection by Six Companies Inc. to be General Superintendent of Hoover Dam. One Boulder City park plaque has: "The Old Man,"

"Frank Trenholm Crowe, Born October 12, 1882, Died February 26, 1946, World's Outstanding Builder of Dams." Crowe worked and supervised construction of nineteen dams from 1905-1946. One plaque had the American Society of Civil Engineers shield and a quote from Frank T. Crowe: "We had 5,000 men in a 4,000 foot canyon. The problem was to set up the right sequence of jobs so they wouldn't kill each other off." Rocca (2007) wrote the extended sentence: "We had 5,000 men in a 4,000 foot canyon. The problem, which was a problem in materials flow, was to set up the right sequence of jobs so they wouldn't kill each other off." "By late summer, 1932....., the New Mexico Construction Company laid out and hooked up the Boulder City sewer and water lines and paved nearly twenty miles of streets" (Stevens, 1988).

Six Companies Inc. set up a separate subsidiary: the Boulder City Company, for providing lodging, commerce, recreation, fire protection, transportation to and from the dam, a laundry service, and boarding (McBride, 1992). Boulder City grew within a few years to 1,000 homes (Reclamation engineers developed standard floor plans for five houses), 12 air-cooled and heated dormitories each with 172 rooms, 4 churches, a 700-seat free air-conditioned theater (the Boulder Theater), an elementary school (with operation/salaries financed by Six Companies Inc.), shops and stores in a Southwest styled shopping plaza (predecessor of today's shopping plazas), garages, restaurants, and the Boulder City Municipal Building (city manager, library and police).

A water system was built and Colorado River water was piped to Boulder City. The water, from twin intakes just below the Nevada diversion tunnel outlets (Stevens, 1988), averaged 6,000 ppm silt and was clarified in settling works a hundred feet above the river, pumped 1,800 feet up to a Boulder City plant that treated the water with hardness of 350 ppm by filtering and chlorination, before pumping 200 feet more up to a 2,000,000 gallon distribution tank on a north hill. Four pumping plants and five water tanks were required. About 200,000 kwh of electricity for Boulder City and the dam came from a 232-mile transmission line operated at 88,000 volts from San Bernardino, California and the Nevada-California Power Company.

Will Rogers performed nightly to packed houses at the Boulder Theater (see FIG. 3). At lunch one day, Will Rogers asked Hoover Dam General Superintendent Frank T. Crowe how many men he had working. Crowe replied about half of them. Will Rogers stated: "Mr. Crowe, I'm supposed to be the comedian."

At the peak of construction, there were 5,000 plus people living in Boulder City from all states of average age 32 with 40% unmarried. Boulder City was the third largest city in Nevada. Today, Boulder City has over 16,000 residents.

There were 5,250 men at construction peak. The workmen ate at a mess hall for 1,300 workers per shift, seven times a day operated by Anderson Brothers Supply Co. Single men were charged $1.60 per day for meals, rooms, and large bus transportation to and from the dam site. Married men rented unfurnished houses for $15 to $50 per month. Six Companies Inc. housing totaled 660 detached dwellings. Private citizens granted land leases could build for a ground rental payable to the U.S. government. Federal Rangers enforced laws.

FIG. 3. Boulder Theater (Free Air-conditioned Theater for Movies for Dam Workers)original photo

OTHER DEVELOPMENTS

In 1931, the Boulder City Hospital (now Sisters of Charity Retreat) was built by the contractor. Also, Ida Browder's Café was completed and is the oldest remaining commercial building. The Boulder Dam Hotel was built in 1933 by Paul "Jim" Webb (McBride, 1992) as a southern colonial-style building. Later, river boating trips from the Boulder Dam Hotel were led by Georgie White. The boating people were called "River Rats."

On September 15, 1935, the dedication of Hoover Dam by President Franklin D. Roosevelt was attended by 20,000 people. (The American Society of Civil Engineers will celebrate the 75th Anniversary of Hoover Dam October 20-22, 2010 with a Hoover Dam History Symposium/Tour.)

In 1956, an annual reunion of dam workers and their families was called: "The 31ers" and held at the Boulder City Hotel (Ferrence, 2008). The hotel is administered by the Boulder City Museum and Historical Association, including the Boulder City Hoover Dam Museum (see FIG. 4). In Boulder City, the 62nd Annual Damboree with Fourth of July Parade is scheduled in 2010.

FIG. 4. Boulder Dam Hotel (Renovated with the Boulder City Dam Museum Inside)original photo

In 1939, the Uptown Hardware Building (with second floor apartments) was finished. In 1939-1940 the Southern California Edison Company built houses for workers, and the Los Angeles Bureau of Power and Light (later the L.A. Department of Water and Power) Building was completed. The Nevada Drug Building (with prefabricated elements) was completed in 1941.

FROM BUREAU OF RECLAMATION TO AN INCORPORATED CITY

The Boulder City Act of 1958 gave the Secretary of Interior the authority to transfer Boulder City. Citizens overwhelmingly approved a charter for incorporation effective October 1959. In January 1960, ownership was moved to the Boulder City government of 33 square miles of land; municipal electric, water, and sewer systems; municipal buildings; streets, sidewalks and curbs; parks and parkways, and other property and equipment. Only the Lower Colorado Regional Office and Hoover Dam operation and maintenance facilities were kept by the Bureau of Reclamation.

BOULDER CITY OFFICES AND OFFICIALS

Early principal Bureau of Reclamation officials for the Boulder City office were: City Manager: Sims Ely (who reported only to the Construction Engineer); Construction Engineer: Walker Rolla Young (M. ASCE) (later Bureau of Reclamation Chief Engineer from 1945-1948); Office Engineer: John C. Page (Hon. M. ASCE 1953) (later Bureau of Reclamation Commissioner from 1936-1943); Field Engineer: Ralph Lowry (M. ASCE) who replaced Walker R. Young as Construction Engineer by 1935-1936; Chief Clerk: E.R. Mills, and Chief Counsel: J.R. Alexander.

For the initial Six Companies Inc. Boulder City office, leaders were: General Superintendent: Frank T. Crowe (Hon. M. ASCE 1943) (able to complete Hoover Dam 22 months ahead of schedule by three continuous shifts of workers, cableway concrete delivery, and lighting at the dam site.); Assistant Superintendent: Benjamin Franklin (B.F.) Williams (M. ASCE); Chief Engineer: Augustine Haines Ayers (M. ASCE); and Administration Manager: J.F. Reis. At the Lodge Street dead-end in Boulder City, the Six Companies built a two-story Executive Lodge and nearby a Spanish Hacienda for Crowe and his family. For an interesting summary of the careers and personalities of Crowe and Williams, please review: Fredrich (1989), *Sons of Martha*. ASCE Press: "Never My Belly to a Desk," "Who the Hell Does He Think He Is?" For stories and details about the dam working conditions and nearby living conditions and some names, please review: (Allen, 1983). For an extensive database of Hoover Dam construction workers and Boulder City families (over 20,000 names), please review: (Irons, 2008).

For the Babcock & Wilcox Company Boulder City office, the Project Superintendent was B.T. Kehoe.

The Lake Mead National Recreation Area has headquarters in Boulder City. Tourism to Hoover Dam (Boulder Dam) and the reservoir became so popular during and after the dam construction that tourism is still a major attraction to the area.

SUMMARY AND CONCLUSIONS

Engineering by the Bureau of Reclamation was completed for Boulder City for the infrastructure of water, sewer, streets, and power for a new town development to house the workers for Hoover Dam. Saco Reink DeBoer was the experienced city planner for Boulder City with a triangular arrangement of streets and superblocks. Wilbur Weed was the landscape architect. Sims Ely was the City Manager overseeing public works, issuing private building permits, collecting revenues, and supervising law enforcement. In 1981, a municipal park in Boulder City was named in the honor of Frank T. Crowe (Hon. M. ASCE 1943), the Six Companies Inc. General Superintendent. The single workers for the Six Companies Inc. lived in air-cooled and heated dormitories, and married workers leased houses/cottages. Administration buildings were planned for the Bureau of Reclamation, power and light companies, and others who built Hoover Dam. The successful new town planning for Boulder City led the way in early American city planning.

ACKNOWLEDGEMENTS

The author thanks the Boulder City Chamber of Commerce, USBR Lower Colorado Regional Office, ASCE's Carol Reese (Aff. M. ASCE) and Donna Dickert (A.M. ASCE), review by Don W. Rogers (M. ASCE) and Thomas C. Piechota (M. ASCE), computer assistance from Donna Rogers, and the following for hosting/participating in the Hoover Dam 75th Anniversary History Symposium/Tour/ASCE Proceedings (October 20-22, 2010): Richard L. Wiltshire (F. ASCE) Symposium Chair and David R. Gilbert (F. ASCE), ASCE History &

Heritage Committee, the ASCE Annual Conference, EWRI History & Heritage Committee, and the Nevada Section-ASCE and Southern Nevada Branch-ASCE.

REFERENCES

Allen, Marion V. (1983). *Hoover Dam & Boulder City*, House of Steno Inc., Redding, CA.

Boulder City Chamber of Commerce (undated). Paper: "BOULDER CITY HISTORIC DISTRICT," 465 Nevada Way, Boulder City, NV 89005 (www.bouldercitychamber.com).

Boulder City Chamber of Commerce (undated). Brochure: "Boulder City: The Town That Built Hoover Dam," 465 Nevada Way, Boulder City, NV 89005 (www.bouldercitychamber.com).

Boulder City History (September 29, 2009). (www.bcnv.org/History.asp).

Ferrence, Cheryl (2008). *AROUND BOULDER CITY: Images of America*, Boulder City Museum and Historical Association, Arcadia Publishing.

Fredrich, Augustine J., (1989). *Sons of Martha*, ASCE Press, "Never My Belly to a Desk," "Who the Hell Does He Think He Is?"

Irons, Judith Sattler (2008). *Construction Workers and Pioneer Boulder City, Nevada Families: 1929 to June 1936*, CD, Hoover Dam Store.

McBride, Dennis (1992). *In the Beginning...A History of Boulder City, Nevada*, Boulder City Hotel/Hoover Dam Museum.

Rocca, Al M. (2007). *AMERICA'S MASTER DAM BUILDER: The Engineering Genius of Frank T. Crowe*, Renown Publishing.

Stevens, Joseph E. (1988). *Hoover Dam: An American Adventure*, University of Oklahoma Press.

U.S. Bureau of Reclamation (1976). *Construction of Hoover Dam*, Thirty-third Printing, KC Publications, Box 94558, Las Vegas, NV 89193.

U.S. Department of Interior (1966). *The Story of Hoover Dam.*

U.S. Department of the Interior (January 2006). *Reclamation- Managing Water in the West: Hoover Dam*, Bureau of Reclamation, Lower Colorado Region.

Reclamation, the Army, and Hoover Dam During World War II

Jim Bailey, M.A., Ph.D.[1]

[1] Historian, Bureau of Reclamation, Technical Service Center, 86-68010, PO Box 25007, Denver, CO 80225-0007; jbailey@usbr.gov

ABSTRACT: Based largely on research into the federal archival record, this paper, an abridged version of the original study, examines the efforts by Reclamation and the Army—who did not always agree—in protecting Hoover Dam, with major events of World War II woven into the narrative to provide global context. This interpretive structure will illustrate how defense priorities shifted as the war evolved, how the Army assigned higher priorities to the U.S. Army Corps of Engineers' hydropower dams than Reclamation's, and how the Army sent its worst soldiers to the desert to protect Hoover Dam. Also mentioned is the sole material remnant of this protection effort, a fortified machine gun bunker that sits above the dam. It not only serves as a symbol of the war effort, it is soon slated to become a historic interpretive site.

INTRODUCTION

In the late 1930s, Americans sensed an ominous, yet vague global threat to their liberties and livelihoods. Although Adolf Hitler's Nazi Germany was aggressively expanding its territory throughout Europe, this threat seemed distant to a country focused on pulling itself out of one of its worst economic depressions in history. But on the clear, brisk Hawaiian morning of December 7, 1941, another threat erased any vague feelings. A surprise air attack by Imperial Japanese forces on Pearl Harbor Naval Base and Hickam Army Air Base, in which nearly 3,000 sailors, soldiers, and civilians died, served as a morbid wake-up call to Americans of this real threat to their domestic security. The attack not only rallied Americans against a tangible enemy, it exposed vulnerabilities in America's critical infrastructure, especially factories, airports, highways, canals, and harbors located in states adjacent or near the Pacific Ocean.

In the wake of Pearl Harbor and America's formal entry into World War II soon after, the defense of America's hydropower infrastructure from sabotage and attack became top priority. As west coast industrial plants ramped up to produce weapons, airplanes, tanks, ships, and other military products for America and its allies, War Department officials realized that although located well inland, the hydroelectric dams and powerhouses that supplied this electricity to major wartime contractors rated protection from potential sabotage and attack.

Among the major western hydroelectric dams that produced most of the cheap electric power needed to fuel wartime industries, and gained the most attention from officials bent on protecting this power, was the Bonneville Power Administration's dams on the Columbia River east of Portland, Oregon, and the Bureau of Reclamation's Grand Coulee Dam on the Columbia River in eastern Washington and Hoover Dam on the Colorado River near Las Vegas. While the Bonneville and Grand Coulee facilities produced most of the electricity consumed by defense plants in the Pacific Northwest, Hoover's wires transmitted nearly half of the electricity consumed by southern California war industries.

Of these high-profile federal hydropower facilities, none captured the imaginations of downtrodden Americans during the Great Depression like Hoover Dam. Over the past seven decades countless historians and scribes have chronicled the human narratives behind one of America's most ambitious public works projects. Electricity and water controlled by the 726–foot-high curved, thick-arch concrete dam not only sparked the enormous growth of southern California's urban areas, it helped develop the state's industrial and agricultural base, and allowed southern Nevada to gain a dependable water supply in Hoover's reservoir Lake Mead—the world's largest man-made body of water (Stevens 1988).

Yet at stake during World War II was protecting Hoover's electrical output. Without this, southern California's aircraft and shipbuilding industries would have been severely hampered. While the stories of Hoover's planning and construction are abundant to the point of saturation, nothing had been written on how Reclamation protected this facility during the war until recently, when Reclamation historian Christine Pfaff forged new narrative paths with an article in the National Archives' quarterly journal of history *Prologue* (Pfaff 2003). For this article, Pfaff examined confidential documents in Reclamation files at the National Archives in Denver and analyzed the policy issues and decisions by the Department of the Interior, Federal Bureau of Investigation, and other agencies as they attempted to devise plausible plans to assess and protect Hoover and its powerhouse from enemy attack, real or perceived.

The value of Pfaff's article is that it establishes the broader wartime policy context behind Hoover's protection. Moreover, her article provides a roadmap upon which researchers can glean compelling narratives underlying the multifaceted dimensions of infrastructure protection. Yet the Army's story was only minimally detailed. For the study this paper evolved from, recently declassified and opened War Department documents at the National Archives in College Park (a location she did not visit for her initial research) provided new, first-hand information regarding the human issues and problems both Interior and War faced while assessing this threat, then protecting the power, before and after Pearl Harbor.

PRE-PEARL ANXIETIES: ASSESSING THE THREAT

Completed in 1936, Hoover not only stood as the world's tallest dam, it was America's most ambitious and challenging public works project to date. Designed as a multipurpose facility, it provides flood control, hydroelectric power, municipal and irrigation water, and water recreational opportunities for southern California and

Nevada. One of America's top tourist attractions, every year hundreds of thousands of visitors from across the globe take time to tour the dam and its powerhouse.

Yet Hoover's concrete had barely set when rumblings of conflict and possible war in Europe and Asia began to unfold. As Adolf Hitler expanded his empire—and the Japanese escalated their campaigns in China and the western Pacific—the federal government realized hydroelectric dams were key infrastructural components that rated protection. By mid-1939, in the wake of Nazi invasions of Czechoslovakia and Poland, Reclamation Commissioner Harry Bashore noted that given the likelihood of war in Europe, there was the possibility that Hoover Dam and its power plant could be closed to the public, and that additional help might be necessary to protect the facility from possible saboteurs.

By November 1939, it became evident that sabotage of the dam was a distinct possibility. That month, a report of possible sabotage reconnaissance by German agents living in Las Vegas heightened security awareness. Filtered through the State Department's Embassy in Mexico City, this report noted that two men, one of them a explosives expert, allegedly planned to sabotage the dam's intake towers by chartering a fishing boat, attaching two heavy bombs to the towers, then detonating the bombs. Their objective: cripple southern California's manufacturing industries by halting the electric power generated by Hoover's two powerhouses.

A couple weeks later, Reclamation initiated new restrictions for visitors, including closing pedestrian access to the intake towers, increasing supervision and inspection of all who enter the facility, including employees, restricting powerhouse tours to five persons per guide, raising the number of Reclamation rangers and patrols, forbidding fishing off the dam, and directing the National Park Service to start 24-hour patrols of Lake Mead, especially near the dam's intake towers.

The War Department, however, thought otherwise. It considered the greatest danger to Hoover were its employees and contractors. War Secretary Harry Hines Woodring stressed to Interior Secretary Harold Ickes that Reclamation should investigate all of its employees to insure their undivided loyalties to America. Ickes replied that Interior was unable to make investigations due to no appropriated funds to hire investigators.

During 1940, as Hitler's unchecked war machine steamrolled through Europe, and as the Japanese empire expanded their aggression in the western Pacific and China, more rumors circulated of plans to sabotage Hoover and other western power facilities. These rumors contributed to the amplified anxieties felt by Interior, War, and the FBI. Maintaining a cool countenance, but obviously concerned, Reclamation Commissioner John Page stressed the facility was perfectly safe because of increased security measures.

The FBI, however, believed that in order to better serve the public and protect the dam, they would have to train Reclamation guards in basic espionage and sabotage recognition and counter-espionage/sabotage techniques. In June 1940, they sent four specialists to Hoover to conduct a training school for Reclamation guards. At the same time, the FBI pointed out to Reclamation officials that any possible sabotage plot(s) would involve stealth, not brute force.   Reclamation also agreed that if tensions escalated into a formal declaration of war, it might be necessary to detail armed military troops to help Reclamation guards protect the facility.

During these anxious times, ideas were tabled as to how Interior might protect the dam. These ideas had one thing in common: all focused on protecting the facility from air attack; none of the proposals considered the FBI's stealth-sabotage option. One letter to the War Department suggested that a steel and concrete canopy be erected over the intake towers, spillway, and the powerhouse, then cap the canopy with native rock and gravel. Another thought the facility should be protected through a series of anchored and horizontally staggered steel cables and shields to deflect and pre-detonate enemy bombs.

It was obvious something needed to be done. By the end of 1940, world events pointed to the increased possibility of American involvement against Germany, Italy, and/or Japan. The June fall of France to the Nazis, coupled with the August start of the Luftwaffe's blitzkrieg on Britain—and Japan's three-year-old theatre of war in China and the western Pacific—made it clear to Americans that what had started as a series of smaller regional hostilities had morphed into a larger global conflict. Actions by the War Department reflected this increasing strife. On December 18, 1940, the Army announced plans to establish a military base for nearly 1,000 soldiers near Boulder City, along with the likelihood of imposing martial law, which chafed local residents.

By March 1941, this military post, initially known as Camp Sibert, was taking shape. With a targeted date of June 1 for arrival of the 524th Military Police Battalion (MPB), the camp included various buildings for enlisted barracks, officer's quarters, mess halls, offices, a fifty-bed hospital, commissary, and recreation facility. These soldiers would be used to patrol critical outlying facilities, including the switchyards, the Boulder City water line, and communications lines. Page also mentioned the military police detachment could supplement the existing force of Reclamation rangers.

On September 17, Secretary Ickes approved the "Regulations Governing the Protection of Structures" regarding not just Hoover, but other major agency projects either completed or under construction such as Washington's Grand Coulee Dam and California's Shasta Dam. Categorically divided into five protection classes, Hoover Dam was rated highest for its irreplaceable importance to national defense by reason of a major power supply, thus rating Class 1 protection, defined as complete protection of all vulnerable features by a sufficient number of armed guards.

Even before Ickes issued this directive, soldiers from Camp Sibert were already on station assisting Reclamation rangers with facility security. Thirteen privates, one corporal, and one sergeant stood on duty on each of the four six-hour daily shifts. Rangers and soldiers also were on duty at the entrance to the switchyards north of the Black Canyon Highway and two lookout perches on the Nevada side, with the fences around the Nevada State, California Electric Power Company, and the Los Angeles switchyards patrolled by three soldiers.

The lives of Reclamation's rangers, the soldiers, and other Americans, however, was about to get more complicated. On December 7, Imperial Japanese forces launched their successful surprise attack upon America's military facilities in Hawaii, shocking Americans to the tune of thousands of military and civilian casualties. For America, that vague threat was no longer far away in the Polish countryside or Malaysian jungle; it hit home with a vengeance. As Congress issued their formal declaration of war the next day, the resolve to protect Hoover's power output became even stronger.

POST-PEARL JITTERS: PROTECT THE POWER

On December 8, 1941, the same day Congress declared war on the Axis, Reclamation announced the closure of Hoover Dam and powerhouse to all visitors. Additionally, Reclamation ordered the extinguishing of project illumination floodlights, inspecting all persons and cars crossing the dam's crest, and prohibiting traffic from stopping atop the dam between the Nevada and Arizona gates. The agency also requested that the War Department dispatch a detachment of fighter planes to McCarran Field in Las Vegas. Traffic would be allowed over the dam only via armed escort convoys. Camp Sibert's troops were placed on high alert, with commanding officers announcing that all guard posts at the dam be expanded to full strength by soldiers from the 524th MPB.

Although Interior seemed worried about the possibility of an aerial attack on Hoover and its other high-profile western hydropower dams, the Army did not view this threat as imminent. Like the FBI, they considered stealth sabotage a greater menace. Secretary of War Stimson, however, reassured Interior officials that he would designate the Hoover facility a prohibited zone that would deny right of access to "enemy aliens" and "other classes of persons" within one mile of the facility's boundary.

America's new war also shifted the priorities of the War Department which, effective December 29, 1941, announced that the 524th MPB would leave Sibert for undisclosed duties. Six weeks later, on February 24, 1942, the War Department announced the 2nd Battalion of the 125th Infantry Division would temporarily relieve the 524th MPB. In addition, the Army activated the 751st MPB under orders of the Commanding General, Western Defense Command and Fourth Army. On July 23, Company C of that MPB, 123 enlisted men and 9 officers, arrived at the camp for anti-sabotage duty.

It seemed, however, that Reclamation had more problems with the military than with guarding against Stimson's alien threat. One incident happened on July 6, 1942, when Maj. General George S. Patton Jr. piloted an Army plane barely 500 feet above the dam, causing enormous concern. Reclamation officials also mentioned serious problems caused by raucous off-duty soldiers from the 524th in Boulder City neighborhoods. Reports included prowlers and peeping toms, attempted forced entry into homes, and assaults and molestations on adults and children, actions the Army assured Interior would be checked (and did so by replacing the all-black 524th with the all-white 751st).

The Army also carried out more passive protection and camouflage studies on non-Reclamation facilities key to the war effort. In summer of 1942, they conducted smoke screen protection studies at Bonneville Dam near Portland, Oregon, the Corps of Engineers-built hydropower facility located closest to the Pacific Ocean. The Western Defense Command concluded that while smoke generating machines called "Langmuir" units would be the optimal choice to smoke-conceal the dam, there was a pronounced shortage of these units, as well as a dearth of qualified Chemical Warfare

Service personnel to safely operate them. This idea was later dismissed by one War Department general as "not practicable."

While the Axis escalated their offensives in the European, African, and Pacific theatres throughout 1942, nervous Reclamation officials wondered where the protection of Hoover Dam fell into the Army's priority list. The agency became increasingly frustrated over War's ambiguity over protection of non-Army facilities—specifically where Reclamation's facilities fit within War's scheme of western coastal infrastructure protection. And despite pleas by Interior Secretary Ickes for the Army to install Langmuir smoke generating units on Hoover, Grand Coulee, and Parker Dams, War balked—but still continued to study protecting hydropower facilities built by the Corps like Bonneville.

In September, Army clarified its position, one not cordial to Reclamation. The WDC/Fourth Army's office of the commanding general informed Reclamation that, after a study by that office on protective measures for Hoover Dam, they concluded aerial bombardment was not an immediate probability, that serious damage was more likely from sabotage of the intake towers and powerhouse. In this report, War echoed what FBI had been telling Reclamation since the start of threat assessments four years previous.

As War and Interior bickered over priorities and the installation of passive protective devices, the new MPB detachment at Camp Sibert started active protection of Hoover Dam and the Basic Magnesium plant in Henderson. In September 1942, Camp Sibert officers released ground and air tactical defense plans for the dam, which included heavily armed escorts and sniper-riflemen armed with Thompson submachine guns. Additionally, officers devised three plans involving tactical air support for ground forces, taking into account the worst case (but most unlikely) scenario—enemy ground attack.

But all was not up to par with Hoover's protectors. In a memo to the Commanding General of the Ninth Service Command at Ft. Douglas, Utah, the 751st MPB's commanding officer, Lt. Colonel Harry Travis, expressed concern about the quality of troops dispatched to his remote and hot desert post. Travis complained his battalion "has been and is still being furnished men far below the standards of ordinary efficiency. Stupid, unreliable, and physical misfits." He requested that Ft. Douglas send him competent soldiers to properly protect this vital link in the war effort.

Another memorandum sent by Travis to the Ninth Service Command a couple weeks previous listed his enlisted men unfit for duty—all 44 of them—which represented nearly 10 percent of the battalion's total of 486 enlisted men. While some soldiers had serious medical problems that kept them from performing their duties, such as arthritic knees, ulcerated stomachs, clubfoots, and heart trouble, many were undependable soldiers deemed unable to carry or fire weapons. According to muster reports, some of these men were considered "radically un-American" and "chronic" alcoholics with "communist tendencies," and "habitual drunkards." Court martials for repeat offenders were common; soldiers who went AWOL and were caught, usually in Las Vegas, were fined and discharged.

Reading reports from where these soldiers were transferred—and Col. Travis's other reports—it appeared Camp Sibert was the Army's last-ditch hope to find assignments for unfit and incompetent men physically and/or mentally unable to soldier in the other two theatres of war. And, the Chief of the Military Police, in a surprise inspection, observed the mentality of some men as being "plainly below that required." A week after his visit, the Army announced the official name change of Camp Sibert to Camp Williston, because of a naming conflict with one Camp Sibert located in Gadsden, Alabama.

By 1943, the tides of war had shifted from Axis to Allied favor. Hitler's poorly executed winter attack on Stalingrad, in which over a quarter-million of his soldiers perished, is considered the war's turning point. Furthermore, British and American troops repelling German and Italian forces out of North Africa would lead to the invasion of Italy and the overthrow of Italian dictator Benito Mussolini; all of these landmark events signaled the beginning of the end for Nazi Germany. Additionally, victories by the American Navy at the battles of Coral Sea and Midway kept Japan from securing key central Pacific strategic locations like the Midway and Hawaiian Islands.

Despite favorable shifts, Interior continued to press War for more passive protection of their dams. Secretary Ickes pleaded to Secretary Stimson that the nearly 1.5 million kilowatt combined electric generation capacities of Parker, Grand Coulee, and Hoover powerhouses rated protection by one agency, the military. A couple weeks later, Ickes received an unfavorable response from the increasingly terse Stimson. As the war shifted to Allied favor, he pointed to the increased capabilities of American and Canadian western coastal defense systems, and believed that any air attack threat against western federal hydropower facilities would be intercepted before reaching their objective.

As 1943 wore on, it became clearer that Reclamation was not going to get further help from the War Department to protect the dam, so again they turned to the FBI. After a scathing, verbose report by Reclamation Ranger George Norman that detailed specific inadequacies and problems with defending the dam made its way to Washington, D.C., in November, new Commissioner Harry Bashore requested the FBI reevaluate its January 1940 assessment of protective measures, which FBI Director J. Edgar Hoover politely declined.

Another War Department memo from late 1943 reinforced this position, but from a military perspective. Assistant Provost Marshall Brigadier General Archer Lerch informed Commissioner Bashore that the Army's current policy was to redirect "all available resources and manpower into an increased effort in direct support of the present offensive phase of the war." Lerch, however, kept open the possibility that War would continue to assess key infrastructural facilities like Hoover and Grand Coulee for "adequate security measures commensurate with the current stage of the war."

Lerch's message of hope, however, was little more than lip service. Despite further pleas by Ickes and other Interior officials—and more reports of possible sabotage plots against Hoover—it was clear Reclamation had to defend the dam on its own. Camp Williston's closure cemented this prospect. On March 9, 1944, all officers and soldiers from the 751st MPB were ordered to Fort Custer, Michigan for training and overseas deployment to Grenock, Scotland, where they arrived on July 31 to support the western offensive front opened by the Normandy invasion. Now, the misfits, alleged communists, chronic alcoholics, mental midgets, and physically questionable soldiers that once helped guard Hoover were good enough to be inserted as fodder into an actual theatre of war.

But the main reason why Hoover's defense became less important was the significant progress Allied forces made against the exhausted Axis war machine in 1944: Italy was out of the picture after its surrender; Allied forces continued to breach Japan's outer and inner defense perimeters; Russian troops repelled German forces from all but two of their cities; and plans for the establishment of a second European front were underway. On June 6, 1944, also known as D-Day, 100,000 Allied forces landed on France's Normandy coast, followed by 2 million troops the following September. It was only a matter of time before the *Reich* Hitler ordained to last a millennium collapsed into the ruins of war.

After Germany's May 1945 surrender, Allied forces broke through Japan's inner defense perimeter. It would only be weeks before Stimson would recommended to President Harry S. Truman the use of atomic weapons to hasten Japan's complete and unconditional surrender, because Stimson felt the alternative, a massive Allied ground and air invasion of Japan, had the potential for dragging out the war and costing hundreds of thousands more Allied lives. Truman concurred: after dropping two atomic bombs on Nagasaki and Hiroshima, on August 14, 1945, Japan unconditionally surrendered.

With this surrender, the Second World War was history, and with it Interior's need to protect its western hydropower dams from enemy attack. With the threat abated and restrictions lifted, on September 2, 1945, Reclamation reopened Hoover Dam and powerhouse to 3,452 eager visitors, and one month later trimmed by almost half its ranger force to reflect peacetime levels.

CONCLUSION: AN ICON OF WAR HYSTERIA

Perched high atop a rocky bluff east-northeast of Hoover Dam on the Arizona side sits the sole remnant of the dam's wartime protection: a machine gun bunker/pillbox/emplacement constructed of steel-reinforced concrete and native rock, with 270 degrees worth of narrow, slit openings barely tall enough to accommodate a machine gun barrel. An icon of war that kept vigil over an icon of power, the stout little pillbox blends in well with the rugged desert landscape, a prerequisite for any fortified gun emplacement designed to repel enemy air attacks.

Yet little information on the pillbox exists in the federal archival record. The only mention of it found in War Department records is the crude facility drawing by Reclamation Ranger George Norman, a hand-drawn cross with "MGN" next to it as he pinpointed perceived deficiencies in the dam's defense plan to the FBI. Reclamation mentions the pillbox's construction in the 1941 *Boulder Canyon Project History*, but only as part of the Army's broader appeal for installation of more "protective features," including sentry lookouts, lights, fences, and gates. Additionally, a photograph dated December 19, 1941, shows the pillbox under construction by what appears to be Reclamation force accounts (employees), more than likely using Army specifications.

In hindsight, it is obvious the Army and Reclamation overestimated the capabilities of long-range Japanese bombers to fly 200 miles inland to attack a hydropower target. Although Western Defense Command forces could have intercepted enemy bombers long before they reached Hoover or any other high-profile hydropower dam, the impact of Pearl Harbor upon the nation at large forced the government to check probable threats by building and manning the bunker as fast as possible.

Not much, however, is known about who manned the bunker and for how long. One clue lies within a human interest piece in the Las Vegas *Review-Journal* (Wagner 1986). The subject, retired Army Lt. Colonel Lincoln Clark, claimed that as a young first lieutenant, he was in the second wave of soldiers detailed to Camp Sibert in February 1942 as part of a heavy weapons company in the 2nd Battalion of the Army's 125th Infantry Division, who relieved the first MPB stationed there (the 524th), the MPB Reclamation officials charged with being the source of alleged criminal mischief problems at the dam and Boulder City. According to Clark, two machine gun squads from his battalion manned the pillbox around the clock, even under darkness of night, during the anxious post-Pearl Harbor period. Yet Clark's weapons battalion only stayed at the dam and Camp Sibert for five months when, in turn, they were relieved by the second MPB (the 751st) in July 1942.

The bunker's rapid construction and Clark's short stay parallels the initial post-Pearl Harbor jitters and anxiety felt by many in the federal government. For a brief period after Pearl Harbor, America felt vulnerable, and rightly so. The deadliest attack on U.S. soil by a hostile foreign force since the War of 1812 reinforced these fears, and it is not surprising that the Army dispatched Col. Clark's heavy weapons battalion to help protect the newly-constructed dam in the war's early stages, when air attack was a distinct, albeit unrealistic, possibility.

But as the war evolved, and as western wartime industries fueled by cheap hydroelectric power produced the materials of war for America and its Allies, this lessened the need for air defense. One reason the Axis dominated early was that Allied forces were not prepared, especially the Navy and Army Air Force after Pearl Harbor. But as more weapons and war materials were produced using western hydroelectric power, the Lend-Lease program matured, and as the Axis failed to match these increased levels of Allied war production (along with their strategic blundering), the tides of war and protection priorities shifted.

As the Army's Western Defense Command strengthened its defenses to prioritize the protection of war facilities located closer to the coast, it was obvious the War Department and Interior did not share views on infrastructural threat levels. Finally, as War prepared for the D-Day invasion, it became evident the soldiers guarding inland hydropower facilities, even documented-as-questionable soldiers, would be better served to support the massive western European offensive that eventually toppled Hitler and his thousand-year *Reich*.

Thus, the protection of Hoover Dam during World War II not only reflected broader wartime policy decisions, the varying degree of security measures initiated at the dam after Pearl Harbor mirrored the progress made by Allied forces as they successfully dismantled the once-powerful global Axis war machine. A material remnant of one of America's most tumultuous periods of history, Hoover Dam's solitary machine gun bunker sits silent as an apposite symbol of a terrified country under the grips of war hysteria.

ACKNOWLEDGEMENTS

The author would like to thank the National Archives and Records Administration, Denver and College Park, for their assistance in locating appropriate federal archival records. The author also acknowledges former Bureau of Reclamation historian Christine Pfaff for "paving the way" with her initial research on Reclamation and Hoover Dam's WWII Defense.

REFERENCES

Nash, Gerald D. (1985). *The American West Transformed: The Impact of the Second World War.* Bloomington: Indiana Univ. Press.

_____. (1990). *World War II and the West: Reshaping the Economy.* Lincoln: Univ. of Nebraska Press.

Pfaff, Christine (2003). "Safeguarding Hoover Dam During World War II." *Prologue* 35: 10-21.

Roberts, Robert B. (1998). *Encyclopedia of Historic Forts: The Military, Pioneer, and Trading Forts of the United States.* New York: Macmillan Publishing Co., s.v. "Camp Williston."

Stevens, Joseph (1988). *Hoover Dam: An American Adventure.* Norman: Univ. of Oklahoma Press.

U.S. Department of the Interior, *Records of the Bureau of Reclamation.* National Archives and Records Administration, Rocky Mountain Region, Denver, CO. RG 115.

U.S. War Department, *Records of the Office of the Secretary of War.* National Archives and Records Administration, College Park, MD. RG 107.

_____. *Records of the War Department General and Special Staffs, Office of the Director of Plans and Operations.* National Archives and Records Administration, College Park, MD. RG 165.

_____. *Records of the Provost Marshall General: Military Police Division.* National Archives and Records Administration, College Park, MD. RG 389.

_____. *Records of the Provost Marshall General: Internal Security Division.* National Archives and Records Administration, College Park, MD. RG 389.

_____. *Records of the Adjutant General's Office: WW II Operations Reports.* National Archives and Records Administration, College Park, MD. RG 407.

Wagner, Christie (1986). "Fortification Hill." *The Nevadan* (January 5, pp. 6-7M).

Advances in Mass Concrete Technology – The Hoover Dam Studies

Timothy P. Dolen, P.E.[1]

[1] Research Civil Engineer and Senior Technical Specialist, U.S. Department of the Interior, Bureau of Reclamation, Materials Engineering and Research Laboratory, P.O. Box 25007, 86-68180, Denver, Colorado 80225-0007: tdolen@usbr.gov

ABSTRACT: The Boulder Canyon Project Final Reports documented the greatest leap forward in concrete technology ever. This included the pioneering work in cement chemistry, mixture proportioning, compressive strength and elastic properties, permeability, and thermal properties of mass concrete. The scientific methodology followed by numerous civil engineers and researchers provided the foundation for future mass concrete investigations for the next half century.

INTRODUCTION

The planned construction of Hoover Dam required the most comprehensive concrete design and materials investigation in history. All facets of the mass concrete mixture were investigated in detail, from the 8-inch-size[1] cobbles to the finest cement particles. The unprecedented size of the structure required extensive efforts to mitigate and control thermal cracking within the mass concrete placed in the dam and appurtenant works. How can one summarize the processes and tests leading to construction of Hoover Dam? There were thousands of tests on hundreds of combinations of the materials that when combined, determined the end product of the 4.4 million cubic yards of mass concrete placed in the dam and appurtenant works. This paper will attempt to highlight some key processes leading to the "recipe" for successful mass concrete production: about 1 part cement to 2.45 parts sand to 7.05 parts gravel and cobbles, mixed with about 0.5 part water. This paper draws on several volumes of detailed information found in the Boulder Canyon Project Final Reports published in 1949 (U.S. Department of the Interior, Bureau of Reclamation, 1949). The reports total more than 1000 pages of information on thousands of tests. The shear size of test specimens themselves is mind boggling. A standard concrete

---

[1] Note: there is some discrepancy in the "nominal" maximum size aggregate (NMSA) for Hoover Dam. The 9-inch diameter NMSA is in reference to round holed screen size. This is roughly equivalent to today's 8-inch NMSA using a square holed screen which is used in this paper.

test specimen weighs about 30 pounds. This program tested thousands of standard specimens and hundreds of large size specimens, some over three *tons* each.

The Problem – Strength versus Temperature Rise

It should be first noted that concrete strength gain is a chemical reaction between water and the cement in the paste binder; water is necessary to progress the reactions. Concrete does not "dry", it hydrates, and the mixture has sufficient water to complete the reaction over time. Mass concrete is referred to as such due to the nature of the concrete itself, large sections where cracking due to temperature rise may be a concern after the initial volumetric expansion and later contraction. In the first stages of mass concrete life, about the first year of age, heat is generated from the exothermic chemical reaction of cement "hydration." However, concrete and aggregates have very little capacity to expel the heat, which then builds up in the dam. Without measures to mitigate these reactions, the interior temperature rises to 125 to 150 degrees Fahrenheit (F), creating initial expansion of the structure and producing a temperature gradient between the cooling exterior surface and the heating interior mass. Long-term thermal contraction of the dam would be in the order of about 6 inches over decades. In addition, if the temperature gradient exceeds about 35 degrees F, unreinforced mass concrete cracks. These cracks must be controlled to avoid structural damage, leakage, and durability concerns.

Since the strength of concrete is a function of the ratio of water to cement, higher strength concretes typically also have higher cement contents, and the accompanying higher temperature gain. Thus, the mass concrete mixture investigations were a balancing act to provide sufficient strength for structural performance with the need for minimizing the temperature rise of the mixture to accommodate long-term volumetric contraction.

The Solution – Low Heat Cement and Artificial Cooling

The concrete materials engineer was faced with these several conflicting issues. Compared to previous dams, a higher strength concrete was necessary for the highest and largest dam in history, pushing the cement content up. Secondly, the need to control cracking requires less cement. Thirdly, the mass concrete mixture had to have sufficient workability in the fresh state to be transported and placed into a homogeneous mass in the desert environment. A multiple phased investigation program provided the ultimate solution.

- The *chemistry* of cement and the manufacturing process was examined to find the key parameters governing strength gain and heat generation.
- Detailed *mixture proportioning investigations* were performed to optimize the combinations of aggregates for optimum workability at the lowest possible water and cement contents.
- Extensive laboratory testing was performed to provide the necessary *hardened properties* and understand the *thermal behavior* of mass concrete.

- Lastly, *artificial cooling* was proposed to withdraw the heat from the mixture after placement during the peak strength and heat generating phases of the mass concrete setting.

At the time of the authorization of the Boulder Canyon Project, the nature of cement and concrete were not fully understood. The understanding of the basic mineralogical composition and chemical reactions responsible for strength, heat generation, and durability were in the formative stage. The solutions found during the Hoover Dam concrete investigations provided the fundamental building blocks of how these chemical reactions affected the fresh and hardened properties of concrete. By the end of these investigations, the cement manufacturing process was now capable of being altered to suit the needs for both structural and mass concrete construction. The first Federal Specifications for the Manufacture of Cement, and ultimately the five basic cement types specified by ASTM quickly followed this program, a benefit to all concrete construction projects from then on. These studies also shed light upon the factors governing long-term durability of concrete in severe environments.

CEMENT AND CONCRETE INVESTIGATIONS – INVESTIGATIONS OF PORTLAND CEMENTS

The investigations of Portland cement were conducted by a collaboration of federal agencies, universities, cement manufacturers, and trade organizations. Some of these included The Bureau of Reclamation's research laboratory, the National Bureau of Standards (later renamed the National Institute of Standards and Technology), the University of California at Berkley Engineering Materials Laboratory, the Portland Cement Association (PCA), and many cement manufacturing companies including the Riverside Cement, California Portland Cement, and Southwestern Portland Cement Companies. Not only were the properties of Portland cement investigated, but also the manufacturing processes themselves on both a laboratory and commercial scale. Even before the construction of Hoover Dam, some commercially available cement met desirable characteristics for mass concrete. C. P. Williams, formerly an assistant chief engineer for Reclamation, recognized the value of low heat cement during construction of Rodriquez Dam in Mexico in the early 1930s (U.S. Department of the Interior, Bureau of Reclamation, 1949). This trial cement was manufactured by the Riverside Cement Company, one of the suppliers of low heat cement for Hoover Dam. However, the processes responsible for these desirable characteristics were not fully understood.

The "hydration" of Portland cement is a chemical reaction between water and four principal chemical compounds to form calcium silicate hydrates, the building blocks of hardened cement paste. The reaction also produces calcium hydroxide *plus heat*. It is the relative proportions of these four compounds that determine the reaction products, the rate of strength gain, *and the amount of heat evolved*. The four basic chemical compounds found in Portland cement are commonly known as follows:

- Tricalcium silicate (abbreviated as C_3S)
- Dicalcium silicate (abbreviated as C_2S)

- Tricalcium aluminate (abbreviated as C_3A), and
- Tetracalcium aluminoferite (abbreviated as C_4AF)

These abbreviated forms tend to simplify the incredibly complex chemical reactions that occur. The silicates and aluminates dictate most of the strength and heat generating characteristics of cement. The chemical compounds form during the manufacturing process. Raw calcareous, alumina, and siliceous materials, such as obtained from limestone and clay or shale, respectively, are combined and fed into in a rotary kiln and burned at temperatures of about 1450 $^{\circ}$C (2600 $^{\circ}$F) until they sinter and fuse together into a "clinker." The clinker cools and is finely ground in ball mills, and combined with gypsum (to control setting) to form the end product. *The chemical composition of the cement is thus a function of the relative proportions of raw materials used and the manufacturing process itself.* About five principal oxides; calcium (CaO / lime), silica (SiO_2), alumina (Al_2O_3), iron (Fe_2O_3), sodium (SO_3), are found in Portland cement, as well as several minor compounds. The chemical oxide analysis provides the clue to determining the ultimate compound composition and the properties of the resulting concrete.

One of the more fortuitous discoveries concerning the nature of Portland cement occurred just at the time of the authorization of the Boulder Canyon Project. R. H. Bogue published a methodology for describing and calculating relative compound compositions of cement based on the chemical oxide analysis (Bogue, 1929). Once the strength and heat contributions of the principal chemical compounds were understood, the processes could be varied to optimize their relative proportions and "customize" the cement for its desired purpose.

A systematic approach was necessary to understand both the strength and heat evolution of cement. Each chemical compound was evaluated for strength characteristics by physical testing and heat evolution through calorimetry; essentially, taking its temperature over time in insulated containers. Although this seems relatively straight forward, the various combinations of each compound return different results through complex interactions. The manufacturing process and physical characteristics of the cement were evaluated, such as calcining temperatures, rate of cooling, fineness, and setting time of the cement and concrete. Many minor constituents are also found in cement which had to be evaluated for their contributions to strength and long-term durability. Not every critical parameter was solved in the Hoover Dam investigations. However, these same participants uncovered solutions to sulfate attack and alkali aggregate reaction through the process of identifying the chemical composition and reaction products of cement and concrete within about seven years of the completion of Hoover Dam.

The primary contributions of the chemical compounds found in cement resulted in some the following conclusions:

Strength Development

- C_3S and C_2S are the primary strength gaining compounds in cement.
- C_3S contributes to early strength gain; primarily during the first week of hydration and not beyond one month.

- C_2S contributes very little to early strength gain but is responsible for most beyond about one week.
- Increasing the percentages of C_3S and C_2S increases overall strength.
- C_3A contributes primarily to very early age strength. Increasing C_3A and C_4AF decreases long-term strength.
- Strength gains are also affected by other processes, such as cement fineness and curing temperature. Concretes cured at higher early temperatures have lower strengths than those with lower temperatures.

Heat of Hydration and Temperature Rise

- The various compounds in cement hydrate and liberate heat at different rates. C_3A, hydrates chiefly during the first day; C_3S, chiefly during the first week; and C_2S, chiefly after the first week." C_4AF contributes little to heat of hydration.
- C_3A contributes about twice as much heat as C_3S, and C_3S contributes more heat than C_2S.
- A reduction of heat of hydration can be obtained by a reduction in C_3A with an accompanying increase in C_4AF.
- A reduction of heat of hydration can be obtained with a decrease in C_3S and accompanying increase of C_2S.
- The heat of hydration and rate of hydration are also affected by cement fineness, curing temperature, and water to cement ratio, but to a lesser degree.
- Cements high in C_2S were more resistant to grinding than cements high in C_3S.

Through these studies, the desirable properties of cement could be obtained. The optimum cement combination is one that meets construction requirements as well as thermal requirements. Early strength gain is necessary for stripping forms which allowed for the rapid rates of concrete production necessary for Hoover Dam. Long-term strength gain produces higher ultimate strengths and low heat of hydration is necessary to decrease cooling requirements and control cracking. Long-term heat gain must be anticipated well after the concrete has gained sufficient strength for structural and volumetric stability. The concrete had to be placed in hot summer temperatures as well as cool winter temperatures.

Two primary cements were chosen for the dam; standard cement used during early construction and winter months and low heat cement for the remainder of the dam. During the winter, about 40 percent standard cement and 60 percent low heat cement was used to provide early strength for form removal. Results for the low heat cement compared to typical standard cement and the "modified" cement chosen for Grand Coulee Dam are shown in Table 1 (Savage, 1936).

Table 1. Properties of Cements for Mass Concrete. (U.S. Department of the Interior, Bureau of Reclamation)

	Chemical Compound Composition percent				Mortar Compressive Strength – lb/in^2		Heat of Hydration cal/g	
	C_3S	C_2S	C_3A	C_4AF	7 days	28 days	7 days	28 days
Low Heat	23	50	5	14	1,770	3,760	55	64
Typical Standard	50	25	10	8	2,660	3,350	85	97
Modified GCD	46	30	5	13	2,720	5,030		

THE MASS CONCRETE "RECIPE" FOR SUCCESS

0.5 Part Water - One Part Cement - 2.45 Parts Sand - 7.05 Parts Gravel

The mass concrete mixture design and proportioning investigation looked at all combinations of water, cements, and aggregates to produce concrete with the necessary properties in both the fresh and hardened states. In one of the pioneering investigations in concrete technology in the 20th Century, Abrams established the basic relationships for concrete strength, quality, and a "volumetric" proportioning method for standard concrete mixtures using the results of typical 6-inch-diameter by 12-inch-high test cylinders (Abrams, 1919). The Hoover Dam mixtures were anything but "standard." The mass concretes had to have specific strength and thermal properties for performance in such a huge structure. Most of the mass concrete tests were performed at the Welton Street concrete laboratory in downtown Denver, Colorado (U.S. Department of the Interior, Bureau of Reclamation, 1949).

A comprehensive mixture proportioning investigation is one that optimizes all ingredients to suit both the fresh and hardened states. The dam was to be constructed in a desert environment under harsh conditions and the mixture had to be effectively consolidated under these conditions. At the time of initial construction, methods of consolidating concrete still relied on physical "tramping, ramming, and spading." This constrained the initial water content to a lower bound to have the necessary slump of about 3 to 4 inches. All mixtures had to be proportioned to this consistency for effective consolidation. The introduction of the mechanical internal vibrator allowed a lowering of the water content of concrete and slumps of 2 to 3 inches. Up until this time, mixture proportioning was by volumetric methods. The original 94 lb bag of Portland cement equates to 1 loose cubic foot, that is, including voids. The original water to cement ratio was by volume; a W/C ratio of 1.0 by volume equates to about 62.3/94 by weight, or 0.66. Previously, aggregates were measured in calibrated boxes before entering the mixer. This proportioning method was particularly affected by the moisture content of the sand. Sand "bulking" decreased the amount of the sand in the measures at moisture contents within the normal range expected in construction. The Hoover Dam mixture proportioning program was the transition to computing proportions by weight and the associated solid volumes of ingredients. The W/C ratio was expressed by weight; however, sometimes the weight of water included the water of absorption by aggregates.

When proportioning mass concrete, the nominal maximum size aggregate or NMSA determines to a large degree the cement content of the mixture. Increasing the NMSA decreases the void space between the largest through the smallest particles and thus reduces the volume of voids that must be filled with cement paste. The lowest volume of cement paste reduces the internal heat of hydration that must be expelled and reduces the overall cost of the concrete due to less cement. The largest practical NMSA is based not only on what is available in the borrow areas, but also on the ability to effectively mix, transport, and place the concrete. Early Reclamation dams used either quarried stone blocks or embedded "plum stones," up to 3 or 4 feet in diameter, surrounded by 1- to 4-inch NMSA mixtures. These dams had high dimensionally stability due to the high aggregate volume, but were not practicable for Hoover Dam construction. Quarried stone blocks or plum stones were also not readily available, and would slow mass concrete placing considerably. An 8 –inch NMSA was selected to take full advantage of the grading existing in the "Arizona" borrow area.

Even before mass concrete testing could begin, mass concrete testing equipment had to be manufactured and test procedures had to be established. This includes the testing machine required to crush the test specimens and all methods of fabrication, preparation, and testing. The diameter of a standard compression test cylinder used in the test program was originally fixed at about four times the diameter of the largest size aggregates. The largest sized cylinder tested was 36 inches in diameter by 72 inches high with a volume of about 1.6 cubic yards each. From the engineering properties perspective, large test specimens better represented the in-situ mass concrete. From the construction quality control perspective, casting large cylinders every day in the field was next to impossible. Tests were compared to establish the relationships between the mass concrete using the 8-inch, 6-inch, 3- inch, and finally the 1-1/2-inch NMSA sample "wet screened" from the mass mixture and placed in the standard sized test cylinders as shown in Figure 1. After initially utilizing the largest size test specimens, most mass concrete mixtures were limited to that representing the minus 6-inch-size aggregate and could be cast in 24 – by 48 – inch specimens. Curing procedures were also compared. A mass concrete test specimen will generate additional heat compared to the standard fog-cured specimen, accelerating early strength gain. Fog-cured test specimens were compared to sealed cured specimens. Strength gain relationships were continued for two or more years using both standard and low heat cements. The elastic properties of mass concrete were critical to the analysis of the dam. Precision comparators, accurate to 0.0001 inch, were fabricated to measure the deformation of compressive strength cylinders under load; a large cylinder might deform about 0.1 inch before failure. Intermediate readings were normally taken to develop the stress versus strain curve, usually to no more than about 40 percent of the failure strength to avoid damaging equipment *and* laboratory technicians!

Strength and Elastic Properties of Mass Concrete

Three primary test series were conducted during the mass concrete proportioning stage and summarized in Bulletin 4 of the Boulder Canyon Project Final Reports. The first two series used the Arizona aggregates borrow area and the last series used the "Brett" source from the proposed Grand Coulee Dam. From the strength standpoint, one of the first tasks was to evaluate the effects of test specimen size for each NMSA. The same "basic" mix was tested in different size test cylinders, first with the full mass mixture with 8 – inch NMSA. Then, larger sized aggregates were progressively removed and the strength was determined on the remaining concrete in smaller sized cylinders while maintaining a cylinder to NMSA ratio of at least 4 to 1. This also has the effect of varying the gravel to sand ratio for a constant W/C ratio. Once the specimen size relationships were established, the task of evaluating key parameters contributing to workability and strength followed. The W/C ratio series, summarized in Figure 2, varied the cement content from 0.8 to 1.2 barrels[2] per cubic yard (300 to 450 lb per cubic yard) while maintaining the same slump within a narrow range. Again, the full mass concrete mixture and the "wet-screened" size fractions were tested, the cylinder size varied with the NMSA.

FIG. 1. Cylinder molds for casting concrete and curing room filled with various size test specimens (Photographs courtesy of U.S. Department of the Interior, Bureau of Reclamation, about 1933)

---

[2] One barrel of cement is about four sacks or 376 pounds of cement.

Table 2. Mixture Proportions and Properties of Fresh Concrete of Hoover Dam Laboratory Mixture B-1. (U.S. Department of the Interior, Bureau of Reclamation)

C:S:G	G/S	W/C ratio	NMSA	Slump	Unit Weight	Paste Volume	Cement Content	Water Content
ratio by weight		net	in	in	lb/ft^3	percent volume	barrels per cubic yard	lb/yd^3
			8	3	155.6	0.19	1.01	205
1 : 2.45 : 7.05	2.88	0.54	3	3	153.3	0.24	1.23	250
			1.5	3	150.5	0.28	1.49	297

Notes: C - Cement; S - Sand; G – Gravel; Net water to cement ratio includes the water of absorption in aggregates; the W/C ratio excluding absorbed water is about 0.48 to 0.50.

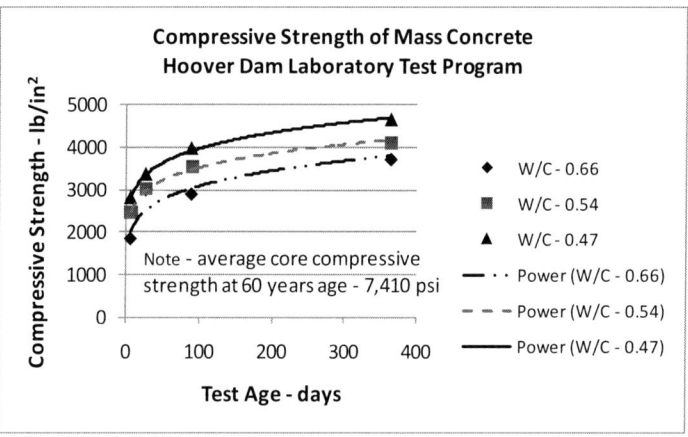

FIG. 2. Compressive strength trends for mass concrete, the effect of W/C ratio on compressive strength. (U.S. Department of the Interior, Bureau of Reclamation)

Table 3. Compressive Strength, Modulus of Elasticity, and Poisson's ratio of Hoover Dam Laboratory Mixture B-1 - Mass Concrete Test Cylinders. (U.S. Department of the Interior, Bureau of Reclamation)

Cylinder Size - inches	7 – day tests			28 – day tests		
	Compressive Strength	Modulus of Elasticity	Poisson's ratio	Compressive Strength	Modulus of Elasticity	Poisson's ratio
	lb/in^2	10^6 lb/in^2		lb/in^2	10^6 lb/in^2	
36- by 72				3,090	5.10	0.18
8- by 16	2,370	4.90	0.19	3,490	5.40	0.19
6- by 12	2,990	4.90	0.20	3,970	5.40	0.22
	90 – day tests			1 year tests[3]		
36- by 72	3,680	5.50	0.20	4,110	6.20	0.18
8- by 16	4,430	6.10	0.19	4,960	6.60	0.20
6- by 12	4,700	5.90	0.21	5,720	6.70	0.23

Based on the reported mixture proportions from Hoover Dam, laboratory mixture B-1 comes closest to the in-situ dam mixtures. Comparisons of the concrete properties of the full mass and smaller sized NMSA mixtures obtained from the mass mixture are shown in Tables 2 and 3 and Figure 3 (U.S. Department of the Interior, Bureau of Reclamation, 1949).

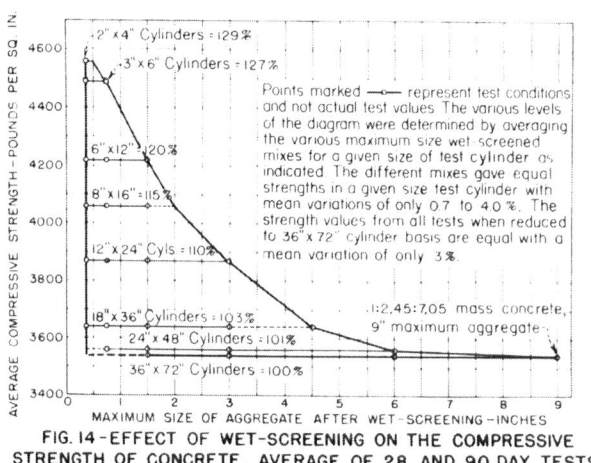

FIG. 3. Sample figure summarizing laboratory testing data, effects of cylinder size on compressive strength of mass concrete. (U.S. Department of the Interior, Bureau of Reclamation)

---

[3] 1-year compressive strength and elastic properties from laboratory mixture B-2 with same W/C ratio.

Permeability of Mass Concrete

The permeability of mass concrete was of great interest for such a high dam. Development of the permeability test and size of the test specimens again provided a challenge. The equipment settled on is shown in Figure 4. The test pressures were up to about 430 lb/in^2, equivalent to about 1000 ft of hydraulic head. The permeability of concrete is largely through the pores of the cement paste. Thus, increasing the aggregate volume again decreases the net permeability of the concrete. As the concrete hydrates, the permeability will also decrease. Permeability was determined for an equivalent test age of 60 days and is expressed as cubic feet per second per square foot of concrete 12 inches thick. The concrete was first sealed in a steel container and tested until there was a uniform inflow and outflow. Permeability test values ranged from about 10 to 20 x 10^{-12} ft/sec for mixtures with 1-1/2-inch and 8-inch NMSA at similar W/C ratios; incredibly small permeability rates. This is attributed to the fact that well-hydrated cement paste at these W/C ratios has few interconnected pores large enough to transmit water flow. The permeability increases about 10 times if the W/C ratio is increased from about 0.55 to 0.7 and another 10 times for a W/C of about 1.0. The permeability of concrete is roughly 10^6 times less permeable than the un-cemented sand and gravel itself.

FIG. 4. Permeability testing apparatus for mass concrete. (Photograph courtesy of U.S. Department of the Interior, Bureau of Reclamation)

THERMAL PROPERTIES OF MASS CONCRETE

Because artificial cooling would be necessary to cool down the dam in a timely manner, the thermal properties of mass concrete played a critical role in the mass concrete mixture investigations. According to the Boulder Canyon Project Final Reports (U.S. Department of the Interior, Bureau of Reclamation, 1949):

> *"preliminary calculations indicated that, without some artificial means of removing the excess heat ... more than 100 years would be required for the concrete in the dam of this size to attain thermal equilibrium and, consequently, volume equilibrium."*

Custom equipment was developed and manufactured by Reclamation in the laboratory. Tests were performed to determine the conductivity, specific heat, diffusivity, thermal expansion, density, and adiabatic temperature rise of mass concrete. The adiabatic calorimeter, Figure 5, was essential to determine the maximum potential temperature rise that could be expected from different cement types. Adiabatic means "no heat loss or gain." Samples of mass concrete were cast, sealed in steel containers, wrapped in insulating material, and placed in closely monitored environmental chambers to record the temperature rise. As the temperature of the concrete increased, the chamber temperature was changed to mirror the test sample to prevent loss or gain of heat from the specimen. This test normally was conducted for about one month, the equipment monitored 24 hours per day for the first week or two. Other equipment was fabricated to determine the thermal conductivity, the rate of heat flow in concrete, Figure 5 and the specific heat, the amount of heat required to change the temperature of concrete one degree.

As with all previously discussed tests, a systematic approach was used to analyze all mixture parameters affecting the thermal properties of mass concrete. This included test variables such as aggregate type, volume of aggregate (and NMSA), water content, cement content, hydration of cement, type of cement, and initial and final temperatures. Mixtures with aggregates from other dams, such as Grand Coulee Dam were included in the test program. The Hoover Dam thermal properties studies formed many of the fundamental relationships and test methodologies used to analyze mass concrete to this date. Derivation of mathematical formulas, design of test equipment and laboratory procedures, and evaluation of results and possible sources of errors were analyzed during the test program. The results of the tests were verified in laboratory scale heating and cooling tests and again in Reclamation's recently built large dams with embedded resistance thermometers.

FIG. 5. Adiabatic calorimeter test chamber (left) for temperature rise and thermal conductivity (right) for determining thermal properties of mass concrete. (Photographs courtesy of U.S. Department of the Interior, Bureau of Reclamation)

Table 4, (from Bulletin 1, Table 1) summarizes the key relationships between the material constituents and thermal properties of mass concrete (Department of the Interior, Bureau of Reclamation, 1949). The influence of aggregate type and volume were key to understanding the thermal behavior of mass concrete. Cooling tests were used to verify experimental results in the laboratory, followed by comparing calculated versus observed temperature histories in concrete dams. The maximum percentage difference between observed and calculated temperatures in mass concrete for Ariel, Bull Run, Gibson, and Owyhee Dams ranged from about 1 to 7 percent, and the average difference between observed and calculated temperatures for three of these dams was less than 3 percent. A section at Owyhee Dam served as the first test by Reclamation to embed pipes and cool mass concrete. This system was adopted for Hoover Dam in its entirety. From these tests, the cooling needs of mass concrete at Hoover Dam were calculated. Two comparisons of Hoover Dam data were documented in this bulletin. Based on two weeks of cooling in two blocks of mass concrete, the average differences between calculated (73.2 and 76.8 °F) versus observed (73.4 and 76.8 °F) temperatures, respectively, were negligible. The accurate determination of thermal properties of mass concrete at Hoover Dam and subsequent cooling system implemented were an outstanding achievement. By combining the strength results from the mixture proportioning program with the adiabatic temperature rise test results, strength to heat efficiency was obtained.

Table 4. Relative Effects of Major Mixture Variables on Thermal Properties of Concrete. (U.S. Department of the Interior, Bureau of Reclamation)

Test Conditions	Thermal Properties		
	Conductivity (Btu/(ft-hr-°F)	Specific Heat (Btu/(lb-°F)	Diffusivity (ft²/hr)
Type of coarse aggregate varied through range of tests.	1.2 to 2.0 Varied as much as 42 per cent	0.23 to 0.245 Varied as much as 8 per cent	0.032 to 0.058 Varied as much as 47 per cent
Water content increased from 4 to 8 per cent of the concrete by weight.	(1.7 to 1.55) Decreased as much as 10 percent	(0.22 to 0.24) Increased as much as 12 percent	(0.049 to 0.042) Decreased as much as 16 per cent
Mean temperature of concrete increased from 50 to 150 ° F.	Increased as much as 12 per cent and decreased as much as 6 percent	Increased as much as 24 percent	Decreased as much as 21 per cent

BOND STRENGTH OF MASS CONCRETE "LIFT LINES"

To assure monolithic behavior in the dam, tests were conducted to evaluate the preparation methods and strength of lift lines, the intermediate horizontal joints between subsequent mass concrete placements. The specification required a 72 – hour delay between each 5-foot-high placement and that the difference in height between adjacent blocks should not exceed 35 ft in height (U.S. Department of the Interior, Bureau of Reclamation, 1949). Preparation of the lift surface before resuming mass concrete placement was primarily accomplished by scarifying the lift surface after appropriate time delays. The typical lift line was cleaned by pressurized air-water jetting and timed to remove the surface layer of cement paste from the exposed aggregates. Studies were conducted at the engineering and materials laboratories of the University of California at Berkley under the direction of Professor Raymond E. Davis (Davis, 1932).

The early construction of Hoover Dam was somewhat of a transition between traditional and "modern" methods of concrete placing. Wire brushing had been the traditional means of cleaning lift lines and tamping and spading the traditional method of concrete consolidation. The Hoover Dam construction process included systematic air-water "green – cutting," eventually followed by wet sand blasting of lift lines. The tramping and spading method of consolidation was succeeded by internal vibration.

This test program focused on (1) wire brushing versus air-water jetting, (2) rodding, internal, and surface vibration, (3) normal and hot-air exposure during the lift line cure period, (4) the time delay between lift surface cleanup and the time of placing the subsequent lift, and finally (5) using bonding mortar broomed onto lift surfaces just prior to placing the following lift. Tests included flexural beam tests to determine the modulus of rupture, with and without lift lines, and permeability tests in the center of the lift line at about 100 lb/in^2 pressure. Although not stated, the flexural beam tests were assumed to be single point loading at the construction joint. Tests were performed on 30 – by 30 – by 60-inch-long and 18 – by 18- by 36-inch-long beams. The first half of the specimens were cast vertically with a time delay and surface treatment performed on the lift line before casting the remaining half of the prism. Test specimens included monolithic beams (no lift line) and lift lines with and without a mortar layer. The results of these tests formed the basis of construction procedures used on the dam. Some conclusions include the following:

- Consolidation of test specimens by internal vibration was superior to rodding.
- Air-water jetting was superior to wire brushing and to surface roughening.
- The best treated lift lines were nearly as strong as monolithic concrete.
- The optimum time delay (after casting the underlying lift) for air-water jetting cleanup was about 6 hours for low-heat cement mixtures. The lift line strength decreased with increasing time delays using the air-water jetting method. Air-water jetting 20 hours after placement was about 80 percent as strong as with 6 hour preparation.
- Specimens with a mortar layer averaged about 35 percent higher strength than specimens without mortar. Specimens with a mortar layer were more impermeable than without a mortar layer.
- The general relationships between W/C ratio and the modulus of rupture test for bond strength followed similar trends as with compressive strength.
- Moist curing was essential to maintain lift line strength during time delays between placing.
- The permeability tests generally followed the same trends as flexural strength.

Based on these tests, the procedures for lift surface preparation at Hoover Dam were established. Generally, this included the following: (1) between 8 and 20 hours after casting the surfaces were cleaned by air-water jetting, (2) lift surfaces were sprinkled with water between initial and final cleaning, generally for about 72 to 96 hours, (3) surfaces were again cleaned by air-water jetting 4 to 12 hours before placing the next lift, (4) sand was introduced through the air-water jet for final cleanup, if necessary, (5) all lift surfaces were covered with a ½-inch-thick layer of mortar ahead of placing the next mass concrete lift, and (6) mass concrete was placed in about 12-inch-thick layers and consolidated by tramping, spading, and internal vibration. The critical time delay between placing and air-water jetting was adjusted with seasonal ambient air temperatures changes, more time during the winter and less during the summer.

CONCLUSIONS

Suffice to say, Hoover Dam was the engineering marvel of it's time. In 2004, the author toured the lower galleries of Hoover Dam where concrete was placed 1933. Hairline cracks were carefully noted and dated in 1934, the handwriting is still visible today. There is not even a hint of leakage. This kind of performance is a part of the legacy from the dedication and hard work employed by the Denver concrete laboratories.

ACKNOWLEDGEMENTS

The author wishes to thank all of the engineers, scientists, and especially the engineering technicians who carried out Reclamation's mass concrete mixture proportioning programs past and present. The difficult work was carried out with precision and accuracy. The shear size of the test specimens used in these comprehensive test programs required great care and patience to achieve the test results. The author would like to thank Mr. Edward Harboe who was his early-career mentor. The author worked closely in the lab with engineering technicians Neil Johnson, Stephen Reo, and the late Kurt Mitchell. Messrs. Charles Prusia, Gary Stevens, and Thomas Strickland were the field laboratory chiefs that carried out much of the field investigations and quality assurance testing on some of our large scale construction projects during the author's career.

REFERENCES

Abrams, Duff, A., "Design of Concrete Mixtures," Structural Materials Research

Bogue, R.,H., "A Calculation of the Compounds in Portland Cement," *Industrial and Engineering Chemical Analysis*, Vol. 1, No. 4, October, 1929.

Davis, Raymond E., Davis, H.E., and Kelly, J.W., "Bonding of New Concrete to Old at Horizontal Construction Joints," Concrete Laboratory file 390.7.5, date about 1932.

Savage, J.L., *Special Cements for Mass Concrete,* Prepared for the Second Congress of the International Commission on Large Dams World Power Conference, Washington D.C., U.S. Department of the Interior, Bureau of Reclamation, Denver, Colorado, 1936, pp. 37 – 53.

U.S. Department of the Interior, Bureau of Reclamation, *Boulder Canyon Project Final Reports, Part VII – Cement and Concrete Investigations, Bulletin* 2, *Investigations of Portland Cements,* Denver, Colorado, 1949.

Op cit, *Bulletin 4, Mass Concrete Investigations,* 1949.

Op cit, *Bulletin 1, Thermal Properties of Concrete*, 1940.

U.S. Department of the Interior, Bureau of Reclamation, *Boulder Canyon Project Final Reports, Part IV, Design and Construction, Bulletin 4, Concrete Manufacture, Handling, and Control*, Denver, Colorado, 1947. *Historic Photos, about 1933.*

Long-Term Properties of Hoover Dam Mass Concrete

Katie Bartojay[1], P.E. and Westin Joy[2]

[1]Civil Engineer, Bureau of Reclamation, Denver Federal Center, Bldg 56 (86-68180), Denver, CO 80225-0007; kbartojay@usbr.gov
[2]Civil Engineer, Bureau of Reclamation, Denver Federal Center, Bldg 56 (86-68180), Denver, CO 80225-0007; wjoy@usbr.gov

ABSTRACT: Hoover Dam, at its original construction in 1931-1936, was constructed with over 3.25 million cubic yards of concrete in the dam alone. Most of the concrete for the dam was mass concrete made with up to 9-inch maximum size aggregate (MSA). The original 28-day compressive strength tests performed for quality assurance during construction were found to be in the 3,500 lb/in^2 range. In 1995, the Bureau of Reclamation's Materials Engineering and Research Laboratory completed a testing program to evaluate the long term properties of the Hoover Dam concrete. Compressive strengths at this time were found to average 7,230 lb/in^2. Discussions will include the core sample locations and compressive, tensile and shear strength test results from this study, along with a discussion of determining the need for the core testing program.

INTRODUCTION

In 1992, potential dam safety deficiencies were identified during a Safety Evaluation of Existing Dams (SEED) investigation of Hoover Dam (Reclamation, 1992). The concern was that at a dam height of 726-feet potential earthquake motions from faults in the proximity of the dam could be magnified in the upper portion of the dam. The SEED investigation prompted a Modification Decision Analysis (MDA) on seismotectonic issues which included direction to core, test, and evaluate the concrete in the top portion of the dam. The coring program was initiated to obtain updated concrete material properties for the dam that could be used in subsequent analyses.

CONCRETE CORING PROGRAM

In December 1994 and January 1995, concrete coring was performed by Reclamation's Lower Colorado, Phoenix Area Office, Exploration Section (Reclamation, 1995a). Five vertical holes were drilled along the dam crest and four horizontal holes were drilled into the upstream face of the dam from a floating barge. The exact locations for these drill holes were not found in the documentation and no precise locations could be determined by the authors. Over 140 feet of six-inch-diameter concrete core was extracted from the dam.

The concrete cores were wrapped in cellophane to preserve the moisture and packed in pine boxes filled with moist saw dust for shipment. Thirty boxes of concrete core were shipped to Reclamation's Technical Service Center in February of 1995 for testing by what is known today as its Materials Engineering and Research Laboratories (MERL) group.

CORE DIAMETER AND LOCATION SELECTION

The nominal maximum size aggregate (NMSA) in Hoover Dam is nine inches, while the concrete core extracted from the dam was only 6 inches in diameter. Typically, concrete cores for compression strength testing are two to three times the NMSA (ASTM, 2009). This would have resulted in a minimum core diameter of eighteen inches. Although Reclamation has taken large diameter cores from projects, they are extremely costly and difficult to extract. This factor was considered when establishing the concrete testing program and the following was concluded:

> The decision to use 6-inch core was based on the concept that we would have plenty of specimens and we would discard and retest results considered invalid due to aggregate size influence...The specimen selection process should try to avoid testing core which contains [an] unreasonably high percentage of aggregate surface area to paste surface area (i.e. a core containing a large aggregate which is not surrounded by paste) (Reclamation, no date).

The decision to use the smaller diameter cores can be substantiated by the conclusions drawn by Mather (1961) that only the precision of compressive strength is affected by the size of a drilled core specimen, and accuracy is not significantly affected by the type of specimen. The precision was found to increase with core diameter. Price (1951) found that 10-inch and 22-inch-diameter concrete cores extracted from Shasta and Friant Dams at an age of 5 years were practically the same. This suggests that age may reduce the influence of specimen diameter to representative compressive strength.

The 1992 Hoover Dam SEED report indicates that the top portion of the dam was the main concern in an earthquake event (Reclamation, 1992). This explains the lack of concrete cores extracted from the remaining 700 feet of dam. Prior to this testing program there were no indications that the concrete in Hoover Dam was performing less than satisfactory and there were no visual signs of increased or excessive deterioration. This likely contributed to the limited scope of the 1995 coring program.

TESTING PROGRAM

The Hoover Dam concrete cores received by MERL included five vertical holes from the dam crest, three cored down contraction joints, and two cored in unjointed (parent) concrete. In addition, four horizontal holes were cored from the upstream face of the dam, three along lift lines (LL) and one into unjointed (parent) concrete.

Test results were summarized in a draft MERL memorandum (Reclamation, 1995a). Tests performed on the concrete specimens included:

Compressive strength testing
Modulus of Elasticity and Poisson's ratio
Splitting tensile strength testing
Direct tensile strength testing
Direct shear strength and sliding friction testing

The compressive and direct tensile strength testing was performed with strain controlled loading in an attempt to obtain post-peak behavior.

All cores were inspected and logged upon receipt. Drawings, pictures, and notes describing the cores were obtained and placed in log books for future reference. All cores were maintained in a moist condition throughout the logging process. Examples of photos taken during the logging process are shown in the figures below. After logging, test specimens were identified and cores were cut using a diamond saw. Once cores were cut for testing, they were placed in a 100-percent relative humidity (fog) room until testing, at which point they were sealed in plastic to prevent moisture loss. All test specimens, except those used for shear testing, were cylinders 6 inches in diameter and 12 inches in length. Bureau of Reclamation, American Society for Testing and Materials (ASTM) and International Society for Rock Mechanics (ISRM) standard test methods were used to perform the various tests.

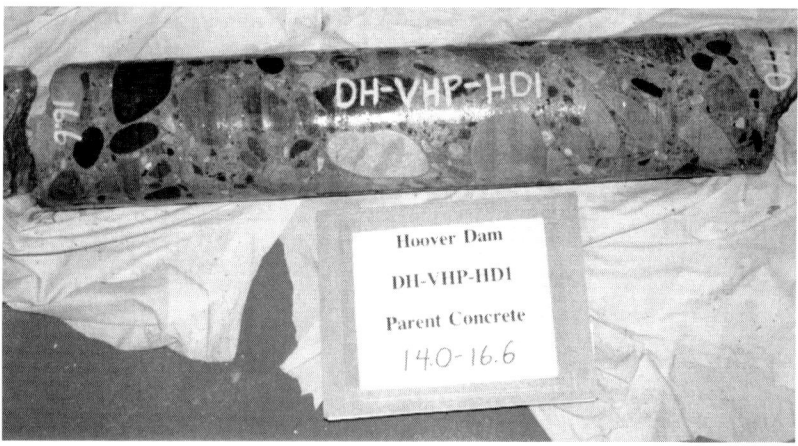

FIG. 1. Hoover Dam vertical parent concrete core from 14.0 to 16.6 feet deep (Reclamation, 1995a)

FIG. 2. Hoover Dam horizontal concrete core along a lift line (LL) at a depth of 2.7 to 4.8 feet from the upstream face of the dam (Reclamation, 1995a)

Compressive Strength and Elastic Properties Testing

The average compressive strength of the 6-inch diameter specimens was found to be about 7,230 lb/in^2, significantly higher than is typical for mass concrete. The compressive strength ranged from 5,120 to 9,230 lb/in^2. The elastic properties were also indicative of normal mass concrete of this strength. The average modulus of elasticity and Poisson's ratio of the specimens were 6.59 x 10^6 lb/in^2 and 0.21, respectively. A summary of the testing results for compressive strength and elastic properties are presented in Table 1.

Table 1. Average Compressive Strength and Elastic Properties (original work)

Drill Hole Type	Compressive Strength (lb/in^2)	Modulus of Elasticity (x10^6 lb/in^2)	Poisson's Ratio
Horizontal	6,690	6.25	-
Vertical	7,410	6.65	0.21
Average of All	7,230	6.59	0.21

Tensile Strength Testing

Both splitting tension and direct tension testing were performed in the investigation. Unjointed concrete specimens and concrete containing a horizontal lift line were tested. The splitting tensile strength of the 6-inch diameter unjointed (parent) concrete specimens in both the vertical and horizontal direction was about 600 lb/in^2. This is around 8-percent of the compressive strength, which is consistent with published values of 8- to 14-percent for all concrete (Mindess and Young, 1981).

Horizontal concrete specimens containing a horizontal lift line were also tested for splitting tensile strength with the lift line centered along the splitting plane to the degree possible. The splitting tensile strength of these specimens averaged 550 lb/in^2, about 92-percent of the unjointed splitting tensile strength.

Direct tension testing results of the unjointed concrete specimens averaged 285 lb/in^2 in the vertical core and 185 lb/in^2 in the horizontal core. These results were more variable than the splitting tension test results with a range of 60 to 420 lb/in^2. Concrete specimens containing a horizontal lift line achieved similar average tensile strengths of 290 lb/in^2. In fact, the only specimen that broke at the lift line had a direct tensile strength of 385 lb/in^2. Taking into account the variables induced by the small diameter core and the small number of tests, the results show that the joints are at least as strong as the parent concrete in direct tension. Typically, the direct tensile strength at the lift lines is found to be less than the unjointed concrete and the greater strength of the samples containing lift lines is likely attributed to the bias associated with a small sample size. Overall the direct tension strength was around 3- to 4-percent of the compressive strength, which is consistent with the average direct tension to compressive strength ratios for Reclamation structures reported by Dolen (2005).

Splitting tensile strength is typically about twice the value of direct tensile strength for mass concrete. However, the splitting tensile strength of small diameter mass concrete cores (relative to the NMSA) appears to test higher than larger diameter cores (Reclamation, 1998). The results from the Hoover Dam core testing program follow this trend, and the comparison between splitting tensile strength and direct tensile strength for the vertical cores can be seen in Figure 3.

In 1995, both splitting and direct tension test results indicated that the intact horizontal lift lines at Hoover Dam were well bonded after 60 years. A summary of the testing results for splitting and direct tensile strength are presented in Table 2.

Table 2. Average Tensile Strength Results (original work)

Specimen Type	Vertical Cores		Horizontal Cores	
	Split Tensile Strength (lb/in^2)	Direct Tensile Strength (lb/in^2)	Split Tensile Strength (lb/in^2)	Direct Tensile Strength (lb/in^2)
Lift Line	N/A	290	550	N/A
No Lift Line	605	285	590	185

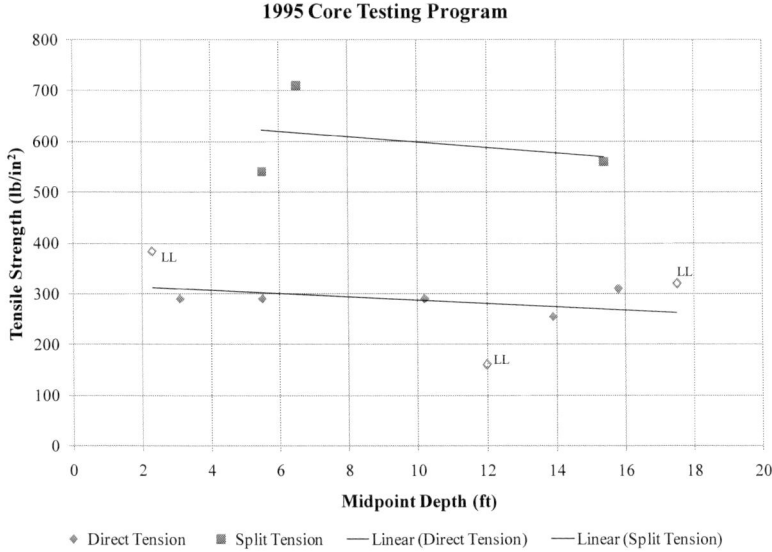

FIG. 3. Graph comparing direct tensile strength to splitting tensile strength (original work)

Direct Shear Strength Testing and Sliding Friction Testing

Direct shear strength test specimens selected from the vertical concrete cores included either unjointed (parent) concrete or a horizontal lift line. All horizontal concrete cores tested contained a horizontal lift line. Bonded lift line locations were not always visible and samples were selected based on measurements and most probable locations. Two specimens containing unbonded lift lines were also tested for comparison.

Direct shear strength analysis was performed using Coulomb-Navier failure criterion. Either 7-inch or 10-inch holding rings were used to encapsulate the specimens. Results are presented in Figure 4 (Reclamation, 1996). The direct shear and sliding friction test results are normal for mass concrete. Figures 5 through 8 show photographs of direct shear specimens in both pre-test and post-failure configurations (Reclamation, 1996).

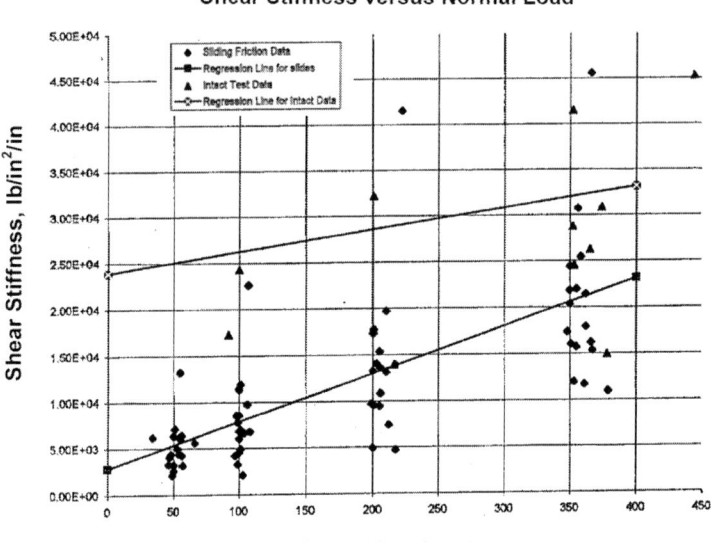

FIG. 4. Plot of stiffness data from direct shear tests (Reclamation, 1996)

FIG. 5. Direct shear parent concrete specimen (Reclamation, 1996)

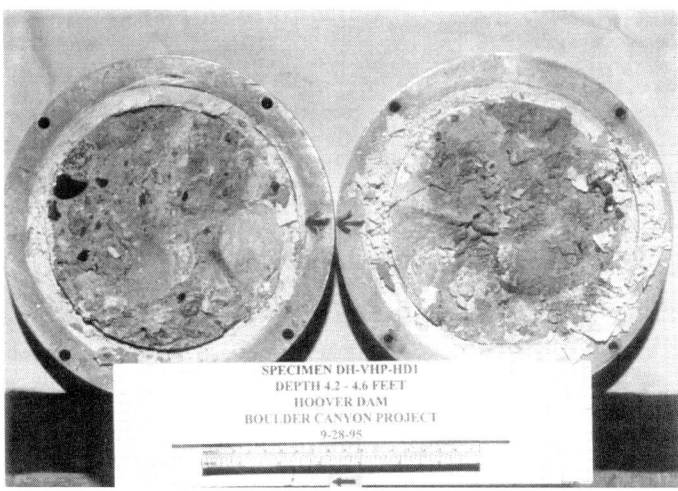

FIG. 6. Direct shear parent concrete specimen post-failure (Reclamation, 1996)

FIG. 7. Direct shear lift line specimen (Reclamation, 1996)

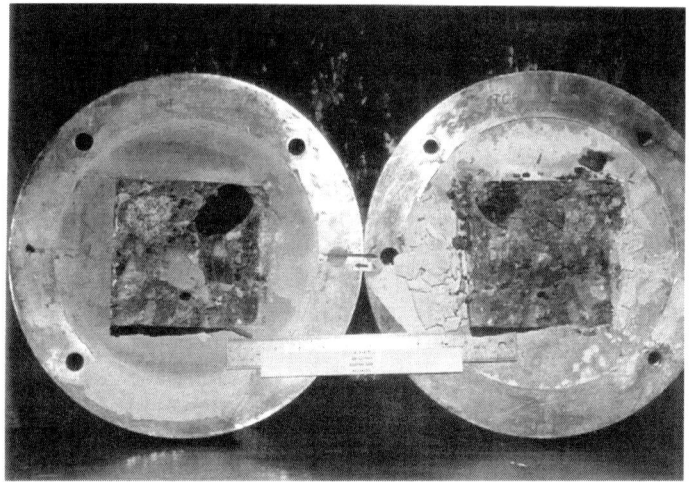

FIG. 8. Direct shear lift line specimen post-failure (Reclamation, 1996)

COMPARISON WITH OTHER MASS CONCRETE DAMS

The testing results from the 1995 Hoover Dam concrete core testing program are quite impressive. Additionally, no deterioration has been identified in recent years. Hoover Dam's concrete has continued to slowly gain strength over the past 75 years, something that is typical of normal concrete without sources of deterioration. In contrast, Seminoe Dam in Wyoming is roughly the same age and is experiencing multiple types of deterioration, causing the strength of the concrete in the upper portion of the dam to decline with age. Causes of deterioration at Seminoe Dam include freeze-thaw as well as the progression of Alkali-Silica Reaction (ASR).

Parker Dam, situated approximately 155 miles downstream of Hoover Dam, is also experiencing ASR damage. The development of ASR in Parker Dam is due to the combination of the reactive aggregates and the high alkali content of the cement used. Hoover Dam was constructed with the same cement, but contains non-reactive aggregates from a different source than the aggregates used in Parker Dam. During the first few years following construction, the two dams exhibited roughly the same concrete strength development. However, due to the deterioration caused by ASR, the concrete strength at Parker Dam is now considerably lower than that at Hoover Dam.

The average compressive strength of mass concrete cores taken from Parker Dam in 2005 is about 4,390 lb/in^2, slightly less than two-thirds the average compressive strength of the Hoover Dam mass concrete cores (Reclamation, 2005a). The development and subsequent difference in concrete strength for Parker Dam and Hoover Dam is shown in Figure 9 below.

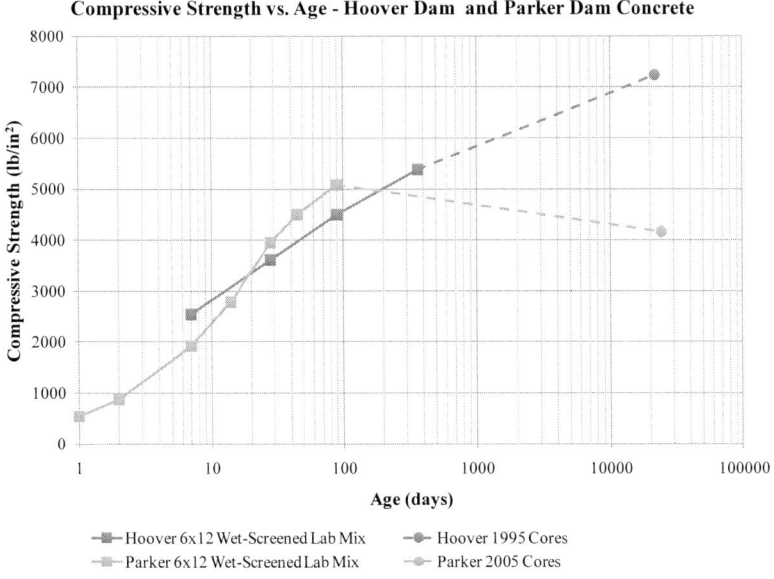

FIG. 9. Comparison of concrete strength from Hoover Dam and Parker Dam (adapted from Reclamation, 2005)

CONCLUSIONS

The overall quality of the concrete cores from Hoover Dam indicates a durable concrete having a compressive strength exceeding the range typically found in normal mass concrete. As of 1995, the intact horizontal lift lines at Hoover Dam remain well bonded after 60 years and are acting essentially the same as the unjointed (parent) concrete. There is no major deterioration mechanism at work within the concrete in the upper portion of the dam that would cause a decrease in strength or elastic properties.

The 1995 recommendations from the Second Consultant Review Board for Hoover Dam (Reclamation, 1995b) concluded that it was not necessary to obtain additional concrete cores to further investigate concrete properties.

ACKNOWLEDGMENTS

The authors appreciate the work performed by Reclamation's Dr. William Kepler who was instrumental in the completion of the 1995 core testing program. Historical information and data provided by Mr. Tim Dolen was also valuable in the writing of this paper.

REFERENCES

ASTM International, (2009). Annual Book of ASTM Standards, West Conshohocken, PA.

Dolen, Timothy P. (2005). Materials Properties Model of Aging Concrete, DSO-05-05, Dam Safety Technology Development Program, Bureau of Reclamation, 2005.

Mather, Bryant and Tynes, William O. (1961). "Investigation of Compressive Strength of Molded Cylinders and Drilled Cores of Concrete" *Journal of American Concrete Institute* JL57-37, American Concrete Institute, Vol. 57, Issue 1.

Mindess and Young (1981), "Concrete" Prentice-Hall Inc., Englewood Cliffs, New Jersey, 1981.

Price, Walter H. (1951). "Factors Influencing Concrete Strength" *Journal of American Concrete Institute* JL57-37, American Concrete Institute, Vol. 47, Issue 2: 417-432.

Reclamation (no date). Factors Considered in Establishing the Concrete Testing Program, Modification Decision Analysis of Hoover Dam, DRAFT Date Unknown Bureau of Reclamation.

Reclamation (1992). Safety Evaluation of Existing Dams Analysis Summary Hoover Dam, Bureau of Reclamation, June 30, 1992.

Reclamation (1995a). Report of Mass Concrete Core Testing, Hoover Dam, Boulder Canyon Project. Materials Engineering Branch Referral Memorandum No. MERL-1995-DRAFT. Bureau of Reclamation, 1995.

Reclamation (1995b). Second Report of Board of Consultants on Hoover Dam Modification Decision Analysis, Boulder Canyon Project, Nevada. Bureau of Reclamation, 1995.

Reclamation (1996). Results of Direct Shear Testing of Concrete Core Specimens, Hoover Dam, Boulder Canyon Project. Earth Sciences and Research Laboratory Referral Number 8340-96-03, Bureau of Reclamation, 1996.

Reclamation (1998). Summary of Material Characteristics, Monticello Dam Core Testing Program, Solano Project, California, Bureau of Reclamation, 1998.

Reclamation (2005). Parker Dam 2005 Concrete Coring – Laboratory Testing Program – Parker-Davis Project, California-Arizona. Materials Engineering and Research Laboratory Report no. MERL-2005-20. Bureau of Reclamation, 2005.

Weaver, Jeff (1995). Memorandum Transmittal of 30 Boxes of Concrete Core Samples from Hoover Dam for Laboratory Analysis Per Modification Decision Analysis (MDA), Bureau of Reclamation, Lower Colorado Regional Office.

Hoover Dam: Evolution of the Dam's Design

J. David Rogers[1], F. ASCE, D. GE, P.E., P.G.

[1]K.F. Hasslemann Chair in Geological Engineering, Missouri University of Science & Technology, Rolla, MO 65409; rogersda@mst.edu

ABSTRACT: Hoover Dam was a monumental accomplishment for its era which set new standards for feasibility studies, structural analysis and behavior, quality control during construction, and post-construction performance evaluations. One of the most important departures was the congressional mandate placed upon the U.S. Bureau of Reclamation (Reclamation) to employ an independent Colorado River Board to perform a detailed review of the agency's design and issue recommendations that significantly affected the project's eventual form and placement. Of its own accord Reclamation also employed an independent board of consultants which convened twice yearly several years prior to and during construction of the project, between 1928 and 1935. Reclamation also appointed a special board of consultants on mass concrete issues, which had never been previously convened. Many additional landmark studies were undertaken which shaped the future of dam building. Some of these included: the employment of terrestrial photogrammetry to map the dam site and validate material quantities; insitu instrumentation of the dam's concrete; and consensus surveys of all previous high dams to compare their physical, geologic, and hydrologic features with those proposed at Hoover Dam. The project was also unique because the federal government provided of all materials, except the concrete aggregate, to minimize risk of construction claims and delays.

EARLY INVESTIGATIONS

Background

Investigations along the lower Colorado River which eventually led to the construction of Hoover Dam were initiated by the U.S. Geological Survey's Hydrology Branch in 1901-02 when hydrologist J. B. Lippincott identified dozens of potential dam sites along the Colorado River, including the bedrock narrows in Black and Boulder Canyons. In 1904 the newly formed U.S. Reclamation Service began evaluating potential dam sites along the Colorado River. The seminal event that eventually led to the dam's construction began with the unintentional flooding of the

Imperial Valley that commenced in March 1905 when the headworks of a privately constructed diversion canal were overwhelmed along the lower Colorado River, about four miles south of the Mexico-California border (Davis, 1907; Grunsky, 1907; Sykes, 1937). The Southern Pacific Railroad attacked the break throughout 1906, eventually closing off the channel at midnight on February 10/11, 1907, after the Colorado River had discharged its waters into the enclosed basin for just under two years, creating the Salton Sea, with a surface area of 500 square miles (Orsi, 2005).

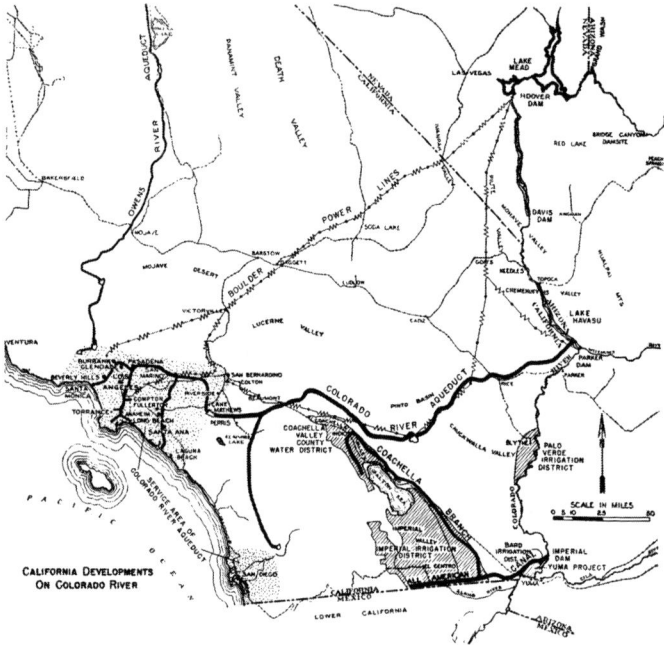

FIG. 1. Location of Hoover Dam, agricultural areas that received irrigation water from the dam, the Colorado River Aqueduct, and the Boulder Canyon Project transmission lines serving southern California (USBR).

The disastrous flooding of the Imperial Valley bankrupted the massive commercial scheme for reclaiming this inland basin of southeastern California. During the winter of 1909-10 the Colorado River once again jumped its banks, this time filling Volcano Lake south of the border and, once again, threatening crops in the Imperial Valley. Congress appropriated $1 million to provide additional flood control along the lower Colorado River.

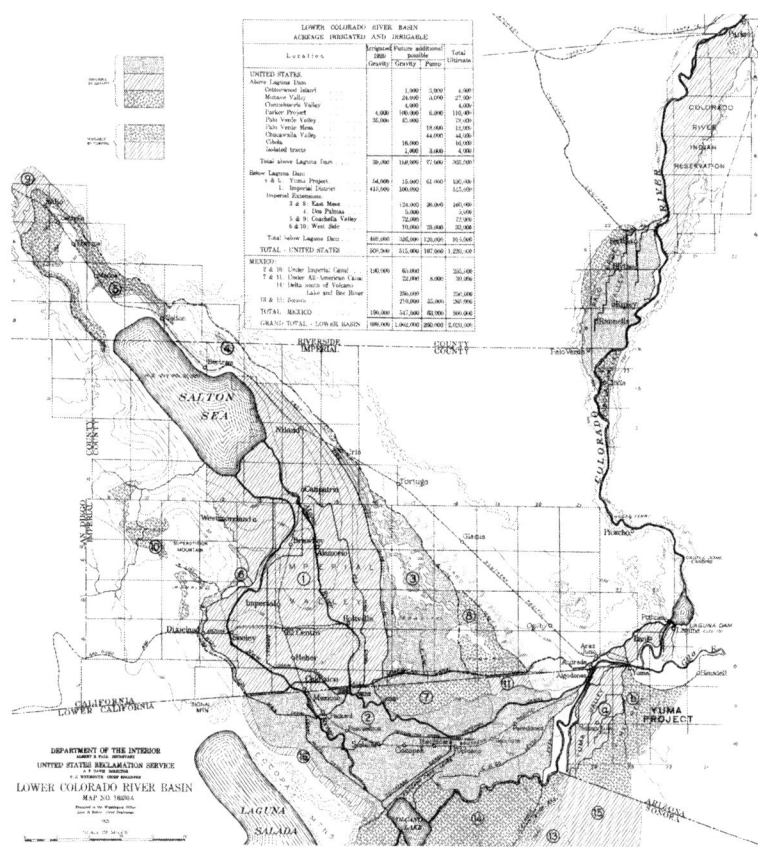

FIG. 2. Map from the 1922 Fall-Davis report showing the non-alkali valleys within and adjacent to the Lower Colorado River Basin that could be irrigated by a dam in Boulder Canyon. These included the Palo Verde, Yuma, Mexicali, Imperial, and Coachella Valleys. At that time about 700,000 acres were under cultivation using irrigation over an area farmed by almost 100,000 people. The reservoir in Boulder Canyon had the potential to bring 2,020,000 acres under irrigation, including areas irrigated by gravity flow and those that would require pumping. (USBR)

This second brush with disaster ignited a growing desire for federal assistance, principally through the U.S. Reclamation Service, which had been established in 1902 to develop irrigation and water conservation projects across the semi-arid west. Political representatives from southern California began lobbying for a Reclamation scheme that would provide reliable sources of irrigation as well as dependable flood control.

In 1914 Congress authorized the Reclamation Service to make an inventory of possible dam sites and irrigation project that would benefit thereof in the Colorado River Basin. In January 1916 a fierce flood swept down the Gila River, inundating the Yuma Valley. This flood damage broadened support for a larger scheme that would address the needs of the lower Colorado River Basin as well as the Salton Sink, which drained the Mexicali, Imperial, and Coachella Valleys. In 1918 Reclamation Service Director and Chief Engineer Arthur Powell Davis proposed that the Colorado River be controlled by a dam of unprecedented height in the granite narrows of Boulder Canyon, east of Las Vegas Junction, which were summarized in the Whistler Report of March 1919. Surveys of the Colorado River by the Reclamation Service through contracts with the U.S. Geological Survey began in 1916 and continued through 1925.

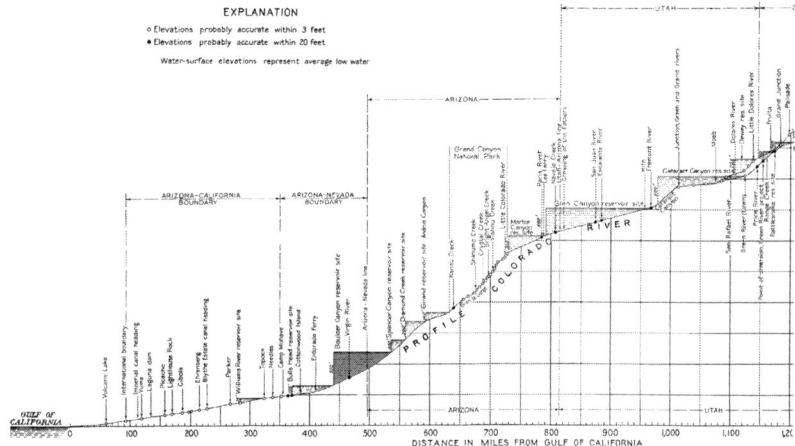

FIG. 3. Stair-stepping series of dams and reservoirs proposed along the Colorado River below its confluence with the Green River by the Reclamation Service in the 1922 Fall-Davis report. The shaded reservoir is the high dam in Boulder Canyon. Only four of the 12 dams shown here were eventually constructed: Hoover (1935), Parker (1938), Davis (1953), and Glen Canyon (1964).

The Imperial Valley interests and their southern California investors vigorously lobbied for the Reclamation Service's Boulder Canyon scheme, which envisioned a substantial reservoir that would provide flood control for the lower Colorado River Basin and supply irrigation water for the Palo Verde and Yuma Valleys of the Colorado River, as well as the Imperial and Coachella Valleys, west of the Colorado River (Figures. 1 and 2).

In 1919 the All-American Canal Board recommended construction of a canal on the American side of the international border that could convey water from the Colorado River to the Imperial Valley. This was intended to circumvent the loss of 50% of the water then being conveyed through the Mexicali Valley within Mexico before it reached the Imperial Valley. That same year Congressman Phil Swing introduced a bill seeking to authorize construction of the All-American Canal (Moeller, 1971).

The Reclamation Service was reorganized into the Bureau of Reclamation (Reclamation) in 1923 and the newly empowered agency began developing grandiose plans to construct a string of dams along the Grand, Green, and Colorado Rivers, from the headwaters down to the Gulf of California (Figure 3).

Boulder versus Black Canyons

Working for the Reclamation Service as a consulting engineer, former Los Angeles City Engineer Homer Hamlin supervised the first surveys of dam sites in Boulder Canyon and Black Canyon in March-April 1920. Educated as both a civil engineer and geologist, Hamlin was the first to designate the site in Black Canyon that was eventually chosen for Hoover Dam in 1928. He died in May 1920 while attending the congressional hearings accompanying passage of the Kinkaid Act which authorized the Secretary of the Interior to begin studying problems with flooding in Imperial Valley and report on possible solutions. At that time Reclamation envisioned the dam at the head of Boulder Canyon where granite outcrops would provide firm abutments (Figure 4) with a minimal cross section for either an earth-rockfill dam (Figure 5) or a masonry gravity-arch dam (Figure 6).

The Kinkaid Act provided funds for a 1-1/2-year-long feasibility study that examined the entire Colorado River watershed with specific emphasis on the development of irrigation of the Imperial Valley region. Reclamation crews began drilling the three principal dam sits in Boulder Canyon between January and May of 1921, under the direction of Walker R. Young. Their camps were destroyed that spring by unusually high flows, which reached about 210,000 cubic feet per second (cfs) in Boulder Canyon. Young marveled at how *"such a great volume of water"* could pass through such a dry thirsty desert without being harnessed for any *"worldly benefit to mankind."*

The Reclamation Service's preliminary design team was comprised of A.J. Wiley, James Munn, John L. "Jack" Savage, and Walker Young. They envisioned that water storage and flood control would originate from a massive dam as much as 740 ft high rising to an elevation of 1310 feet above sea level in Boulder Canyon, about 33 air miles due east of Las Vegas, Nevada.

The feasibility study was unveiled at a conference titled *"Construction of Boulder Dam"* convened in San Diego in November 1921. The Reclamation findings and the proceedings of the San Diego Conference were jointly published by the Government Printing Office in March 1922. It was thereafter referred to as the "Fall-Davis Report," because it was submitted by Albert B. Fall, Secretary of the Interior and

FIG 4. Between 1916 and 1929 most everyone assumed that the Granite Narrows shown here at the head of Boulder Canyon would be the logical site for a mighty dam controlling the lower Colorado River. This was how the scheme came to known as the Boulder Canyon Project (USBR).

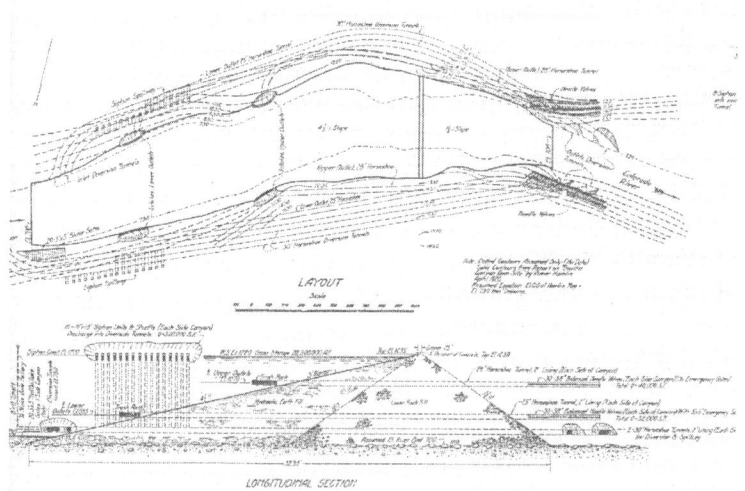

FIG. 5. Schematic plan and section view through a combination earth-rockfill embankment dam in Boulder Canyon, rising 535 feet above river level, as envisioned by the Reclamation Service in 1920. Note that the channel gravels would have been left in-place and the absence of seepage filters between the upstream hydraulic fill and the downstream rockfill (USBR).

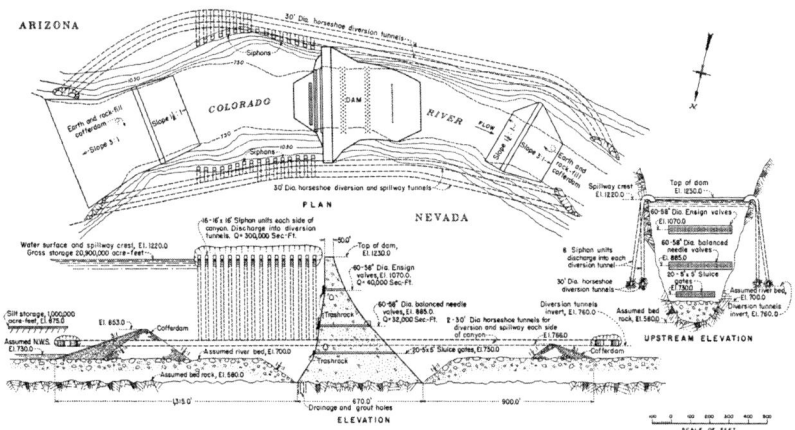

FIG. 6. Schematic plan and section view through a 650 feet high concrete gravity masonry dam in Boulder Canyon, 23 river miles upstream of Black Canyon, as envisioned in 1920. The additional height was necessary to remove all the channel gravels to secure a foundation capable of bearing 35 tons per square foot bearing pressure (Fall and Davis, 1922).

Arthur Powell Davis, Director of the Reclamation Service (USBR, 1922). The study examined four possible dam sites in Boulder Canyon with crest elevations between 1230 and 1310 feet. The largest alternative could store 34 million acre-feet of water, about 26 times the largest reservoir (California's Lake Spaulding) in the United States at the time. These developments were so ominous for the upper basin states that the Colorado Legislature voted to change the name of the Grand to the Colorado River in 1922, as a ploy to bolster their claim to the river's waters.

In January 1922 the Reclamation Service began examining alternative sites for a high dam about 23 miles downstream, in Black Canyon (Figure 7). A survey camp was established at the head of Black Canyon on a sandspit known as "Cape Horn," about 1-1/2 miles below the ferry operated by Murl Emery, who provided boat access to Black Canyon throughout the 1920s. This camp was vacated in May to avoid the summer heat, then re-occupied between September 1922 and late April 1923. During those encampments dozens of diamond drill holes were made from barges anchored to cables attached to the sheer canyon walls (Figure 8), in order to ascertain how much gravel filled the bedrock channel. To their dismay they learned that the channel gravels in Black Canyon extended to an average depth of 120 feet, because of a deeper "inner gorge" caved out of the andesite bedrock. Much of this inner gorge was filled with coarse angular boulders, from rockfalls and debris flows. The depth to granite bedrock in Boulder Canyon had varied from 25 to 150 ft, but the average depth at the favored dam site was only about 65 feet.

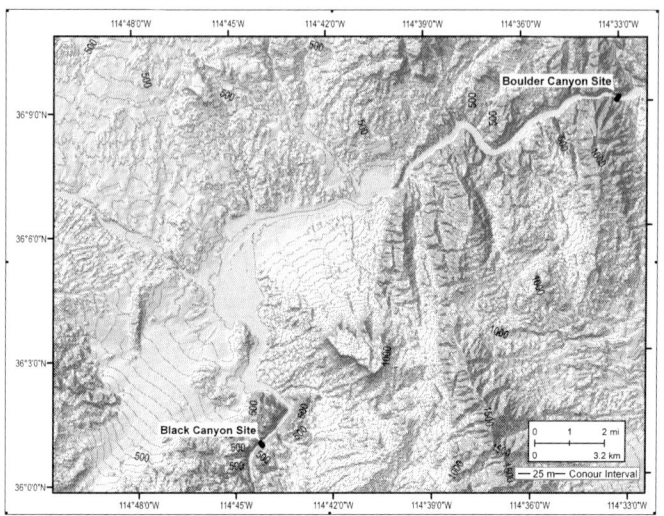

FIG. 7. The geomorphic settings of Boulder and Black Canyons were similar in that both of them were tectonically-controlled by geologically recent faulting, juxtaposing blocks of dramatically different age against one another. The dam sites downstream of Grand Canyon lie within bedrock narrows rapidly excavated by paleo outbreak floods which drained large bodies of water (data supplied by USBR staff).

FIG 8. Drilling a line of three AX-size diamond drill holes in upper Black Canyon on October 31, 1922. This image shows Munns' Cave, developed within a basaltic intrusion of the Dam Breccia, which weathered more easily (USBR).

Colorado River Compact

While the Reclamation Service was studying dam sites in Boulder and Black Canyons, the Colorado Basin states began fearing that California would succeed in laying a legal claim to the river's waters by the doctrine of prior appropriation by swallowing the river's average annual flow for three years behind the massive dam being proposed in Boulder Canyon. Everyone recognized that it would be the kingpin structure of any water resources development with the seven-state basin.

In 1921 Congress authorized the seven basin states to enter into a compact for allocation of the river resources. The acrimonious negotiations were chaired by Commerce Secretary Herbert Hoover, who was appointed as the federal representative to the Colorado River Commission by President Harding. Hoover exercised his considerable political skills to force compromises that eventually resulted in the partitioning of the river's watershed into Upper and Lower Basins, with each basin having the right to develop half (7.5 million ac-ft) of the river's 15 million acre-feet average annual flow volume (this figure was overly optimistic, the average flow being closer to ~13.5 million ac-ft/yr after 1930). The upper basin supplied the majority of the water, but the greatest demand was in the lower basin, where the river flowed through an arid landscape.

The resulting agreement signed in November 1922 was the Colorado River Compact, which came to be known as the "Law of the River," upon which all subsequent water resources development in the western USA would be based (Anderson, 2004; Hobbs, 2008). The compact did not become law until June 1929, when 6 of the 7 basin states ratified the agreement, six months after passage of the Boulder Canyon Project Act (Arizona was forced to accede to the agreement after the Supreme Court dismissed their case in May 1931). The Compact has been amended on numerous occasions since that time, most notably, in 1931, 1944, 1948, 1956, 1964, 1968, 1970, 1973, and 1974.

1924 Weymouth Report

Surveys, foundation investigations, and preliminary structural and hydraulic analyses of the entire Colorado River Basin were carried out by the Reclamation Service throughout 1922-23 and were summarized in a massive document known as the "Weymouth Report" (Weymouth, 1924) because it was submitted by Reclamation Chief Engineer Frank E. Weymouth to the Secretary of the Interior in February 1924. The Weymouth Report contained a preliminary design for an arched concrete gravity dam in Black Canyon, which utilized concrete cofferdams that became integral portions of the completed dam, involving about 235,000 yd^3 of concrete. Outlet works were all run through the main dam structure, as shown in Figure 9.

This preliminary design was subsequently adopted as representative of the yardage and material costs for estimates so crucial to the project in securing approval and appropriations from Congress. The maximum bearing stress of 41.3 tons per square foot (tsf) beneath the upstream heel at maximum section was about twice as high as any dam then in existence or even contemplated. At this stage in the design process the concrete dam was also designed to survive flood overtopping 21 feet deep! This is why both ends of the dam crest were sloped 25 feet upward, to train the overflow over the center of the dam, sparing erosion of the rock abutments.

When the 1924 design was released the issue of generating hydroelectric power had not been decided, so was altogether omitted. All of the Bureau designs for this stage forward were made with an eye towards their being retrofitted to accommodate hydroelectric generation, even if these were not shown on the plans (Raphael, 1977).

Between the fall of 1922 and spring of 1928 California Congressman Phil Swing (representing the Imperial Valley) and Senator Hiram Johnson jointly sponsored Boulder Canyon Project bills at each congressional session (twice per calendar year), but without success (Moeller, 1971). The proposed project cost of $125 million was the largest federal appropriation ever approved up to that time and it was difficult to marshal sufficient political support by the other 47 states, being viewed as a project seeking to benefit one particular area of southeastern California.

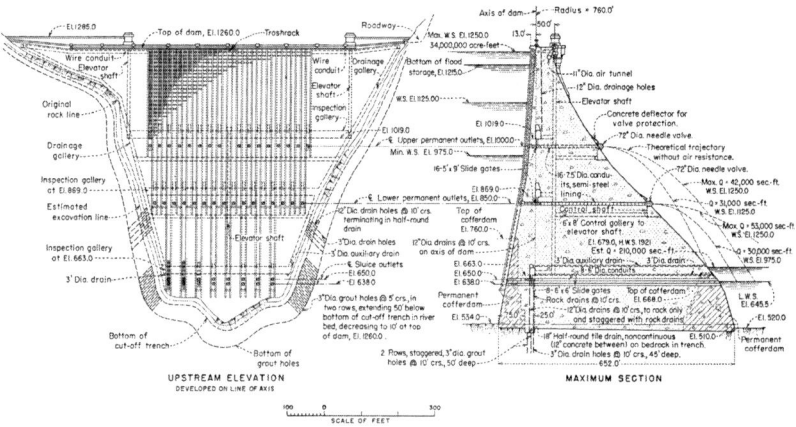

FIG. 9. Plan for a 740 ft high arched concrete dam in Black Canyon from the Weymouth Report in 1924. This design assumed a crest elevation of 1260 ft, 28 ft higher than Hoover Dam. Note the dipped crest to accommodate overflowage at left and the hachured zones in the elevation at right, which were concrete cofferdams wrapped into the dam's heel and toe. (USBR)

THE BOULDER CANYON PROJECT

Congress Debates the Boulder Canyon Project Act

The California delegation labored diligently over the next six years. Spurred by a water crisis Los Angeles entered the fray in October 1923 after its voters approved a bond measure to perform feasibility studies for a Colorado River Aqueduct. In July 1924 Los Angeles filed for rights to one million ac-ft of Colorado River water.

In September 1924 ranchers in the Imperial Valley watched their crops wither as near-record low flows of the Colorado River precluded the necessary diversions for irrigation. The following month President Coolidge announced his support of the Boulder Canyon Project during his election campaign. In April 1925 Los Angeles concluded that a 268-mile-long Colorado River Aqueduct was technically feasible. That fall voters approved a $2 million construction bond for a new aqueduct. In April 1926 San Diego filed for 110,000 ac-ft per year of Colorado River water.

That same month (April 1926) Reclamation Commissioner Elwood Mead issued a statement asserting that the unprecedented price tag associated with the proposed Boulder Canyon Project could be financed through a bond issue that would not interfere with the operations of the federal government, but would generate income through the sale of electricity. These funds would pay the interest on all the monies advanced by the government for construction, and provide a sinking fund for repayment of project costs. This would eventually become the model for countless Reclamation projects for years to come.

Each year the proposal evolved into a larger and more multi-faceted project, providing irrigation to an enlarged area while providing flood control and electricity, but the direct benefits were still limited to the Colorado River Valley and southern California, and were vigorously opposed by Arizona and Utah.

In March 1927 Interior Secretary Hubert Work appointed a Board of Advisors to make an extended survey of the Colorado River watershed to study the problems that might be associated with its development, posing five specific questions that the administration wanted answered because so many criticisms and alternative schemes were being proposed. This board was chaired by former Interior Secretary James R. Garfield, Wyoming Governor F. C. Emerson, former Nevada Governor and State Engineer James G. Scrugham, Colorado Senator Charles W. Waterman, and the board's secretary, Professor W.F. Durand of Stanford University. Durand's name had been suggested to Secretary Work by Stanford President Ray Lyman Wilbur (who succeeded Work as Interior Secretary in March 1929, when Herbert Hoover became President), because he had served on the Consulting Board of Engineers appointed to review the expansive system of aqueducts, reservoirs, and hydroelectric powerplants constructed by the City of Los Angeles from 1906 onward, and had authored an objective article about the various engineering problems posed by development of the lower Colorado River (Durand, 1925; 1953).

The Board of Advisors made a reconnaissance of the Colorado River between Lee's Ferry and the Gulf of California, with particular attention to the potential dam sites in Boulder and Black Canyons (Emerson et al., 1928). Each of the five members wrote their own separate reports, which were compiled together and published in January 1928. Durand's portion was the most comprehensive, occupying 56 of the report's 70 pages.

The political situation changed with the record flooding that struck the lower Mississippi River Valley in the spring and summer of 1927, which wrecked havoc on all the states adjoining the great watercourse, which drains 41% of the continental United States. Commerce Secretary Hoover provided stirring testimony before the House Committee on Irrigation during the 69th Congress (fall 1927) urging them to approve the Boulder Canyon Project (Committee of the Irrigation Division, 1929). The

1927 flood enabled Congressman Swing to forge new alliances and political partnerships with his Midwestern and Southern colleagues, and he brought sweeping support from the California delegation for the precedent-setting federal Flood Control Act approved in May 1928 by the 70th Congress (Moeller, 1971, Nadeau, 1974).

With these new political allies, Swing's Boulder Canyon Project Act bill passed through the House on May 15th, but failed to make it through the Senate before the session concluded. It's likely passage was temporarily derailed by the collapse of the St. Francis Dam on March 12-13, 1928, which killed upwards of 435+ people, making it worst American civil engineering failure in the 20th Century (Rogers, 1995). St. Francis Dam had been built in 1924-26 by the Los Angeles Department of Water & Power, one of the most vociferous proponents of the Boulder Canyon Project. The Arizona and Utah delegations were quick to point out that St. Francis was the same sort of concrete gravity arch dam being proposed at Boulder Canyon (although much smaller, about 200 ft high). Boulder Canyon Dam would be 3-1/2 times higher and would contain 26 times more concrete than the ill-fated St. Francis Dam. The senators from Utah and Arizona succeeded in filibustering passage and the question of safety had to be answered to most everyone's satisfaction before the Senate would approve the measure.

Colorado River (Siebert) Board

As a compromise measure Congress passed House Resolution 5773/Senate Resolution 65 on May 29, 1928, directing Interior Secretary Ray Lyman Wilbur to appoint "a board of five eminent engineers and geologists," of which at least one was to be an Army Engineer, to examine the proposed sites for the Boulder Canyon Dam, and report on "the safety, economic, and engineering feasibility, and the adequacy of the proposed structure and incidental works," such as all the electrical generating equipment, the structures associated thereof, and the proposed scheme for selling electrical power to pay off the project with interest over a term of 50 years (a major criticism of the proposed project by some members of Congress).

The review also included a critical examination of "appurtenant structures," such as the Imperial Dam about 300 miles downstream of Hoover (completed in 1940), the All American Canal (completed in 1940) to be built along the international boundary, and the 123-mile- long Coachella Canal, north of the Salton Sea (completed in 1948). They were given exactly six months to perform these reviews, with a written report due no later than December 1, 1928.

This panel was officially named the Colorado River Board (CRB), but was referred to at the time as the "Siebert Board," because it was chaired by retired Corps of Engineers Major General William L. Siebert, famous for his role in constructing the Panama Canal and as the first chief of the Army's Chemical Warfare Branch during the First World War (Clark, 1930). Board member Robert W. Ridgway had worked on New York's New Croton and Catskill Aqueducts and as chief engineer of New York's Board of Transportation. He had recently served as the president of ASCE in 1925. The other engineer was Professor Daniel W. Mead of the University of Wisconsin, a respected expert in hydraulics who had considerable experience with dams. He subsequently served as ASCE president in 1936.

FIG. 10. The 'Siebert Board' appointed by Congressional Resolution in May 1928 to review the Bureau of Reclamation design of Boulder Canyon Dam. From left: William L. Siebert, Elwood Mead, Warren J. Mead, Charles Berkey, Daniel W. Mead, and Robert Ridgway. The three Meads were unrelated. (USBR)

The CRB was unusual in that it also included two geologists, because the St. Francis Dam failure had been blamed on faulty foundation conditions, not on the engineering of the structure itself (Committee Report for the State, 1928). The two geologists were Columbia University Geology Professor Charles P. Berkey and University of Wisconsin Professor Warren J. Mead.

Reclamation Commissioner Dr. Elwood Mead served as the board's technical advisor and principal liaison with the Bureau of Reclamation. The board members are shown in Figure 10. They were given six months to review Reclamation's designs and make recommendations on where to site the great dam along the Colorado River.

On December 3, 1928 The Colorado River Board issued its 15-page report (CRB, 1928). Under the section titled "Review of Plans and Estimates; The Dam and Incidental Works," the Board's initial comments were blunt and succinct:

The board is of the opinion that it is feasible from an engineering standpoint to build a dam across the Colorado River at Black Canyon that will safely impound water to an elevation of 580 feet above low water...

The proposed dam would be by far the highest yet constructed and would impound 26,000,000 acre feet of water. If it should fail, the flood created would probably destroy Needles, Topock, Parker, Blythe, Yuma, and permanently destroy the levees of the Imperial District, creating a channel into the Salton Sea which would probably be so deep that it would be impracticable to reestablish the Colorado River in its normal course. To avoid such possibilities the proposed dam should be constructed on conservative if not ultra-conservative lines.

The Board felt that Reclamation should place the dam in upper Black Canyon, not in Boulder Canyon. The reasons given were: 1) superior geology; 2) narrower canyon, steeper walls; 3) site more accessible from existing rail lines and highways near Las Vegas; 4) river channel not as deep, less volume of excavation to competent rock [even though deeper]; 5) a dam of equal height would cost less and store greater volume of water than at Boulder Canyon; 6) rock less jointed; 7) few open fractures; 8) rock appears less pervious than at Boulder Canyon site; and, 9) rock easier to drill and excavate than at Boulder Canyon (one deep hole was advanced to a depth of 557 feet below the low water surface at the Black Canyon site which only encountered andesite). The Board also pointed out that the Black Canyon site was also much more favorably situated for the development of rail and highway connections to the job site, which were crucial to the project's estimated costs, as the government would be supplying all of the construction materials, with the exception of the concrete aggregate.

The Board also recommended a series of important changes in the dam's design and construction. They recommended that the Bureau of Reclamation reduce foundation contact pressure from 40 to 30 tsf; doubling the design capacity of river bypass diversion tunnels from 100,000 cfs to at least 200,000 cfs (25 yr flood); increasing the spillway capacity from 110,000 cfs to something greater than 160,000 cfs; and increasing the volume of flood storage to 9.5 million ac-ft; increasing the reservoir's maximum storage from 26 million to 30.5 million acre-feet (an increase of 31%). They also felt that the depth of water behind the upper cofferdam should be limited to no more than 55 ft (elevation 700 ft). The Board also affirmed their confidence that an All-American Canal could be built north of the Mexican border and that the electricity generated by dam could be absorbed by the expanding market of greater Los Angeles. The proposed changes increased the estimated cost of the project by 32%, from $125 million to $165 million.

Passage of the Boulder Canyon Project Act

The Boulder Canyon Project name was retained because Swing and Johnson had been trying to push the bill through Congress every year since 1922. The favorable and thorough report by the CRB resulted in rapid approval of the fourth version of the Boulder Canyon Project Act by a vote of 63-11 in the Senate on December 14th, 1928. The House approved a similar version, but with specific amendments requested by Utah on December 18th, and President Coolidge approved the act on December 21st. It was the largest government appropriation ever approved up to that point and its successful prosecution initiated a string of significant appropriations to the Bureau of Reclamation for western water projects from the general funds of the United States over the succeeding three decades.

Appropriations were delayed until a hard fought debate about the pros and cons of publicly generated versus privately generated electrical power, and setting the prices the Bureau of Reclamation would charge for water and electricity generated by the project. These agreements and the power allotments were not agreed upon until late March 1930. The sale of electricity would be the impetus for economic justification of every major Reclamation project (for dams over 200 feet high) for the next half century. The first appropriation of $10.66 million was authorized on July 3, 1930, which was earmarked for the construction of the government rail spurs and highways to the dam site.

A contract was also let for detailed surveying of the Black Canyon dam site and surrounding area, at a scale of 50 ft to an inch with 5-ft contour intervals. Final surveying of dam site continued throughout 1929-30 and into the spring of 1931, when Claude Birdseye (Chief Geographer of the USGS) and Heinz Gruner used terrestrial photogrammetry to construct a detailed topographic map of the dam site in Black Canyon, which was of unprecedented severity of slope (Gruner, 1972). They took stereopair photographs from ground stations tied in by triangulation. The Boulder City town site was laid out by Brock & Weymouth of Philadelphia.

Walker R. Young was appointed to be the Construction Engineer for the Bureau of Reclamation and he would remain on-site as the senor government representative throughout the project. Young was a logical choice, as he had been one of the principal authors of the 1922 Fall-Davis Report and he had supervised the Bureau's surveys and subsurface explorations carried out in Boulder and Black Canyons since 1921.

PERFECTING THE DAM'S DESIGN

Responses to expert advice (1928-31)

The Colorado River Board (CRB) continued to review the various design amendments made by Reclamation before the project went to construction. In April 1930 the CRB recommended that Reclamation increase the height of Boulder Dam by 25 ft, from 557 to 582 feet, with crest elevation increasing from 1207 to 1232 ft. The purpose was to provide 4,500,000 ac-ft of additional flood storage with a minimum freeboard of three feet, increasing the maximum seasonal flood storage to 9,500,000 ac-ft. This decision was influenced in great part by the 1884 flood, a recurrence of which would require Hoover Dam to spill 160,000 cfs downstream, even with maximum flood storage (ENR, 1930). The CRB recognized that this capacity should likely diminish to 4,000,000 ac-ft by 1988, if or when additional reservoirs were emplaced upstream. This decision increased the combined reservoir storage capacity to 30.5 million ac-ft.

The board also warned that flows over 75,000 cfs could cause considerable damage to training and control structures downstream. They recommended that the dam's radius of curvature be reduced from 740 to just 505 ft, and that Reclamation construct physical models of the arched gravity dam to test the new theories of stress distribution using the Trial Load Method of analysis, which Reclamation was using (USBR, 1940).

The board felt that the downstream water demand could be met with a minimum discharge of 8,355 cfs at the dam, apportioned as follows: 55 cfs to southern Nevada; irrigation needs between Black Canyon and Yuma 2,800 cfs; Los Angeles Aqueduct 1,500 cfs; All American Canal 2,000 cfs; and base flow into Mexico 2,000 cfs. The entire system could be self sustaining in terms of power demands and regulation by constructing additional dams with hydroelectric generation at Parker (Parker Dam) and Bulls Head (Davis Dam).

On September 17, 1930, Herbert Hoover's Secretary of the Interior Ray Lyman Wilbur, went to southern Nevada to presides over the project's initial construction at what came to be known as the "Silver Spur" ceremony, because he drove a silver spike at the foot of the Union Pacific Railroad spur that would provide a strategic connection with the dam site (Stevens, 1988). In his dedication speech, Wilbur announced that the dam would be named Hoover Dam, in honor of President Herbert Hoover.

In accordance with the recommendations of their Colorado River Board, in December 1930 the Bureau of Reclamation issued Specifications No. 519, which fixed the dam's crest elevation at 1232 feet, with the maximum reservoir level approximately 586 feet above expected tailwater elevation at the powerhouses. The dam's crest would be about 730 feet above the lowest point in the foundation (this figure turned out to be 726.4 ft). They abandoned plans for glory hole inlet spillways and chose to employ side channel spillways on either abutment (above the outboard diversion tunnels), controlled by 50 ft square Stoney Gates, and four dual bank valve houses on opposing canyon walls, one above another.

Reclamation's Consulting Boards

On January 10, 1931 Reclamation solicited bids for the construction of Hoover Dam and Power Plant. The Bureau also appointed a special Concrete Research Board in November 1930, which began meeting in January 1931. This board was comprised of P.H. Bares, W.K. Hatt, H.J. Gilkey, F.R. McMillan, and R.E. Davis. On March 11th (the same day the construction contract was awarded to Six Companies, Inc.) the Bureau of Reclamation's own Board of Consulting Engineers (BCE) approved its plans and specifications for Hoover Dam. The original board was comprised of: David C. Henny of Portland, Louis C. Hill of Los Angeles, and A. J. Wiley of Boise. All three were former Reclamation engineers actively engaged in private consulting. A.J. Wiley passed away on October 8, 1931 and Dr. William F. Durand of Stanford University was appointed to fill his slot. Reclamation's BCE remained in-place for approximately 10 years thereafter, providing peer review of subsequent projects, including Grand Coulee, Shasta, and Friant Dams, as well as many smaller projects in the American Southwest (subsequent members included geologist Charles P. Berkey, and engineers Joseph Jacobs and C. H. Paul).

ASSUMING RISKS TO REDUCE PROJECT COSTS

Hoover Dam was the first federal project to assume a broad spectrum of risks that had previously been the burden of contractors. These factors, known in the construction industry as "contingencies," were the principal unknowns that drove bid prices up or down. Many concerns were voiced about problems that would be associated with a project of such unprecedented size and duration (up to that point in time most dam construction projects had only taken between one and three years to construct). Risk of flooding might cause great uncertainties, while unforeseen changes in materials and labor costs during the unusually long term of the contract (seven years) was a very real proposition during the Great Depression, which had witnessed unprecedented fluctuations in materials costs.

The government's solution to these concerns was innovative and unprecedented:
1. The bid bond and construction surety to be advanced by the winning bidder were fixed at the moderate amounts of $2 million and $5 million, respectively. An incredible bargain for a $50 million project, but it was difficult for anyone to come up with $5 million dollars in early 1931.
2. Once the cofferdams were accepted by Reclamation, it accepted responsibility for flood damage to all property, except the contractor's plants.
3. Military veterans and U.S. citizens would have preference to being employed on the project.
4. Government was to supply ALL materials, except the concrete aggregate.

The federal government also contracted directly to provide:
1. Construction of a Union Pacific Railroad spur line 22.6 miles long from Las Vegas to Boulder City, then the U.S. Construction Railroad to the rim of Black Canyon overlooking the Nevada powerhouse, at an elevation just 138 feet above the dam crest (elevation 1370 feet).
2. Contracts with Southern Sierra Power Co., to deliver electricity to the dam site no later than June 25, 1931 via a 222-mile-long power transmission line from San Bernardino, California.
3. Awarded a contract to construct a paved highway to the dam site from Boulder City.
4. Awarded separate contacts for construction of Boulder City, to be administered by the Department of Interior.

Government Administration of Housing

It was initially envisioned that Six Companies would house 2,000 workers and their families in a temporary work camp that the government named "Boulder City," nine miles from the dam site. This number eventually swelled to 5,200, although the average number was closer to 3,500. Six Companies began by constructing wooden dormitories for 500 workers at a "river camp" at the head of Black Canyon, on the Nevada side of the river. Additional "family housing" was badly needed to replace the derelict abodes that characterized the unofficial settlement of "Ragtown," along the Colorado River in this same area. The town that became Boulder City was built in just 15 months to house 5,000 government employees and contract workers and their families. Boulder City remained a federal reservation under control of the U.S. Department of Interior until 1959, when it became self-governing. No alcohol was sold within the city until 1969 and no gambling has ever been allowed within the city limits.

HYDRAULICS AND HYDROLOGY ISSUES
Estimating Flood Frequencies with a Paucity of Hydrologic Data

Prior to the construction of federal dams and reservoirs, the Colorado was a river of extremes like no other in the United States. The flood-of-record along the lower Colorado River was 384,000 cfs measured at Topock, Arizona in February 1884, and the historic low flow about 500 cfs in 1911 at Lava Falls, in the western Grand Canyon. The 2.5 yr flood was believed to be about 120,000 cfs, while scour lines in Boulder Canyon suggest floods as great as ~500,000 cfs in prehistoric past.

FIG. 11. The Colorado River Board exercised considerable alarm upon their observance of high water marks 80 feet above low water level at the Boulder Canyon dam site (shown above behind rafts supporting three exploratory drilling rigs). This site was near the head of Boulder Canyon. These water marks dropped to about 40 feet above low water by the time the river passed out of the mouth of Boulder Canyon, about five miles downstream.

Table 1. Probable Frequency of Flood Discharges at Black Canyon Assumed in 1930

Discharge, cubic feet/second	Frequency With Which Discharge May be Equaled or Exceeded
130,000	Once in 5 years
160,000	Once in 10 years
190,000	Once in 20 years
230,000	Once in 50 years
260,000	Once in 100 years

320,000 cfs Once in 500 years
450,000 cfs Once in 10,000 years

When Reclamation revised their designs in 1928 the available data consisted of just 26 years of flow volumes recorded at Yuma, Arizona, 312 river miles downstream of Hoover Dam. Flow heights had been recorded at Yuma since 1878, but no velocity measurements had been recorded until 1902. Reclamation had only six years of flow measurements at Lee's Ferry, (346.6 river miles upstream), where a gauging station had been established in 1922; five years of flow volumes at Bright Angel (260 miles upstream, in Grand Canyon), and five years of reliable data at Topock (111 miles downstream). The largest dam in history was being sized with less than 10 years of reliable flow records, a very low figure considering the magnitude of the project.

During their review in 1928 the CRB exercised concern about how large the maximum probable flood might be after observing high water marks 80 feet above low water level at the head of Boulder Canyon (Figure 11). They concluded that a flow of 320,000 cfs every 500 years and 450,000 cfs every 10,000 years were altogether likely, and the frequencies shown in Table 1 were adopted for the project in 1930.

The board recommended that Reclamation increase the spillway capacity from 110,000 cfs to something greater than 160,000 cfs. At that time (1929) the general assumption employed by most designers was to build dams strong enough to withstand *double the largest flood that ever been observed.* The highest recorded flow Reclamation had to work with was 200,000 cfs at Yuma in 1902 (they were not apparently aware of, or did not trust, the estimate of 384,000 cfs at Topock, Arizona in January-February 1884 by W. A. Drake, Chief Engineer of the Atlantic & Pacific Railroad), so they doubled this figure and assumed this to be a conservative estimate (Debler, 1930).

After publication of Gumbal's flood probability triparte diagram in 1941 Reclamation appears to have used that plot to justify their estimates. By using Yuma gage flows for the period 1878-1929 (51 yrs) they used their design flood of 400,000 cfs to back out a recurrence frequency of once-every-3,950 years (Rogers, 2008).

In 1990 Reclamation retained Morrison-Knudsen Engineers (1990) to perform meteorological studies of the Colorado River Basin above Hoover Dam. Just prior to letting this contract the Bureau of Reclamation and the U.S. Geological Survey performed a paleoflood analysis of the 1884 high water marks in the Black Canyon, using flood observations at Lees Ferry; and gage observations at Grand Junction, Colorado and Yuma, Arizona. They estimated that this flood reached a peak flow of about 300,000 cfs in Black Canyon (Hoover Dam) in July 1884 (Swain, 2008).

Estimates of Annual Flow and Flood Storage

In their November 1928 report the Colorado River Board expressed these concerns:
"The information on which this flow has been estimated is inadequate to furnish an accurate or sound estimate on which to base an important project without using factors of safety sufficiently great to make such estimate conservative and safe."

The Board also noted problems with using Yuma gage readings taken between 1902 and 1922, bereft of reliable volume calculations. Reclamation had extrapolated these data to estimate the average annual flow volume of 16,200,000 ac-ft at the dam site (Debler, 1930). USGS estimated an average annual flow of just 13,600,000 ac-ft for the period 1878-1922, a much longer, and, therefore, more reliable sampling. The Board then made their own estimates of the average annual flow, based on the available data. They concluded that for the period 1887-1904 the average annual flow may only have been 9,360,000 ac-ft/yr, not the 15,000,000 ac-ft/yr assumed in the Colorado River Compact. 75 years later the Board's skepticism was somewhat justified insofar that the actual annual runoff since 1930 has been about 13,500,000 ac-ft/yr (Anderson, 2004).

The Board also recommended that the volume of flood storage of the Boulder Canyon Project reservoir be increased to 9.5 million acre-feet of the total reservoir capacity of 30.5 million acre-feet, a startling 31% of the total storage. This was an unprecedented conservative figure for that time, which drew considerable ire on the part of some Reclamation engineers.

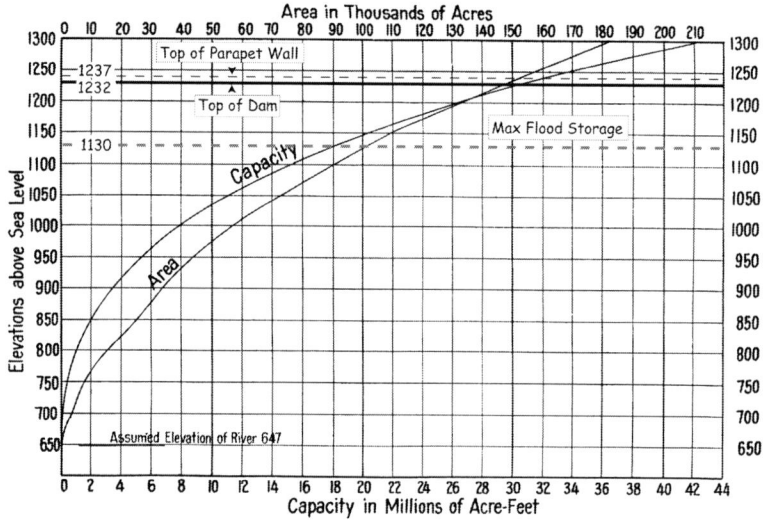

FIG. 12. Reservoir area and capacity curves for a dam in upper Black Canyon, prepared by the Bureau of Reclamation in 1928. The respective levels for Hoover Dam, as constructed, are indicted (USBR).

In April 1930 the decision was made to raise the dam 25 feet to increase flood storage (Debler, 1930). The dam's height was increased to 730 ft above the assumed deepest point of the foundation, with a crest elevation of 1232 ft. This was intended to provide 30,500,000 ac-ft of total reservoir storage, equal to two years cumulative flow of the Colorado River. The reservoir area and capacity curves constructed by the Bureau of Reclamation in 1928 are presented in Figure 12. Of that total, 9,500,000 ac-ft of seasonal flood storage between April 1st and September 1st of each year was intended to accommodate spring runoff from the mountainous interior (in accordance with the recommendations of the Colorado River Board). June and July typically see the greatest volume of runoff, while Reclamation normally targets July 1st as the optimal date for "topping off" their storage reservoirs. Over the years the flood storage has gradually been reduced to just 1,500,000 ac-ft between January 1st and August 31st as more reservoirs have been completed upstream of Lake Mead (Swain, 2008).

In 1930 Reclamation assumed a design flood inflow of 300,000 cfs with a 60-day volume of 23,200,000 ac-ft (Debler, 1930). In 1990 Morrison-Knudsen Engineers (1990) re-examined the hydrologic models for the Colorado River Basin, utilizing a much greater volume of data than was available in 1930, along with considerations of the various reservoirs that now dot the basin. It assumes that most critical situation could occur in August, when a Pine and Cedar Mountains centered storm follows a San Juan Mountains centered storm by seven days. This storm sequence would produce a probable maximum flood (PMF) at Hoover Dam with a peak discharge of 1,130,000 cfs and a 60-day volume of 9,300,000 ac-ft. This peak discharge is nearly four times that assumed in 1930, with a flood volume of about 40% that estimated in 1930 (Swain, 2008).

Spillway Design Evolution

The spillways at Hoover Dam went through a remarkable design metamorphosis. In 1920 the Reclamation Service envisioned the use of 32 siphon spillways, each 16 x 16 ft, splitting these with 16 on each side of the canyon, each group connecting to two 30-foot-diameter horseshoe-shaped diversion tunnels. Their aggregate capacity was 300,000 cfs.

The 1924 Weymouth Report presented conceptual designs for a concrete gravity arch dam (Figure 9) capable of passing 80,000 cfs through interior outlet works, with any excess passing over the dam's crest to a maximum depth of 21 ft. The rockfill dam alternative (Figure 5) envisioned seven drum gates on the Arizona abutment connected to seven tunnels along with an excavated channel that conveyed spillage back into the Colorado River about 1.5 miles downstream. The bypass flow capacity of this scheme was between 180,000 and 300,000 cfs.

Between 1924 and 1928 there was a turnover in staff at Reclamation because of political battles over funding and Chief Engineer Frank Weymouth and a number of his key subordinates, such as Julian Hinds, resigned. The new staff began concentrating on how large bypass flows could be safely passed around a dam of unprecedented height. They chose massive glory hole spillways similar to that which was then being constructed at Owyhee Dam in Idaho (which employed a 60-foot diameter glory hole spillway necking down to 28 ft. With a design capacity of 41,730 cfs, it was the largest of its kind till 1957, when a slightly larger version of the same design was built for California's Monticello Dam). Massive 50-foot-diameter shafts would be connected to two of the four the diversion tunnels, as shown in Figure 13.

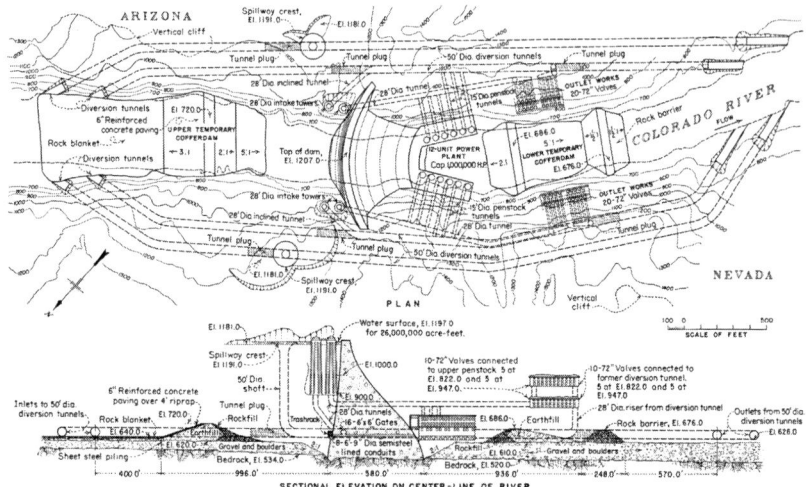

FIG. 13. 1928 design for Boulder Dam at the Black Canyon site, which employed 50-foot-diameter glory hole spillway inlets and a dam with a radius of curvature of 740 ft. This was the design reviewed by the Colorado River Board, which was extensively revised before actual construction began in 1931.

These could accommodate spillage of up to 200,000 cfs. The problem with these was the flow velocities approached 180 ft/sec (fps), far above any other glory hole spillways then in existence. Reclamation engineers brought this figure down to 100 fps by employing risers at the downstream end of the diversion tunnels, but this greatly reduced the outflow capacity.

Reclamation undertook extensive 1:60 scale model studies with the University of Colorado to examine the velocity issues and ended up concluding (erroneously) that smooth high strength concrete could be employed at the elbow transitions between the glory hole shaft and the diversion tunnel to avoid any problems with "erosive action of water" (cavitation). The crest of the spillway was surrounded by an 11-foot-high Ogee crested weir, 234 ft in diameter, in the familiar morning glory shape. Further model studies which included the impacts of the sloping canyon walls showed that the massive glory hole spillways would fill asymmetrically, causing undesirable suction and pulsation when the depth of water approached design levels. For these reasons it was decided to employ open side-channel spillways.

In the fall of 1930 Reclamation altered the spillway design to employ massive side channel, or trough, spillways on either abutment, connected to the two outboard diversion tunnels (Figure 14). These spillways were to be controlled by using 50 x 50 ft steel Stoney Gates at either end (entry and point of discharge), similar to those seen today at the mouths of the two inboard diversion tunnels. The Stoney Gates at the upstream entries allowed the reservoir to be controlled between elevations of 1173.6 and 1223.6 ft, and could theoretically spill 270,000 cfs at a maximum reservoir level of 1232 ft (the top of the dam, without considering the parapet wall, at el. 1237 ft). This is the spillway design that was included in the original plans and specifications bid in early 1931.

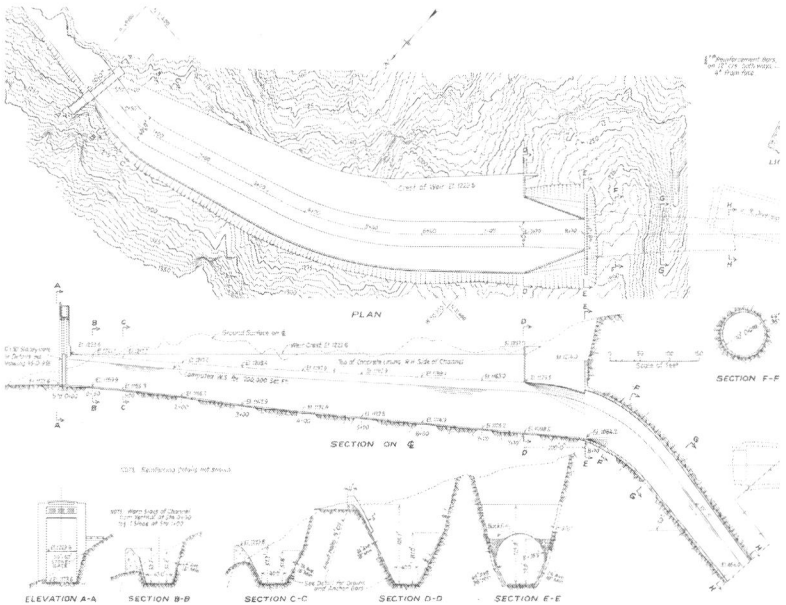

FIG. 14. Reclamation's December 1930 design for side-channel spillways controlled by 50 x 50 ft Stoney Gates. These massive spillway troughs were 870 ft long, passing into 70-foot-diameter throated inlets (Wilbur and Mead, 1933).

The Decision to Employ Drum Gates on the Side Channel Spillways

Model tests on the free crested side-channel spillways showed that they exhibited considerable loss of economy due to turbulent mixing and that an extensive array of training vanes would be needed to alleviate the turbulence, which would entrain air and thereby increase the risks of cavitation. It was then discovered that in order to pass 200,000 cfs at acceptable velocities, they would need a net spillway crest length of about 3,000 feet, which was out of the question because each spillway trough was 650 ft long.

After the winning bid had been submitted, but before the contract was let (mid-1931) Reclamation changed the design of the spillways to full gate control, using floating drum gates. These would be the largest drum gates in the world, allowing spillage of 63,000 cfs with the reservoir at an elevation of 1229 ft, to a maximum discharge of 400,000 cfs with the reservoir at elevation 1232 ft (top of the dam). These side channel spillway troughs are 650 feet long, 150 feet wide, and 170 feet deep on each canyon wall. More than 600,000 cubic yards of rock were excavated for the spillways. The troughs led into inclined shafts 50 feet in diameter and 600 feet long.

The spillway crest elevation is controlled by four 100 x 16-foot hollow drum gates on each spillway, each drum weighing 250 tons. By allowing water into their nested chambers, the hollow drum gates are lifted upward, to a maximum height of 17 ft above the spillway sill. Maximum discharge velocity in the voluminous spillway shafts is about 175 feet per second, or 120 miles per hour. The flow over each spillway would be about the same as the flow over Niagara Falls, and the drop from the top of the raised spillway gates to the river level would be approximately three times as great. The general layout of the main spillways, outlet valves, and penstocks feeding into the powerhouses are shown in Figure 15.

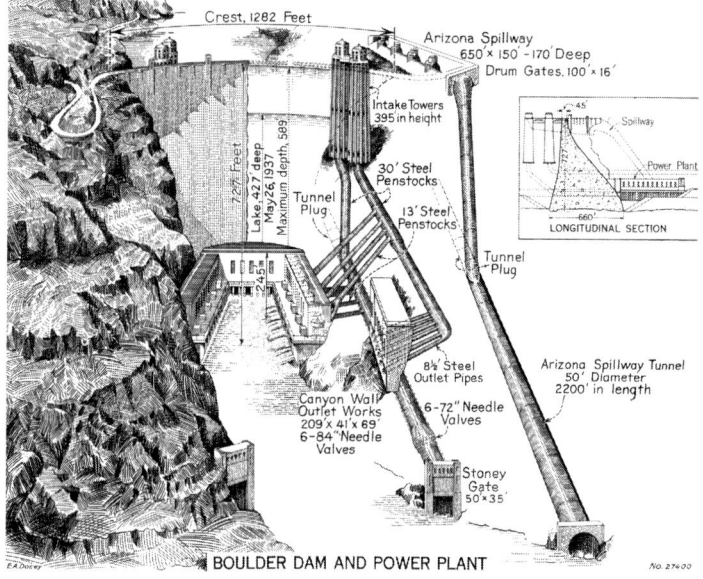

FIG. 15. Cutaway view showing the layout of the main side channel spillway, canyon outlet works, needle valve outlets in the inboard diversion tunnel, and penstocks leading to the powerhouse, all on the Arizona abutment. A similar array exists on the Nevada side (USBR).

Construction access to the higher elevations on the dam's left abutment were accommodated by constructing high catwalks (Figure 16) with the workmen using equipment that had to be driven 237 miles, south through Searchlight, Nevada to Needles, California, and another 10 miles to cross the Colorado River at Topock (about 10 miles downstream of Needles, California). From Topock workers followed old Route 66 through Oatman to Kingman, Arizona, before turning northwest, along the old mining road connecting Kingman to the mining town of Chloride. From Chloride an unimproved road extended up the Detrital Valley to Householder Pass.

From here a new road had to be blazed across White Rock Canyon to the dam's Arizona abutment. A new high-speed highway was graded a year or so later between Kingman and the dam, which became U.S. Highway 93.

As completed in 1935 the total spillage capacity of the dam was about 491,200 cfs. This was broken down as 400,000 cfs in the main side-channel spillways; another 48,000 cfs through the canyon wall outlet works; and 43,200 cfs through the diversion tunnel plug outlet works. An additional 28,800 cfs capacity was gradually absorbed as the powerhouse turbines were added between 1938 and 1961. The canyon wall outlet works were removed in 1954. Today up to 50,000 cfs can be passed through the powerhouses, so the aggregate spillage capacity is assumed to be 493,000 cfs.

FIG. 16. This daring catwalk suspension bridge situated 650 feet above the river allowed workers to cross from the Nevada side to the opposite side to work on the Arizona spillway (USBR).

Valve House Outlet Works

When Hoover Dam was designed valve houses (Figure 17) were originally situated 180 feet above the river on both canyon walls. The first steel penstocks and outlet works pipes began to be placed in 1934. There were 4,700 feet of 30-foot-diameter pipe and 2,000 feet of 8 1/2-foot-diameter pipe. The maximum thickness of the largest pipe was 2-3/4 inches. Each of these canyon-side valve houses was originally configured with six 72-inch-diameter needle valves.

At this same time four 68-inch jet flow gates were installed as "plug outlets" within the inner diversion tunnel plugs. The gates are designed to bypass water around the dam under emergency or flood conditions, or to empty the penstocks for maintenance work.

Between 1956 and 1964 Glen Canyon Dam was constructed along the Colorado River 370 river miles upstream of Hoover Dam and began storing water in March 1963. The erection of Glen Canyon Dam provided significant flood control benefits for Hoover Dam and greatly diminished the rate of sediment accumulation in Lake Mead. In 1979 Reclamation removed two needle valves from each of the tunnel plug outlet works and replaced by a pair of 90-inch- diameter jet flow gates (total of four), which discharge into the river. They also removed six needle valves from each of the canyon wall outlet structures. The abandoned outlets were permanently sealed with high-pressure steel bulkheads and new valves replaced the worn valves in the remaining outlet conduits.

FIG. 17. Water being discharged from the Nevada and Arizona valve houses during the spillway tests in the late summer of 1941, looking downstream (USBR).

STRUCTURAL DESIGN ISSUES

Design of the Mass Concrete Mix

Reclamation's Concrete Research Board (P.H. Bates, W.K. Hatt, H.J. Gilkey, F.R. McMillan, and R.E. Davis) met throughout 1931-34 to provide advice on the many challenging issues posed by the proposed construction, which involved considerable research and innovation. Reclamation's Chief Designing Engineer John L. "Jack" Savage specified four sacks of cement per cubic yard for the mass concrete in all of his dams (Raphael, 1977). Each sack weighed 96 lbs, so 376 pounds of cement were included in each cubic yard, which was the weight of one barrel of cement. The cement used in the early 1930s was much coarser than that used today and the 1931 mix employed more water than would be used today. For these reasons engineers of that era were obliged to use more cement to obtain the desired strength (this was an era before the use of internal vibrators and air entrainment).

Low heat cement was used for the main dam after a small portion of the base was poured (low heat cement not yet being available in large quantities). During the winter months they used a blend of 60% low heat and 40% standard Portland cement.

The mass concrete allowed rock aggregate up to 9 inches in diameter, which was unusually large for mass concrete at that time (this would be equivalent to an 8-inch size using modern aggregate screens). A fairly 'dry mix' was specified, allowing for a 3-inch slump (Figure 18). Standard 6 by 12-inch test cylinders, removing the aggregate greater than 1.5 inches in diameter, were used.

FIG. 18. Slump test on dam concrete, using 4-sack per cubic yard mix (USBR, 1947).

Higher strength structural concrete (with a greater proportion of cement) was used in the powerhouses, inlet towers, and tunnel linings (all with steel reinforcement). The daily cement demand during construction of the dam was from 7,500 to 10,800 barrels per day. Reclamation had used only 5,862,000 barrels in its 27 years of construction activity prior to June 30, 1932.

Between 1932 and 1935 10,000 concrete test specimens were made and tested for Hoover Dam. The materials and included cements represented a wide range of chemical composition and fineness. These specimens included: (a) cement paste, (b) mortar, and (c) mass concrete. For economy, most of the concrete specimens were 4- by -8 inch test cylinders with ¾-inch maximum size aggregate. Some large diameter cylinders were also poured to evaluate other properties, such as modulus of elasticity and Poisson's Ratio, and long-term creep of the actual mix using 36 x 72-inch test cylinders, which allowed the full range of aggregate sizes (Blanks and McNamara, 1935; USBR, 1947).

Chilled Water to Absorb Concrete Heat of Hydration

Hoover Dam was of unprecedented height and volume. Jack Savage and his design team at Reclamation determined that the concrete would remain warm for hundreds of years due to the cement's heat of hydration (Savage, 1936). This was the first time that internal heating of a concrete dam was viewed as a major design problem. At first, it was thought that the solution would be to develop cement with sufficiently low heat liberation that the increased temperatures would be tolerable. Low-heat cement was indeed developed and used in the dam, but even with the low heat cement, the internal temperature still reached 150 degrees F in Hoover Dam.

In 1931 Professor Raymond E. Davis, a member of Reclamation's Concrete Research Board and Director of the Engineering Materials Laboratory at the University of California (Berkeley) began a concerted research program funded by Reclamation to examine the cement chemistry, heat of hydration, maximum aggregate size, most favorable mix proportions, design of an artificial cooling system, the most favorable dimensions of monolithic pour blocks comprising the dam, grouting methods and materials, and the methods employed in the manufacture and handling of the mass concrete (Davis, 1932).

The first step of his research was to devise a cement composition that would produce a low heat of hydration because Reclamation had recently measured the heat of hydration produced by curing concrete on Gibson (1929) and Ariel (1930) Dams (Townsend, 1981). There had also been widely reported problems with heat of hydration causing undesirable tensile stresses and cracking of Rodriguez Dam near Tijuana, Mexico (Noetzli, 1934), described later.

As the dams grew increasingly higher in the late 1920s to early 1930s engineers demanded higher strength concrete, which required additional cement (mixes using 4 or more sacks per cubic yard). The increased heat of hydration was recognized as a potentially troubling aspect of employing higher strength mass concrete. The higher strength concrete was deemed necessary to handle such factors as long-term loading (creep), natural variations in strength, and the requirement that design strength of test cylinders be significantly higher than the maximum design stress, and so forth. No corresponding compensations were made for favorable factors, such as the automatic transfer of stress from highly stressed regions to those adjoining regions of lower stress.

Tests of arch dam models suggested that large stresses develop at the downstream haunch which diminishes to near-zero at the upstream face. Failure only ensued when the average stress over a section reached the compressive strength of the concrete. When the elastic limit was exceeded, a redistribution of stress occurred, transferring stress away from highly stressed regions. Unfortunately, such favorable transfer did not appear to occur in the case of tensile stress. For these reasons, the most noticeable adjustments in Hoover Dam's final design were made along the upstream face, which was sloped noticeably upstream, to reduce the likelihood of the dam developing tensile stress at its upstream toe.

The calculated heat of hydration for the final design of Hoover Dam was 40 degrees F with 125 years to cure and cool, absent any artificial cooling. The Concrete Research Board felt that this volume of concrete would set off thermal stresses that would certainly crack the dam. It was therefore determined that various measures should be employed to reduce the heat of hydration, which were being tested on Owyhee Dam, then under construction near Boise, Idaho (Scott, Nuss, and LaBoon, 2008).

At that time it had been determined that cement contained four principal compounds: tricalcium silicate, dicalcium silicate, tricalcium aluminate, and tetra-calcium aluminoferite. For convenience, these were referred to as C_3 S, ~ C_2 S, ~C_3 A, and C_4 AF. The two silicates produced most of the cement's strength. Cement with a low heat of hydration could be fabricated by increasing the amount of ~ C_2 S and decreasing the amount of ~C_3 A. This became the formula used in producing "Modified Low Heat Cement" for Hoover Dam (Davis et al., 1933; Savage, 1936). The heat of hydration of normal Portland Cement is between 85 and 100 calories per gram, while that for the low-heat cement at Hoover Dam was 65 (at 7 days) to 75 (28 days) calories per gram, with an average 28-day strength $f_c' = 2,000$ pounds/square inch (psi) (Savage, 1936). In retrospect, there was actually no need for low-heat cement because the internal heat of hydration was removed by circulating cooling water in embedded pipes.

In addition to using low-heat cement, two other measures were adopted to help alleviate problems with internal heating of the mass concrete during hydration. One was to cast the concrete in blocks small enough so that they would shrink as a monolithic block, and thereby avoid development of uncontrolled shrinkage cracks. In 1931 Reclamation let a three-year contract with the University of California to carry out extensive tests to ascertain the heat of hydration from cement curing (Davis and Troxell, 1931; Davis 1932; Davis et al., 1933). These tests found that about 90% of the heat is generated in the first 28 days, but that this heat could not be dissipated adequately if the concrete was insulated by warmer concrete above and around it. Hoover was the first mass concrete dam to receive this level of analysis, although research and adjustments to mass concrete mixes for dams continued for many years thereafter (Liel and Billington, 2008).

Dry mixes were specified to reduce shrinkage from moisture change. The dam was built in blocks or vertical columns varying in size from about 60 feet square at the upstream face of the dam to about 25 feet square at the downstream face, using steel forms. Adjacent columns were locked together by a system of vertical keys on the radial joints and horizontal keys on the circumferential joints (Figures 19 and 20). Lift heights in each block were limited to five feet in 72 hours, and 35 feet within 30 days. After the concrete was cooled, grout was forced into the spaces created between the columns by the contraction of the cooled concrete to form a monolithic (one piece) structure. Shrinkage was about 0.5%.

Water stops were employed near the up and downstream faces of block joints. Vertically serrated joints were used between blocks in the dam. These joints were grouted after the blocks had shrunk. Horizontally serrated joints used against the abutments. All joints between blocks were to be grouted in 100-foot lifts, after cooling occurred. All cooling pipes were grouted as well, after water circulation ceased.

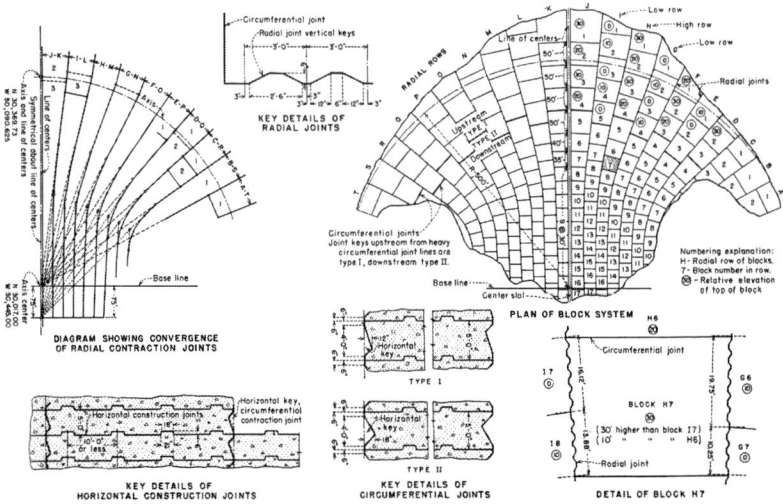

FIG. 19. Plan illustrating the block system used by Reclamation to isolate concrete shrinkage and heat of hydration. The spaces between the blocks were grouted after most of the expected shrinkage and curing had occurred (USBR, 1947).

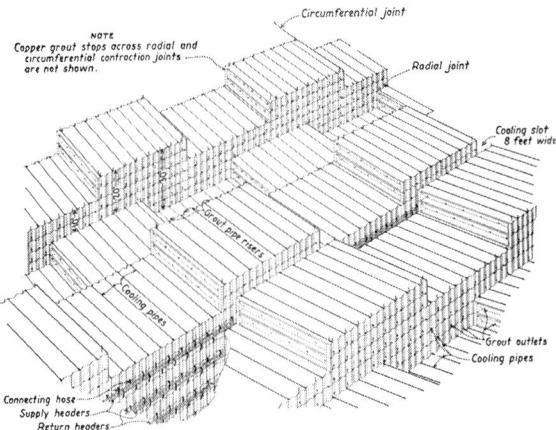

FIG. 20. Detail showing the offset nature of the monolithic blocks comprising the dam, intended to avoid the formation of random shrinkage cracks within the dam. Note beveled shear keys between blocks and the cooling pipes, which were laid parallel to the dam's radius (USBR, 1947).

Designing the Embedded Instrumentation

In 1925, a group of engineers led by Swiss engineer Fred Noetzli conceived the idea of building an experimental arch dam along Stevenson Creek a few miles below Shaver Dam, in the western Sierra Nevada of California (Veltrop, 1988). The test dam was a thin arch 60 ft high and just 7.5 ft wide at its base. The purpose was to test the evolving theories of arch and cantilever loading in a constant radius arch dam. The structure was fitted with every sort of measuring instrument available at that time, including carbon-pile telemeters, which were a kind of strain meter which could be embedded in the wet concrete at the time it was poured. This was the first time that meters were used to measure internal strains of a dam. A young physicist from Caltech named Roy Carlson was placed in charge of the strain measurements, a consultation that proved providential for Carlson as well as dam engineering, world-wide.

The carbon-pile telemeters used at Stevenson Creek did not exhibit long-term stability. After the project concluded Carlson took a position testing materials for the Los Angeles County Flood Control District (LACFCD) began devising a better device for measuring internal strain, and then set about developing a companion device to measure internal stress. Carlson soon determined that the measurement of strain was far easier than measuring stress, and he searched for a suitable sensing element. He found that carbon-steel wire, when drawn down to smaller and smaller diameter kept increasing in tensile strength, reaching up to 700,000 psi. He found that for each one percent change in length, the wire's electrical resistance changed 3.6 percent. This was the physical attribute that allowed the development of the modern strain gage (Davis and Carlson, 1932).

Carlson then experimented with various schemes of mounting the elastic-steel wire to the sensing element, settling on two coils of the steel wire, initially stressed and mounted in such a way that one would increase in length when the ends of the meter were brought closer together and the other would decrease. The beauty of his device lay in its simplicity, which overcame concerns with temperature effects because the ratio of the resistances of the two coils would be affected to double degree when the gage length was increased. The current resistance was independent of temperature change because the temperature effects were compensating. Since the total resistance changed only with temperature, the device could also be employed to measure concrete temperature.

While working at Stevenson Creek in 1926, Carlson observed that concrete increased in temperature while hardening. He set about measuring the temperature rise in concrete poured for the Pacoima and Big Tujunga concrete arch dams built by LACFCD in the San Gabriel Mountains. He used an adiabatic calorimeter to study the heat generated by concrete during its curing.

At the same time (1928-30) a chemical engineer for the Riverside Cement Co. named Hubert Woods was trying to measure the heat of solution of dry cement and compare it with that of the hardened cement paste. Woods succeeded in demonstrating that the difference between the two was the cement's unique heat of hydration. This method proved to be reliable and far more efficient than Carlson's adiabatic calorimeter. However, it was also discovered that the heat-of-solution method was inaccurate when applied to cements containing pozzolan admixtures.

In 1929 Woods began collaborating with Carlson at his LACFCD lab on solving the puzzle posed by fresh cracks that had appeared at Rodriguez Dam near Tijuana, Mexico, where they were using a 5 sack/yd^3 mix (Noetzli, 1934). The cracks appeared within a few days of placement and were being blamed on Riverside Cement, the supplier. Hubert suspected that the cracks were ascribable to the temperature change induced by concrete curing. Carlson had noted significant cracks that developed in the Stevenson Creek Test Dam, which he ascribed to thermal stresses when the concrete warmed significantly during the first 48 hours after placement.

Carlson had been fabricating electrical-resistance thermometers by winding enameled copper wire on insulating spools and dipping them in hot tar to protect the wire from the corrosion (Carlson, 1938a). By embedding some of these thermometers in Rodriguez Dam they recorded a temperature rise of about 120 degrees F. over two days, which then decreased. A 120 degree temperature drop could trigger tensile stresses of more than 300 psi, sufficient to crack the dam's concrete (a five sack mix with 28-day compressive strength $f_c' = 2{,}650$ psi). Although assuming zero stress at maximum temperature initially seemed questionable, it was later found that this was very nearly the case (Davis and Carlson, 1932).

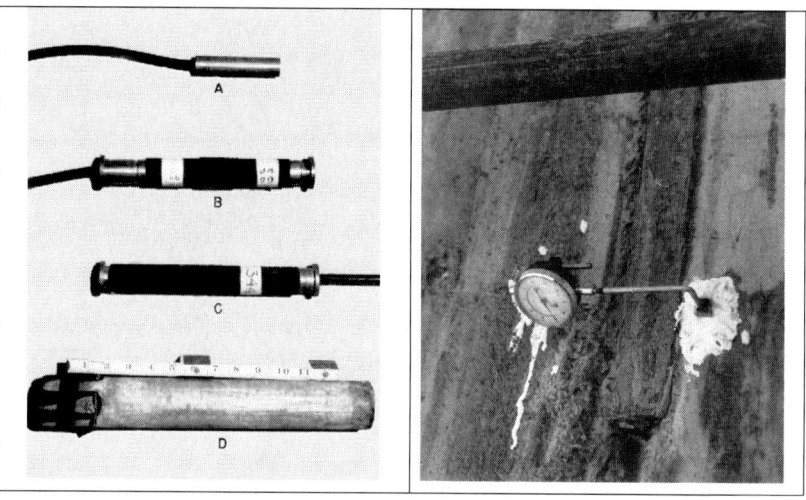

FIG. 21. Left image shows the four basic instruments invented by Roy Carlson that were used to instrument Hoover Dam. These included: A – resistance thermometer; B- joint meter; C – strain meter; and D- strain meter packed in protective tube for embedment. Image at right shows the conventional instrumentation used at exposed locations, in this case a radial dial gage placed across a horizontal joint (USBR).

Carlson's unique experience and his instruments were perceived as vital to Reclamation's needs and interests in developing a state-of-the-art protocol for concrete placement at Hoover Dam, because of its unprecedented scale (Carlson,

1977). Reclamation Chief Designing Engineer Jack Savage tried to hire Carlson to work at their Denver office, but this possibility fizzled when it was learned that Carlson's degrees were in mathematics and physics, not civil engineering (so he was not eligible to take the civil service examination required by Reclamation). In May 1931 the impasse was solved by Professor Davis at Berkeley, who talked Reclamation officials into funneling the concrete research, testing, and instrumentation development aspects of Hoover Dam to Berkeley's Engineering Materials Lab in a three-year contract (June 1931 to June 1934), with the understanding that Davis would hire Carlson as a research engineer (Carlson subsequently earned his master's and Sc.D. degrees in civil engineering, at Berkeley and MIT, respectively).

By the time Six Companies began pouring concrete in June 1933 Carlson had developed and tested the electrical instruments that could be embedded in the dam's concrete to measure strain, joint opening, and temperature (Figure 21). These included 450 of his electrical-resistance joint meters, which provided crucial instrumentation of the dam's expansion joints (although some of the joint meters were rendered inoperative by being carried beyond their design range, as described previously).

Carlson's resistance strain meters (Figure 22) were used to measure the strains engendered by the cement heat of hydration, dead weight accumulation (as the dam rose higher and higher), and, through inference, validate the design assumptions about cantilever and arch stress distributions in the Trial Load Method of analyses employed in the dam's design (Savage et al., 1931; USBR, 1939; 1940).

These measurements were continued on a regular basis up through the end of 1941, when the reservoir filled and the spillways were tested. Instrument readout banks were established inside the dam's galleries, as shown in Figure 23.

FIG. 22. Left image shows an array of resistance strain meters being embedded in a block, just prior to removal of the template and barrier. This array would record the strains in a horizontal plane. Right image shows an 11 strain meter spider array being assembled just prior to embedment. Spider arrays allowed a three dimensional assessment of the stresses developed within a dam for the first time (from Raphael and Carlson, 1965).

FIG. 23. Left image shows a battery of terminal boards in the Arizona Radial Gallery near elevation 705 ft. Each bank contained copper wire leads to 55 strain meters, which could be plugged into a portable Carlson strain meter testing set. Right image shows one of the dam's internal radial inspection galleries (USBR).

Measurements made during construction were used to control the cooling of the mass concrete and grouting of the dam's joints. These data and measurements were also used to validate the loading assumptions used in the design of the dam and to record its structural behavior. The measurements and resulting analyses validated most of the pre-construction modeling (Carlson, 1938b). The dam and abutments were also fitted with 64 triangulation stations to allow precise external monitoring of deflections. The interior of the dam was also fitted with tilt meters, plumb lines hung in special instrument shafts, recording thermometers, and three strong motion accelerographs. A system of pipes was also placed in connection with the base of the dam to measure the hydraulic uplift against the base of the dam.

The results of the instrumentation program were so successful they became standard practice for all of Reclamation's significant concrete dams thereafter, many of which received substantially more instrumentation, as improved devices, such as Carlson's mercury filled pressure meters, became available (Raphael, 1948; 1953; 1955). This 'second generation' of instruments was sufficiently sensitive to actually measure plastic flow of the dam's mass concrete (ENR, 1942).

Roy Carlson's association with Reclamation would continue for years thereafter. After his work on Hoover Dam concluded he accepted a faculty position at MIT in the fall of 1934. While teaching at MIT he recorded numerous patents for his measurement devices, including the elastic-wire strain meter (in 1936) and an electric-resistance pressure meter (in 1939). The strain meter came to be widely used and a modification of the pressure meter called the "pore pressure cell," was used to a lesser degree by Reclamation on a number of projects, moistly notably, on Grand Coulee, Shasta, and Friant Dams (Raphael, 1953). Three decades later Reclamation was still installing his instruments on their largest mass concrete dams. 2,000 Carlson strain meters were used at Glen Canyon Dam in 1958-64 and 1,600 strain meters were installed in Flaming Gorge Dam in 1958-64 (Raphael and Carlson, 1965).

CONCLUSIONS

The scale of the Boulder Canyon Project was so massive that it gave rise to an unprecedented volume of scientific research and engineering analyses, which was of inestimable value to the civil engineering community. Most of the technical aspects of the dam's planning and design were subsequently summarized in what were collectively known as the "Boulder Canyon Project Final Reports," 21 volumes released by Reclamation between 1939 and 1950.

At the time of its design Hoover Dam was already recognized as one of the greatest engineering feats of the 20th Century. The unprecedented size of the dam led to studies in almost every aspect of dam design and construction of mass concrete dams, including concrete composition and cooling, stress analysis, hydraulic design, and hydraulic and structural modeling.

ACKNOWLEDGMENTS

The author is indebted to retired Reclamation engineer Richard Wiltshire, who appreciates civil engineering history and organized this symposium commemorating the 75th anniversary of Hoover Dam's completion. In 2008 the author served as one of the speakers participated in a Conference on The Fate and Future of the Colorado River, sponsored by the Huntington-USC Institute on California & the West and the Water Education Foundation at the Huntington Library. This led to many valuable contacts which led to a myriad of sources for information and photographs relating the Boulder Canyon Project and subsequent water resources development in the Lower Colorado River Basin.

In 2009-10 the author received a Trent Dames Civil Engineering Heritage Fellowship and was a Dibner Research Fellow at the Huntington Library in San Marino, California. This fellowship and the residency it afforded allowed the author to review of numerous serial publications, collections, archives, and ephemera that provide invaluable to understanding the engineering decisions and political pressures influencing those decisions, not just during construction (1931-35) but during the decade preceding and following the dam's completion. Huntington Archivists Dan Lewis and Bill Frank proved to be particularly valuable in ferreting out rare or obscure accounts from the Huntington's civil engineering and scientist manuscript collections, as well as rare map and historic photos. The author is also indebted to engineering geologist James Shuttleworth, a volunteer in the Huntington's manuscripts department, well versed in the history of the lower Colorado River, who provided innumerable suggestions.

Between 1976 and 1988 the author conducted interviews with Roy W. Carlson (1900-1990), Milos Polivka (1917-1987), Jerome M. Raphael (1912-1989), and George E. Troxell (1896-1984), all professors in the civil engineering program at U.C. Berkeley. Carlson had been an ex-officio member of the Hoover Dam Concrete Research Board. Jerry Raphael had worked for the Bureau of Reclamation from 1938 to 1953 and had been trained by Reclamation Chief Designing Engineer Jack L. Savage (1879-1967). He inherited many of Savage's files on mass concrete behavior and had served as head of Reclamation's Structural Behavior Group between 1949 and 1953, when he supervised the evaluations of crustal deformation at Lake Mead and reservoir-triggered seismicity. Troxell, Carlson, Polivka, and Raphael were all protégées of Professor Raymond E. Davis (1885-1970), Director of Berkeley's Engineering Materials Lab between 1928 and 1953, which carried out the concrete research for Hoover Dam during its final design and construction (Raphael had taken Davis' faculty position at Berkeley when he retired in 1953). These professors supplied the author with countless documents relating to the concrete research carried out at Berkeley for Hoover Dam, which continued for many years after the dam's completion (e.g. creep studies).

Sincere appreciation for assistance is also rendered to the following people and their respective organizations: Reclamation GIS specialist Steven Belew; Dr. Brit Allan Storey and Christine Pfaff, historians with the U.S. Bureau of Reclamation at the Denver office, Dianne Powell of the Bureau of Reclamation Library in their Denver office; Reclamation archivists and staff Bonnie Wilson, Andy Pernick, Bill Garrity, and Karen Cowan of Reclamation's Lower Colorado Region (LCR) office in Boulder City; Boulder City Library Special Collections; Eric Bittner at the National Archives and Records Service Rocky Mountain Region depository at Denver Federal Center; Paul Atwood and Linda Vida at the University of California Water Resources Center Archives in Berkeley; Su Kim Chung of the University of Nevada-Las Vegas Libraries; civil engineering historian Edward L. Butts; Tom Northrup of the American Society of Photogrammetry & Remote Sensing; and a special thanks to the author's best friend Steve Tetreault, who accompanied the author of his first tour of the dam in February 1973. Over the last several decades Mr. Tetreault has provided invaluable support and logistical connections in support of the author's numerous visits to the University of Nevada-Las Vegas, Boulder City, and Hoover Dam.

REFERENCES

Anderson, D.L. (2004). History of the Development of the Colorado River and "The Law of the River." In J. R. Rogers, G.O. Brown, and J.D. Garbrecht, eds., *Water Resources and Environmental History*. Env. and Water Res. Inst., ASCE, pp. 75-81.

Blanks, R.F., and McNamara, C.C. (1935). Mass Concrete Test in Large Cylinders. *Proceedings* of the American Concrete Institute, v. 31:280-303.

Carlson, R.W. (1938a). A Simple Method for the Computation of Temperatures in Concrete Structures. *Proceedings American Concrete Institute*, v. 34:89-104.

Carlson, R.W. (1938b). Temperatures and Stresses in Mass Concrete. *Proceedings American Concrete Institute*, v. 34:497-516.

Carlson, R.W. (1966). *Manual for the use of stress meters, strain meters, and joint meters in mass concrete, 3rd Edition*. Gillick Printing, Berkeley, CA, 24 p.

Carlson, R.W. (1977). Personal communication.

Clark, E. B. (1930). *William L. Siebert: The Army Engineer*. Dorrance & Co., Philadelphia.

Colorado River Board. (1928). *Report of Colorado River Board on the Boulder Canyon Project*. Report prepared in accordance with Resolution No. 65, Seventieth

Congress, approved May 29, 1928, and delivered on December 1, 1928, 15 p.
Committee of the Irrigation Division. (1929). "A National Reclamation Policy." *Transactions* ASCE, Vol. 95, pp. 1303-1418.
Committee Report for the State. (1928). *Causes Leading to the Failure of the St. Francis Dam.* Sacramento. California State Printing Office, 77 p.
Davis, Arthur Powell. (1907). The New Inland Sea. *National Geographic*, v. 18, pp. 36-48.
Davis, R.E. (1932). Hoover Dam Concrete Research: Progress report on cement investigations for Hoover Dam. Engineering Materials Laboratory, University of California.
Davis, R.E., and Troxell, G.E. (1931). Temperatures Developed in Mass Concrete and Their Effect on Compressive Strength. *Proceedings American Society for Testing and Materials*, v. 31, Pt II, pp. 576-594.
Davis, R.E., and Carlson, R.W. (1932). The Electric Strain Meter and Its Use in Measuring Internal Strains. *ASTM Proceedings*, v. 32:793 – 801.
Davis, R.E., Carlson, R.W., Troxell, G.E., and Kelly, J.W. (1933). Cement Investigations for the Hoover Dam: *Journal American Concrete Institute*, v. 29:413-431.
Debler, E.B. (1930). Hydrology of the Boulder Canyon Reservoir: with reference especially to the height of the dam to be adopted. U.S. Bureau of Reclamation, U.S. Department of the Interior, Denver, 46 p.
Durand, W. F. (1925). The Problem of the Colorado River. *Mechanical Engineering*, v. 47 (February), p. 79-84.
Durand, W. F. (1953). Adventures in the Navy, in Education, Science, Engineering, and in War: A Life Story W.F. Durand. American Society of Mechanical Engineers and McGraw-Hill, New York.
Emerson, F. C., W.F. Durand, J.G. Scrugham, and J. R. Garfield. (1928). Development of the lower Colorado River: reports by special advisors to the Secretary of the Interior. U.S. Bureau of Reclamation, Government Printing Office, Wash., D.C., pp. 365-435.
Engineering News Record. (1930). Boulder Dam Height Increase of 25 Ft. Advised by Board. v. 104:701 (April 24, 1930).
Engineering News Record. (1942). Strain Meters To Measure Plastic Flow at Shasta Dam. v.128:425 (March 12, 1942).
Gruner, H. (1972). Colonel Claude H. Birdseye: Memorial lecture salutes an American pioneer in photogrammetry. *Photogrammetric Engineering*, Vol. 38, pp. 865-875.
Grunsky, C.E. (1907). The Lower Colorado River and the Salton Basin. *Transactions* ASCE. v. 65: 294.
Hobbs, G. J. (2008). Colorado River Compact Entitlements, Clearing Up Misconceptions. *Journal of Land Resources and Environmental Law*, Vol. 28:83-104.
Liel, A.B., and Billington, D.P. (2008). Engineering Innovation at Bonneville Dam. *Technology and Culture*, v. 49:3, pp. 727-751.
Moeller, B. B. (1971). *Phil Swing and Boulder Dam.* University of California Press, Berkeley.

Morrison-Knudsen Engineers. (1990). Determination of an Upper Limit Design Rainstorm for the Colorado River Basin Above Hoover Dam. *Colorado River Basin Probable Maximum Floods-Hoover and Glen Canyon Dams*. U.S. Dept of Interior, Bureau of Reclamation, Denver.

Noetzli, F.A. (1934). Discussion-Foundation Treatment at Rodriguez Dam. *Transactions* ASCE, v. 99:310.

Orsi, R. J. (2005). *Sunset Limited: the Southern Pacific Railroad and the development of the American West, 1850-1930*. University of California Press, Berkeley.

Nadeau, R.A. (1974). *The Water Seekers, Revised Ed*. Chalfant Press, Bishop.

Raphael, J.M. (1948). Determination from Stress from Measurements in Concrete Dams. *Proceedings Third Congress on Large Dams*, Stockholm, Paper No. R-54.

Raphael, J.M. (1953). The Development of Stresses in Shasta Dam. *Transactions* ASCE, v.118:289.

Raphael, J.M. (1955). Effect of Longitudinal Joints on the Stresses at the Base of an Arch Dam. *Proceedings Symposium on Observation of Structures*, Lisbon.

Raphael, J.M. and Carlson, R.W. (1965). *Measurement of Structural Action in Dams, 3rd Ed*. James J. Gillick & Co., Berkeley.

Raphael, J.M. (1977). Personal communication.

Rogers, J.D. (1995). "A Man, A Dam and A Disaster: Mulholland and the St. Francis Dam." In D. B. Nunis, Jr., ed., "The St. Francis Dam Disaster Revisited." *Southern California Quarterly*, Vol. LXXVII, n. 1-2, pp. 1-110.

Rogers, J.D. (2008). *The Boulder Canyon Project and Hoover Dam: Some of Its Impacts on the Engineering Profession*: Conference on The Fate and Future of the Colorado River, sponsored by the Huntington-USC Institute on California & the West and the Water Education Foundation, Huntington Library, San Marino. Online: http://web.mst.edu/~rogersda/dams/Rogers-Hoover%20Dam%20Impacts.pdf

Savage, J.L., Houk, I.E., Gilkey, H.J., and Vogt, F. (1931). *Tests of Models of Arch Dams and Auxiliary Concrete Tests Conducted by the Bureau of Reclamation at the University of Colorado*. Engineering Foundation Committee on Arch Dam Investigation, Sub-Committee on Model Tests. Denver, Colorado (June 1, 1931), 542 p.

Savage, J.L. (1936). *Special Cements for Mass Concrete*; for the Second on Congress, International Commission on Large Dams, World Power Conference, Washington, DC. U.S. Bureau of Reclamation, Denver, 230 p.

Scott, G.A., Nuss, L.K., and LaBoon, J. (2008). Concrete Dam Evolution: The Bureau of Reclamation's Contributions to 2002. In *Bureau of Reclamation: Historical Essays from the Centennial Symposium*. Vol.1, pp. 1-66.

Stevens, J.E. (1988). *Hoover Dam: An American Adventure*. University of Oklahoma Press, Norman.

Swain, R.E. (2008). Evolution of the Hoover Dam Inflow Design Flood – A Study in Changing Methodologies. In *Bureau of Reclamation: Historical Essays from the Centennial Symposium*. Vol.1, pp. 195-207.

Sykes, Godfrey. (1937). The Colorado River Delta. Carnegie Institute of Washington *Publication 460*. and American Geographical Society of New York, *Special Publication No. 19*.

Townsend, C.L. (1981). Control of Cracking in Mass Concrete Structures. *Engineering Monograph No. 34*, U.S. Bureau of Reclamation, Denver.

U.S. Bureau of Reclamation. (1922). "*Problems of Imperial Valley & vicinity.*" Washington, D.C., U.S. Gov't Printing Office. 326 p., 52 leaves.

U.S. Bureau of Reclamation. (1939). Stress Studies for Boulder Dam, Boulder Canyon Project Final Reports, Part V – Technical Investigations, Bulletin 4. Denver.

U.S. Bureau of Reclamation. (1940). Model Tests of Arch and Cantilever Elements, Boulder Canyon Project Final Reports, Part V – Technical Investigations, Bulletin 6. Denver.

U.S. Bureau of Reclamation. (1947). Concrete Manufacture, Handling, and Control, Boulder Canyon Project Final Reports, Part IV – Design and Construction, Bulletin 1. Denver.

Veltrop, J. A. (1988). Arch Dams, Section 3, In E.B. Kollgaard and W.L. Chadwick, eds., *Development of Dam Engineering in the United States*. Pergamon Press, pp. 219-284.

Weymouth, F.E. (1924). Colorado River Development. Senate Document No. 186, 70th Congress, 2nd Session, 231 p.

Wilbur, R.L., and Mead, E. (1933). The Construction of Hoover Dam; Preliminary Investigations, Design, of Dam, and Progress of Construction. U.S. Dept. of Interior, Gov't Printing Office, Wash., D.C.

Hoover Dam: First Joint Venture and Construction Milestones in Excavation, Geology, Materials Handling, and Aggregates

J. David Rogers[1], F. ASCE, D. GE, P.E., P.G.

[1]K.F. Hasslemann Chair in Geological Engineering, Missouri University of Science & Technology, Rolla, MO 65409; rogersda@mst.edu

ABSTRACT: The size of the Boulder Canyon Project necessitated a broad array of innovations in construction engineering and management which had enormous impacts on all of the large scale projects that followed it. Foremost among these was the employment of a joint venture involving eight different firms, organized into six partners (Six Companies, Incorporated). Many of the techniques employed to construct of Hoover Dam were of a pioneering nature, designed to hasten the construction schedule and maximize profits. These were emulated and perfected by Six Companies and most of their competitors for several decades thereafter. Some of these included: multiple-level rail spurs; temporary trestles and suspension bridges of many sizes, employment of construction access adits to allow multiple headings of underground workings; fully automated concrete batch plants; staging of construction materials on the opposite river bank; government provision of all materials except the concrete aggregate (to minimize risk of construction claims and delays). Major achievements were also made in quality assurance and materials testing, despite the fact that the job proceeded round-the-clock.

FIRST LARGE JOINT VENTURE

Six Companies Incorporated

The Boulder Canyon Project approved by Congress had a budget of $165 million, making it the largest federal contract ever awarded up to that time (March 1931). It called for 4.4 million cubic yards (mcy) of concrete, which was more than all previous Bureau of Reclamation projects during the previous three decades, combined (4.3 mcy). The project was so large that no single company in America had the resources to bid the job alone. The government contract called for bidding of 119 separate items and allowed seven years for completion of the project, commencing on April 20, 1931 and running thru April 11, 1938, with a performance penalty of $3,000/day.

The joint venture of Six Companies, Incorporated was organized by Edmund O. and William H. Wattis of Utah Construction Company. There were actually eight firms that originally comprised the joint venture. Bechtel, Kaiser, and Warren Brothers intended to pool their resources to become the largest partner, but Warren was unable to come up with the half million in cash required to become a partner when Six Companies incorporated in February 1931 (Tassava, 2003; Wolf, 1996). The remaining seven firms pooled their financial resources to capitalize the new corporation with $5 million in capital stock, a government requirement to bid the job. A $2 million dollar bond was also required by Reclamation as part of their actual bid submittal. The seven firms were: W.A. Bechtel Co. of San Francisco, Kaiser Paving Co. Ltd., of Oakland, Utah Construction Co. of Ogden, MacDonald & Kahn Construction Co. of San Francisco, Morrison-Knudsen Co. of Boise, J. F. Shea Co. of Portland, and Pacific Bridge Co. of Portland. This was the first time a joint venture of more than three firms was used in public works construction.

On March 4, 1931 Reclamation opened the bids for the construction of Hoover Dam and Six Companies' bid for $48,890,955 was the lowest of any qualifying firm, nudging out Arundel-Atkinson Corporation of New York ($53.9 million) and Woods Brothers of Lincoln, Nebraska ($58.6 million), the only other serious bidders. The winning bid was just $24,000 more than the engineer's estimate by Reclamation, a difference of just 0.05%. The largest line items in Six Companies bid were: $13,285,000 for 1,563,000 yd^3 of tunnel excavation; $9,180,000 for placement of concrete for the dam; and $3,432,000 for lining the four diversion tunnels with 312,000 yd^3 of concrete. No other items exceeded $1,000,000.

Everyone with the Six Companies management agreed that Frank T. Crowe of Morrison-Knudsen (Figure 1) was the most qualified individual to serve as the General Superintendent for Hoover Dam because he had been Reclamation's General Superintendent of Construction. In June 1925 he left Reclamation to join Morrison & Knudsen (MK). MK had secured a subcontract with Utah Construction Company, the prime contractor building Guernsey Dam in eastern Wyoming. Crowe supervised the construction of Guernsey, followed by Combie Dam in California and Deadwood Dam in Idaho (also with Utah Construction).

Crowe possessed the most experience in concrete dam construction of anyone in the Six Companies consortium, and had won every bid he had prepared for Morrison-Knudsen during the previous five years. Known by the workmen as '*Hurry Up Crowe*,' he had an uncanny ability to attract "*reliable, competent, fast-thinking men*" to his job sites (Rocca, 2001). He also possessed superior skills in appreciating critical path management, which he brought to materials handling, processing, and delivery. Crowe was also a natural leader, although he preferred the moniker "*problem solver*" (Raphael, 1977).

Crowe's uncanny ability to solve problems quickly, around the clock, earned the respect of all who worked for him. Crowe gave much of the credit to the supervisors and foremen he hired, who he expected to solve such problems, not waste his time with complaints. Problem solving, after all, was *the* essential skill in heavy construction, where unforeseen delays can easily snowball into logistical, managerial, and financial catastrophe (Gerwick and Woolery, 1983). His generalship was never in question, and when it was all over (in mid-1935) he had managed to complete the most expensive and complicated engineering project 30 months ahead of schedule, netting record profits for the Six Companies partnership (Hiltzik, 2010).

FIG. 1. Civil engineers Walker R. Young (left) and Frank T. Crowe (right). Young was the senior onsite representative (Construction Engineer) for the Bureau of Reclamation while Crowe was the General Superintendent for Six Companies (USBR photo). Everyone expected Young to succeed Elwood Mead as the next Commissioner of Reclamation, but his subordinate John C. Page was given the position because he was a Democrat. Crowe went on to manage construction of Parker and Shasta Dams before dying of a heart attack at age 63 in February 1946. Young went on to head up the Central Valley Project in California, returning to Reclamation headquarters in 1945, where he served as Assistant Chief and then Chief Engineer until retiring in 1948. He lived to be 97, passing away in 1982.

Frank Crowe's greatest impact on the construction industry was his penchant for crafting what came to be known as "unbalanced bids." The contractor's unit costs were manipulated to seek greater compensation for materials excavation, handling, and placement early in the job, to offset the cash bond that the winning contractor had to post to begin the job. At Hoover Dam Crowe sought compensation of $8.50 per cubic yard for rock excavation, a very high price, but just $2.70 per yard for mass concrete in-place. His concrete unit price was 20% below Reclamation estimates and 35% below their nearest competitor. With 4.4 million yd^3 of concrete, this might seem dangerous to an accountant, but Crowe knew exactly what he was doing; he was gathering the most cash possible at the "front end" (beginning) of the job, which offset the $2 million performance bond. By completing the four massive diversion tunnels a year ahead of schedule Six Companies was able to collect $13.285 million in the first 12 months on the job (May 1931 to May 1932). They received another $3.43 million for lining the diversion tunnels between March 1932 and February 1933.

In the end, Six Companies cleared about $13 million in profits (Hiltzik, 2010) on a $53 million job (including the extras), for a tidy profit of 25%, stellar by anyone's standards in the construction industry. Crowe received an annual salary of $18,000 and received performance bonuses in excess of $250,000 during the course of the job.

Between 1931 and 1935 Six Companies employed 21,000 men on the project, with the workforce averaging about 3,500 employees at any given time. Even with all the housing, there was so much commercial activity in the area the population of Las Vegas doubled between 1930 and 1934, while the dam was under construction. As many as 5,218 men were on the payroll during the height of construction, in June 1934. The average monthly payroll was roughly $500,000. Pay days were on the 10th and 25th of each month. Six Companies completed concrete placement on the dam on May 29, 1935, and all features were completed by March 1, 1936, more than two years ahead of schedule.

The official death toll eventually tallied 96 men. Most of these occurred during the first year and a half, while the diversion tunnels were being excavated. Six Companies also took preventive measures to provide water and emergency medical assistance, to placate the concerns of workmen. They were exempt from having to abide by mining safety ventilation standards in force at the time (1931-32), and by modern standards the paucity of forced ventilation in the massive diversion tunnels was shamefully low.

Site Access Secured by the Government

The greatest construction problem attendant to the construction of Hoover Dam was the difficult site access. The federal government contracted with the Union Pacific Railroad to construct a spur line 22.6 miles long from their Los Angeles & Salt Lake rail line in Las Vegas to Boulder City Summit. This contract was let in May 1930. From Boulder Junction Lewis Construction Co. built the U.S. Construction Railroad line that continued eastward for another 10 miles to a point overlooking the Nevada powerhouse, at an elevation of 1370 feet (138 vertical feet above the dam's crest).

Six Companies then constructed two additional rail lines, one down Hemenway Wash to the aggregate classification plant and across the Colorado River to the aggregate quarry on the Arizona side of the river and the second low level spur serving the lower concrete batch plant near river level, just upstream of the dam, within Black Canyon. These were known as the "Six Companies Railroad" lines and are shown in Figure 2.

Construction of the low level spur involved considerable hard rock tunneling in the andesite breccia, the costs of which had to be absorbed by Six Companies. Six Companies was obliged to use boats to access the dam site before temporary construction roads and the rail line could be extended into Black Canyon along the river.

The federal government hired General Construction Co. of Seattle to build 8.3 miles of highway leading from Boulder City to the steel fabrication plant. At the time General Construction was also building Owyhee Dam in Idaho, the Bureau of Reclamation's highest dam prior to Hoover. They became a Six Companies partner at Bonneville Dam in 1934, cementing a relationship that would continue into the 1950s (Wolf, 1996). All of the transportation links were brutally difficult, passing through extremely rough terrain.

On Kaiser's recommendation, General Construction subcontracted R.G. LeTourneau, Inc. of Stockton, California, an earthmoving firm that had worked with

Kaiser on highways and earth dams in California. They were given the task of grading a 22-foot-wide asphalt-surfaced highway from Boulder City to the staging area above the dam's Nevada abutment, known as the "government highway" (Figure 3). The timeframe for completion of this highway in the spring of 1931 was critical to Six Companies so they could move their workforce, equipment, and materials to the dam site.

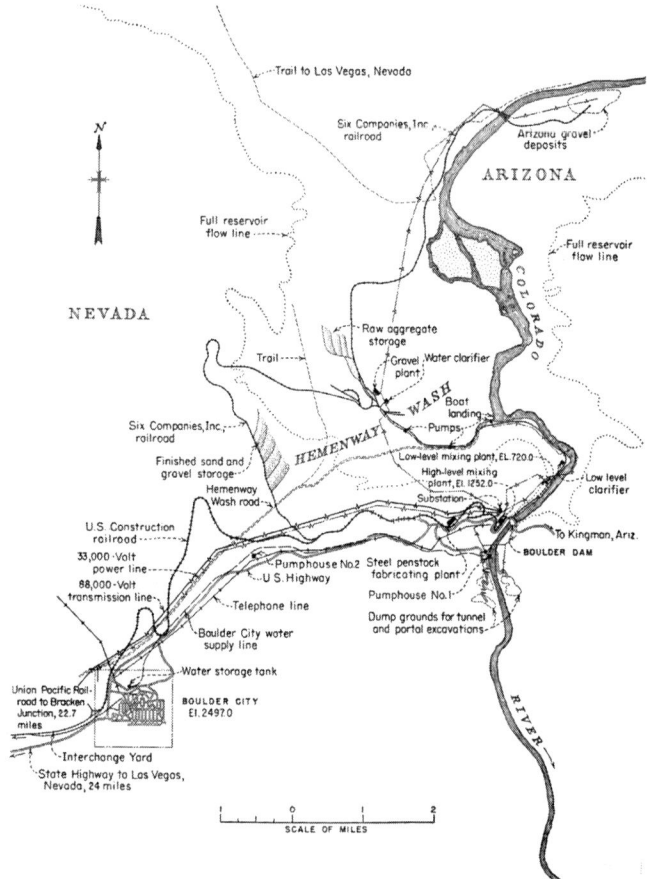

FIG. 2. Map showing Boulder City, the highway, rail spurs, and power lines serving the dam site during construction. The aggregate came from terrace gravels about 11 miles upstream of the dam, on the Arizona side of the river (USBR).

Robert G. "Bob" LeTourneau was respected as the inventor of the tracked bulldozer and the modern articulated scrapper, with a reputation for building the best earthmoving equipment on the planet (he registered over 400 patents and 70% of the

FIG. 3. Upper: The R.G. LeTourneau Co. lost close to $100,000 while grading the government highway between Boulder City and the dam staging area because they encountered resistant volcanic strata that proved much more troublesome than anything they had previously dealt with in California. The lower image shows the completed highway (costing $330,000) with the U.S. Construction Railroad in the background (costing $1 million). (USBR)

FIG. 4. Six Companies quickly built a series of suspended walkways like that shown here under construction to access the Arizona side of the river, to begin excavation of the two diversion tunnels underlying the dam's left abutment (USBR).

FIG. 5. Six Companies employed a variety of bridges to get workers and equipment across the channel. They began by driving timber piles to fashion simple timber trestle bridges such as that shown here, which replaced the span lost in the high water of February 10, 1932 (USBR).

earthmoving equipment used by the Allies during the Second World War were built at five different plants licensed by his firm). LeTourneau's crews battled the lava flows, sandwiched between softer, powdery strata. Every time they drilled holes, loaded them with dynamite, and shot them, the blast would simply shoot up the hole (called "cannon shots"), instead of "springing a pocket" of fractured rock at some depth. They adjusted their stemming and tried drilling deeper holes, but without success. In the end

LeTourneau fell back upon his previous experiences working in a gold mine and had his crews excavate a series of "coyote drifts," or adits, into the tough volcanic strata. These tunnels were then backfilled with "truck loads" of explosives and detonated. This succeeded in keeping the project on schedule, but LeTourneau lost about $100,000 on the contract, which came perilously close to bankrupting his firm (LeTourneau, 1960).

Satisfied that he was the best man for the job, Six Companies let an additional contract to LeTourneau to extend the government highway another 4,000 feet, down to the precipice for the inclined railway serving the main dam site in Black Canyon (seen in Figure 14). This proved tedious, but LeTourneau employed cable-winch-controlled bulldozers, which anchored by cables, were able to winch themselves up and down the precipitous cut slopes, just like the high scalers employed to pluck the loose rocks off the dam's abutments following each blast. This was the route that eventually became U.S. Highway 93, bringing millions of tourists to the dam, and crossed over the dam and continued on to Kingman, Arizona (the highway to Kingman was also graded in 1931, before the dam construction began in earnest).

Six Companies employed numerous subcontractors to help them establish site access throughout the summer and fall of 1931, establishing the roads, rail spurs, and electric power transmission lines crucial to excavating the four massive diversion tunnels around-the-clock. Two levels of access had to be provided; an upper level that serviced the cliffy area above the dam's right (Nevada) abutment, and another at river level, which allowed the river to be crossed by workers and vehicles (Figures 4 and 5). This transportation network extended many miles upstream to the aggregate quarries on the Arizona side, aggregate stockpiles and the main aggregate mixing plant on the alluvial fan in Hemenway Wash, connections to the diversion tunnels (to haul out tunnel muck) and to the two massive concrete batch plants on the Nevada side of Black Canyon.

EXCAVATION MILESTONES

Largest Diversion Tunnels

Reclamation's 1928 plans envisioned two partial excavations of the river bed to construct the upstream heel of the dam to a height of approximately 231 feet above the lowest point of foundation (to an elevation of 760 ft) during the first construction season of nine months, followed the next year by a downstream cofferdam about 153 ft high (to elevation 663 ft), incorporated into the downstream toe of the dam (shown in Weymouth, 1924). This scheme envisioned placement of 235,000 yd^3 of concrete

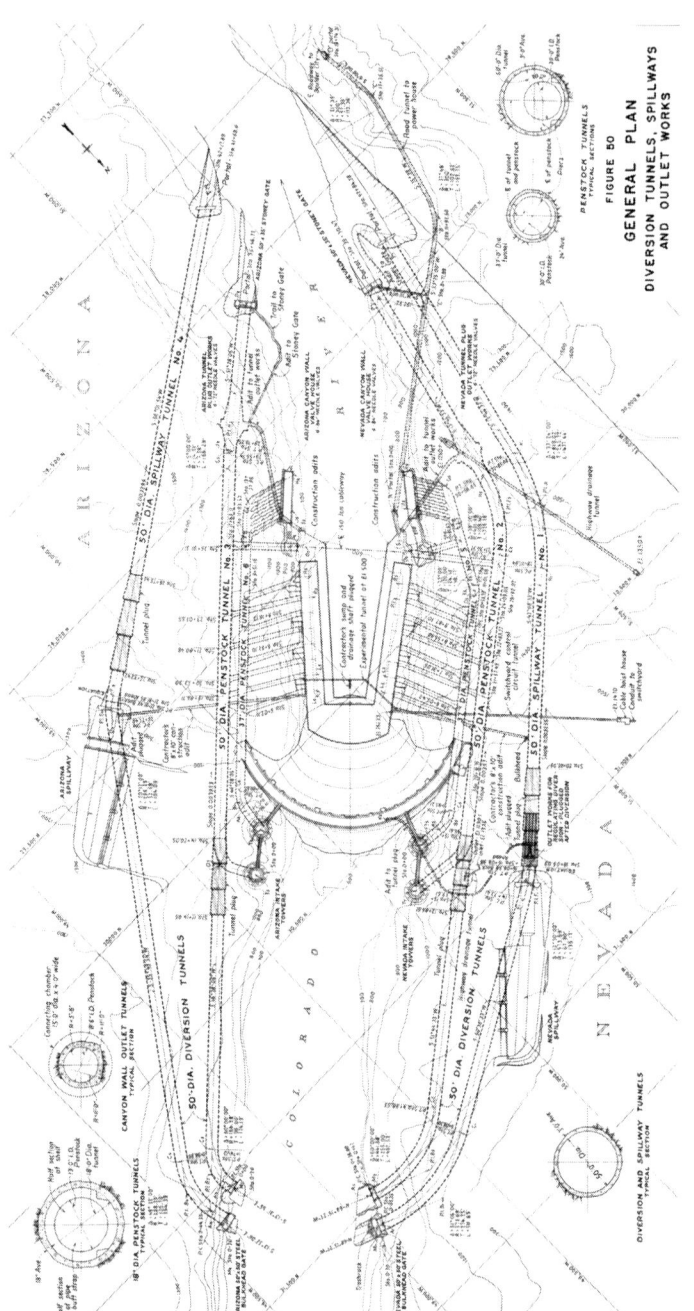

FIG. 6. Plan view of final layout employed for Hoover Dam, diversion tunnels, spillways, outlet works and powerplant. (USBR)

FIG. 7. Suspension bridge built by Six Companies just downstream of the dam site to allow access to 8 x 10 ft cross adits so they could drive multiple headings on the massive diversion tunnels. The portal on the Arizona side is clearly seen, about 50 feet above the bridge deck on the far side. Six Companies was not paid for constructing these additional tunnels, which greatly hastened the construction schedule (USBR).

that would subsequently be incorporated into the main dam, which could retain up to 400,000 ac-ft of flood water. Similar schemes had been employed on several masonry gravity dams built during the previous decade in steep-walled canyons.

The 1928 design only provided sufficient river diversion (80,000 to 100,000 cfs) during the months of low to medium water stages, between the period of high flow, in May-June-July. The Colorado River Board overruled Reclamation on this diversion scheme, deeming it to be too risky to allow flood waters to run over the dam and damage the powerhouse excavations, immediately downstream of the dam (Clark, 1930). They favored the use of separate upstream and downstream embankment cofferdams which would allow the powerhouse excavations to be made "in-the-dry" at the same time as the main dam.

The CRB told Reclamation to double their bypass tunnel capacity from 100,000 to 200,000 cfs. They were therefore obliged to construct four diversion tunnels in lieu of the proposed two. This was the most significant design change proposed by the board, which resulted in significant increased costs and an additional year of anticipated construction (the Board anticipated that the diversion tunnels would have to be at least one mile long; but they averaged 4,000 ft upon completion).

From the project's inception Frank Crowe realized that the most important milestone in constructing Hoover Dam would be how long it took Six Companies to excavate the project's four massive diversion tunnels, which, at 56 ft, were the largest diameter hard rock tunnels ever constructed up to that time. The dam site couldn't be excavated until the Colorado River was safely diverted through these tunnels.

That the Colorado was a fickle river everyone knew. Over the previous half century its flow had ranged between a low value of just 500 cubic feet per second (cfs) in January 1912 to a peak flow of 384,000 cfs in February 1884 (USBR, 1950); a high-to-low ratio of 768:1. The Colorado River is unusual in that almost no accretiary flow joins the stream over its last 400 miles, 80% of the annual flow being derived from the Wind River and Rocky Mountains, far upstream. Late summer thunderstorms in the Colorado Plateau bring occasional flash floods, but these flows mollify rather quickly, downstream.

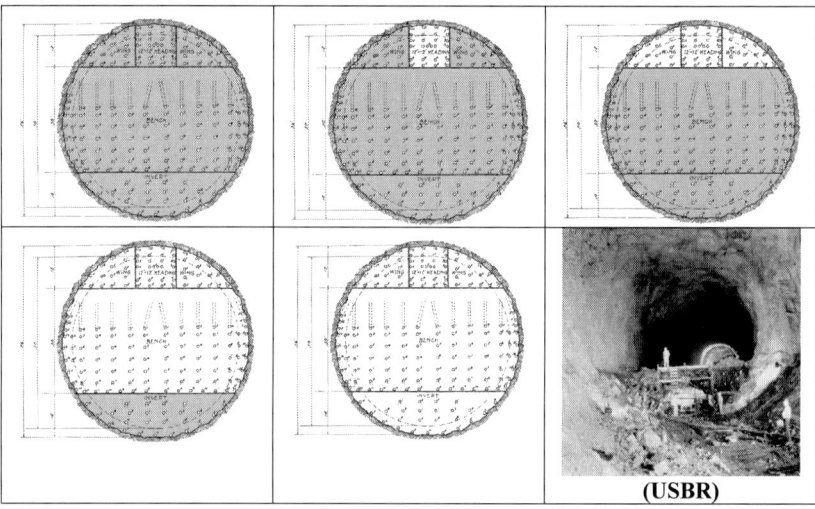

(USBR)

FIG. 8. Sequence employed to excavate the 56-foot-diameter diversion tunnels. Upper left: full face before drilling. Upper middle: 12 x 12 ft pilot bore in tunnel crown. Upper right: Excavation of crown wings. Lower left: seven rows of drill holes were employed with top row drilled upward. Lower middle: Four levels of holes in the invert. 173 drill holes were loaded with each round of blasting. Lower right image shows the completed full-face excavation.

The circular tunnels were to be excavated to a diameter of no less than 56 feet, and then lined with reinforced concrete to a finished diameter of exactly 50 feet. The falsework to support forms for a 3-foot thick-concrete lining 50 feet above the tunnel invert were just one of several unprecedented challenges faced Six Companies.

The tunnels were numbered 1 through 4, beginning with the outboard Nevada tunnel (No. 1) and ending with the outboard Arizona Tunnel (No.4). The tunnels' average length was about 4,000 feet. Their general layout is shown in Figure 6. The contractors would not be able to employ cross-cut adits or additional access shafts to accommodate multiple headings upstream of the dam's axis because these additional openings would lie beneath up to 600 feet of water when the reservoir filled. Multiple headings had been employed since the mid-1860s while driving railroad or highway tunnels of any significant length in order to reduce the duration of excavation.

It was of the utmost importance to divert the river in late fall, or early winter, before spring floods, which were estimated to be about 120,000 cfs every 2.5 years, based on the available records. The aggregate capacity of the four 50 foot diameter tunnels was intended to be 200,000 cfs. If Six Companies could succeed in diverting the river's flow before May 1st, they could accelerate the project by an entire year and thereby save millions of dollars by being able to excavate the channel bed beneath the dam and powerhouses.

FIG. 9. Suspension bridges, roads, and rail spurs accessed the diversion tunnels and dam site at river level. Tunnel muck on the Nevada side was removed by rail while that from the Arizona tunnels was removed by truck. The Arizona tunnels were completed first and lined in late October 1932 (USBR).

On May 21, 1931 Six Companies detonated their first blasts of dynamite on the diversion tunnels. Six Companies began the diversion tunnels by excavating a series of 8 by 10-foot cross adits from just downstream of the dam site (Six Companies was not compensated for these excavations because they were not required by the Reclamation contract). These exploratory adits extended 826 feet in on the Arizona side and 607 feet in on the Nevada side (Figure 7). These allowed access to employ double heading of the 12 by 12 ft pioneer headings at the crown of all four 56-foot-diameter bores (four faces would be worked simultaneously, from the upstream end, the downstream end, and in both ends of the exploratory adits).

The tunnels were then enlarged from the upstream and downstream portals because the drilling jumbo could not fit through the cross adits. The single exception was the outboard Arizona Tunnel No. 4, which employed a full crown heading excavation (12 x 30 ft) proceeding upstream from the downstream portal (because of access problems at the upstream portal and the ease of wasting tunnel muck in side canyons adjacent to both of the downstream portals (see Figure 8). Six Companies erected a suspension work bridge to allow their crews to work on the Arizona diversion tunnels, as shown in Figure 9. During that first summer of 1931 average temperatures reached 120° F in the shade, and 14 workers died of complications of dehydration. Crowe responded proactively by hiring orderlies to deliver ice water to men while they were working, encouraging them to drink, but other hazards manifest themselves as the job progressed (described later).

First Use of Drilling Jumbos

The single greatest problem in excavating such large diameter tunnels was "reach," the slang term used to describe how far a driller could reach with a pneumatically-powered but hand-operated jackhammer, which was referred to as a "jack leg." The maximum reach of a jack leg was something between 6 and 12 feet. Up to that time the only way to accommodate such a large excavation face would have been to employ a pilot bore in the tunnel crown, which would be expanded outward, bit by bit, then dropping down in multiple benches, each about 8 feet high. The problem was how to pass the tunnel muck backward, out of the tunnel, without dropping it on the men working the lower benches.

One of the tunnel shift superintendents named C. T. Hargroves came up with a novel scheme that allowed four levels of drilling to be undertaken simultaneously, which allowed 2/3 of the tunnel face to be drilled by 24 to 30 jack-legs simultaneously. He took an International 10-ton freight truck, stripped it down to its chassis and welded on a semi-circular frame that would support four working platforms, each of which was equipped with two levels of jackhammers. This allowed seven different levels of the tunnel face to be drilled simultaneously, as shown in Figure 10.

By October 1931 Six Companies was employing their first full-face excavations on Diversion Tunnel No 4. By January 1932 they were employing eight Marion 100-ton electric shovels using modified 3.5 cubic yard buckets, removing an astonishing 16,000 cubic yards of rock per day from the four tunnels. The tunnel muck shed by the rapid excavations was removed using gas-powered International 10-ton dump trucks. The trucks were instructed to move as quickly as possible between the tunnel face and the railroad hopper cars lined up just outside the upstream portals, which hauled the muck away. Drivers were required to make a finite number of round-trips during each 8-hr shift, depending on the length of the tunnel in which they were working. The exhaust from the dump trucks caused increasingly poor air quality as the tunnels extended further into the canyon walls, bereft of any forced air ventilation. It's a wonder more weren't killed (96 men were officially recognized as having died during construction of the dam). Six Companies managed to complete excavation of all four diversion tunnels by May 1932, 14 months ahead of schedule.

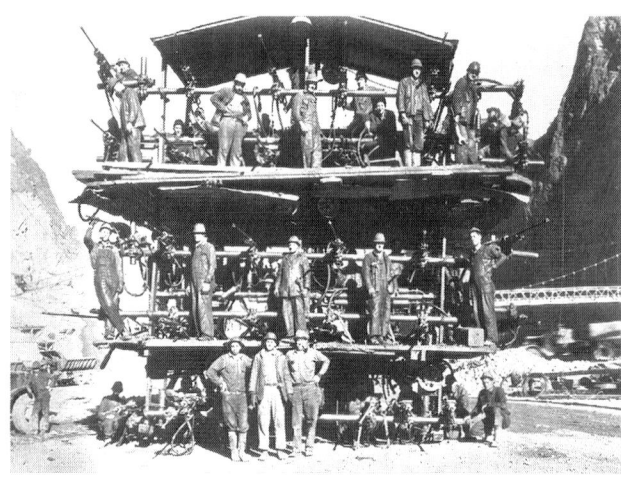

FIG. 10. Upper: Drilling jumbo fashioned upon a 10-ton International freight truck supported 24 to 30 jackhammers to drill 48 holes from four to 20 feet deep on seven levels. Lower: Smaller drilling jumbo designed to drill 24 holes in the semi-circular tunnel invert, on four levels. All of the drills were pneumatically-powered by Ingersoll-Rand compressors (USBR).

Dozens of tunnel muckers and dump truck drivers working inside the tunnels experienced the debilitating effects of what was likely carbon monoxide poisoning. About 50 of these workers brought suit against Six Companies seeking compensation for the loss of their health. The first trials began in late 1932 in Las Vegas courts and Six Companies brought their considerable influence to bear and was able to fend off defense judgments, although several jurors admitted to accepting bribes after one of the trails. In January 1936 an out-of-court settlement was reached by Six Companies with approximately 50 remaining gas-suit plaintiffs and the amounts of their payments were undisclosed.

Largest Traveling Slip-forms

The diversion tunnels were lined with concrete, beginning with their inverts (Figure 11). The tunnel invert was then filled with 15 feet of gravel to serve as a road bed for removing muck at the tunnel face and support the traveling slip-form used for lining the tunnel shoulders, shown in Figure 12.

FIG. 11. Traveling slip-forms were used to pour the concrete invert section of the four diversion tunnels, as shown here (USBR).

The government contract required that the diversions tunnels be completed by October 1, 1933, subject to a $3,000/day fine if they were not finished on time (Figure 13). For 15 continuous hours on November 12-13, 1932 dump trucks bombarded the river channel with load after load of rock riprap; one dump truck dropping its load every 15 seconds. At 11:30 AM on the 13th the outboard Arizona diversion tunnel cofferdam was detonated, allowing the Colorado River to spill into the tunnel. Within 24 hrs entire flow of the Colorado River was diverted through this tunnel, which was running between 3,000 and 6,000 cfs over the next few weeks. The largest river

FIG. 12. Traveling slip-form used to line the crown of the diversion tunnels after pouring the invert and filling it with 15 feet of gravel, to serve as a road bed. This was the largest traveling slip form ever fabricated at the time (USBR).

FIG. 13. Initial diversion of the Colorado River through Diversion Tunnel No. 4 on the Arizona side on November 14, 1932, a full year ahead of schedule. Note the scale of operations by the height of several hundred workers standing at the brow of the fill pile at right center, in front of the dragline (USBR).

diversion in history had been accomplished in just 13 months. The maximum capacity of the four diversion tunnels was 200,000 cfs, about a once-in-25-years event. The government assumed responsibility for flood damage after the cofferdams and diversion tunnels were completed. This opened the way for Six Companies to begin building the cofferdams and start excavating the channel at the dam site.

Abutment Excavations

The most celebrated aspect of the project's early years was the abutment excavations, typified by loud production blasts and swarms of high scalers rappelling from the cliffs, prying off the loose rocks using scaling bars, which looked like 8-ft long javelins with a tapered head, similar to an old Blacksmith's nail (Figure 14). High scalers were the highest paid and the most photographed laborers, earning $3.50/day. High scaling was an art built on teamwork and close coordination, lest a block might fall on another high scaler. High scalers were thus obliged to descend newly blasted faces in a line-abreast fashion, as shown in Figure 15. This allowed ample space between the men, and used adjacent climbers to help pry off the larger blocks, often using wood wedges.

FIG. 14. Noontime production blasts on the Arizona abutment upstream of the dam. Tourists flocked to view the small blasts that accompanied the lunchtime breaks, when workers were cleared from the canyon floor.

FIG. 15. Upper: High scalers working the lower portion of the dam's left abutment, on the Arizona side. Note the raw loose character of the freshly blasted rock and the sheer scale of the operation. Lower: Enlargement of the same image, showing the high scalers worked line abreast of one another (USBR).

High scalers also drilled the shot holes using air-powered jackhammers. These men could be recognized by their bosun's chairs, which were fashioned of a rope wrapped around a small wooden seat, which sailors used to support themselves while painting a ship's hull (Figure 16). Behind these came "powder monkeys," the men who loaded the drill holes with dynamite, stemming, and blasting caps. There were three principal "blasting windows" each day, usually during mealtimes. The greatest audience of gawkers always occurred during the noon hour, when the lunchtime blasts were detonated.

The techniques developed to scale the precipitous cliffs at Hoover Dam became the standard practices employ." Scaling hasn't been improved with time or by employing advanced technologies. High scalers tapped the rock with their scaling bars to listen and feel its "ring." Solid rock tends to elicit a numbing thud, but rock that has separated from the parent cliff reverberates with a sort of hollow clang, alerting the high scaler that it has become loosened or dislodged, even slightly. These same scaling techniques were employed 34 years later, at Glen Canyon Dam.

FIG. 16. High Scalers in Bosun's Chairs operating jackhammers to drill holes for production blasting. This was hot, dirty, rough work, without any chance for rest. On the very steep slopes, like that shown here, scalers would have to stop and help one another 'set up' their 'jacklegs' on a hole and drill it a sufficient distance so that a man could safely control it (USBR).

FIG. 17. A group of 'powder monkeys' loading drill holes with dynamite, stemming, and blasting caps, photographed on August 15, 1932. Note the thin steel tamping bars, which were typically 10 ft long. These were used to tamp the sticks of dynamite and pack the stemming in the upper part of the holes, to force the blast energy into the rock mass (Bechtel Collection, USBR).

ENGINEEERING GEOLOGIC MILESTONES

Geology of Black Canyon Revealed in Excavations

According to Jerome M. Raphael, a Reclamation engineer who worked closely with Frank Crowe at Shasta Dam, Crowe described Hoover Dam as "*nothing more than a three-pronged job; the prongs were just bigger, that's all.*" First was dealing with site access, equipment mobilization, and materials delivery (aggregate, cement, steel, and fuel); Second was excavation of the bypass tunnels, the river channel, and the dam and powerhouse abutments; Third was concrete batching and placement, using an intricate system of overlapping cross-canyon cableways that were Crowe's trademark (Raphael, 1977, 1988; Rocca, 2001). Of these the greatest uncertainties lay in the excavation work because nobody really knew what they would find once they began excavating; that was simply "the nature of subsurface work."

In the end more than 5,500,000 cubic yards of sand, gravel, and rock material were excavated, and another 1,000,000 cubic yards of earth and rockfill were placed (Figure 18). By feature, this included: 1,912,000 yd^3 for the tunnels and shafts; 150,000 yd^3 for the dam's abutments; 1,300,000 yd^3 for the foundation of the dam and powerplants; 750,000 yd^3 for the spillways and inclined tunnels; 410,000 yd^3 for the valve houses and intake towers; 732,000 yd^3 earth and rockfill for the upstream cofferdam; 500,000 yd^3 earth and rockfill for the downstream cofferdam (about 2,000,000 total cubic yards excavated from channel excavations); and 2,300,000 yd^3 for other excavations.

A major limitation was that no tunnel muck or excavation spoil could be dumped in the Colorado River; the channel needed clear for operation of what would soon become the world's largest hydroelectric facility (eclipsing the Vemork hydroelectric power plant at Rjukan, Norway).

The contractor was expected to dispose of six million yd^3 of spoil in the steep ravines lining Black Canyon above and below the dam site. The contractor constructed a network of temporary construction roads to gain access to the diversion tunnel portals and to the dumping spots. These spoil slopes were to be neatly dressed and compacted so as to present a 'neat appearance' and resist surficial rill erosion.

The average depth of channel excavation was between 110 and 130 feet below mean low water level. The deepest excavation was in the upstream cutoff trench, which extended 139 feet. A sawn 2 x 6-inch plank was discovered 50 feet below the low water surface during foundation excavation (Figure 19). This surprised everyone. It was thought to have come from either the June 1921 high flows (170,000 cfs) or during the winter of 1921-22 when a local downpour triggered a debris flow that swept through the old Mormon settlement at the mouth of Callville Wash, about 15 miles upstream of the dam site. A third alternative may have been the high flows of June 1928, which reached 137,000 cfs.

The 'Inner Gorge' of the channel beneath the dam and powerhouses was revealed in borings made ahead of construction, but its physical character was altogether unusual and the subject of considerable fascination with geologists at the time (Ransome, 1931; USBR, 1950).

First Use of Paleoseismology

In his geologic report to Reclamation, California Institute of Technology (Caltech) Geology Professor Leslie Ransome described curious pothole and rill structures in the uppermost reaches of Black Canyon around elevation 1550, about 900 ft above the low flow surface of the river! These potholes were filled with fresh water-rounded cobbles from upper Precambrian and Paleozoic units that outcrop in the Grand Canyon, well east of the Grand Wash Cliffs.

FIG. 18. View looking upstream from the Nevada abutment at the excavation work in December 1932 as the abutments were being trimmed back. The downstream cofferdam is taking shape in the foreground. The cofferdams had to be completed before they could begin excavating the channel gravels. All of the excavated materials had to be transported to disposal sites in steep side canyons upstream or downstream of the dam (USBR).

FIG. 19. Upper image shows excavation of the cobbles, boulders, sand, and gravels of the Colorado River on April 3, 1933 when crews began excavating the "inner gorge," which extended much deeper into the andesite bedrock. This narrow trough was filled with boulders up to 12 feet in diameter. Lower image shows the 2 x 6 inch plank unearthed at a depth of 50 feet below the river bed, thought to have been deposited during the peak flows of June 1921, which reached 170,000 cfs (from USBR, 1950).

FIG. 20. View of the excavated 'inner gorge' of Black Canyon as it appeared on June 3, 1933. This inner gorge was filled with enormous subangular boulders, which had fooled drillers in thinking they had encountered 'bedrock' during the first two seasons of drilling, in 1922-23 (USBR).

FIG. 21. The sides of the inner gorge were carved out into giant pothole-and-rill structures, on a scale not previously observed. These suggested an intensely turbid flow had rapidly cut the channel. These overhangs were chipped off to avoid the problems with bearing and arching over irregularities (USBR).

The potholes were locally surrounded and intermittently filled with older terrace gravels, which Ransome (1931) mapped as "Qtg." Ransome recognized that that these gravels were from a much older river channel, approximately 950 feet above the present level of the Colorado River. One of the potholes was cut by mapped fault A-25, which clearly cuts the biotite latite bedrock, but without any apparent offset of the pothole.

Leslie Ransome correctly deduced that the absence of offset of the old potholes cut by the canyon's faults suggested that they had been inactive since before the Colorado River had excavated the 900 ft deep gorge through Black Canyon. He then reasoned that the other faults, of similar style and inclination, most likely predated the incision of Black Canyon. The concern over state-of-activity of these faults was one of the prime reasons Reclamation had engaged Ransome's services (in 1923 he had been an employee of the USGS, but was teaching at Caltech when he prepared his second report in 1931). Ransome's reasoning appears to have been one of the first applications of in assessing seismic risk to critical engineering structures; the field we now refer to as paleoseismology (McAlpine, 1996).

80 years would pass before the puzzle of the paleochannel 950 ft above Black Canyon was more or less solved. Howard et al. (2008) have assembled data from numerous outcrops between the Grand Wash Cliffs (mouth of Grand Canyon) and the Gulf of California which consistently record major swings in river elevation over the last five million years (5 Ma), since the last basalt flow congealed along the margins of what is now Black Canyon (Howard and Bohannon, 2000). These sediments infilling more than 20 stranded paleovalleys of the ancestral Colorado River record alternating periods of aggradation (filling) and degradation (erosive downcutting). These cycles appear to have resulted in channel incisions of as much as 1,150 ft. 80 years later House, Pearthree and Perkins (2008) examined the Quaternary sediments preserved in proximity to the bedrock narrows of the lower Colorado River and made a convincing case for periodic breakout floods that carved the bedrock narrows that formed favorable dam sites for by Hoover, Davis, and Parker Dams.

MILESTONES IN COFFERDAM DESIGN

Largest Cofferdams

To isolate the dam and powerhouses foundation excavations and prevent their being flooded during construction cofferdams of unprecedented scale were constructed upstream and downstream of the dam site. As mentioned previously, the design capacity of river diversion was a once-in-25-years event. The government assumed responsibility for flood damage after the cofferdams and diversion tunnels were completed. The specifications for the cofferdams were for them to be at least 100 feet high (above the low flow level of the river), which required 732,000 cubic yards in the upstream embankment and 500,000 cubic yards in the downstream embankment. The upstream cofferdam was located about 600 feet downstream of the diversion tunnel inlets. The general layout of the cofferdams and protective rock dikes with respect to other project works are summarized in Figure 22.

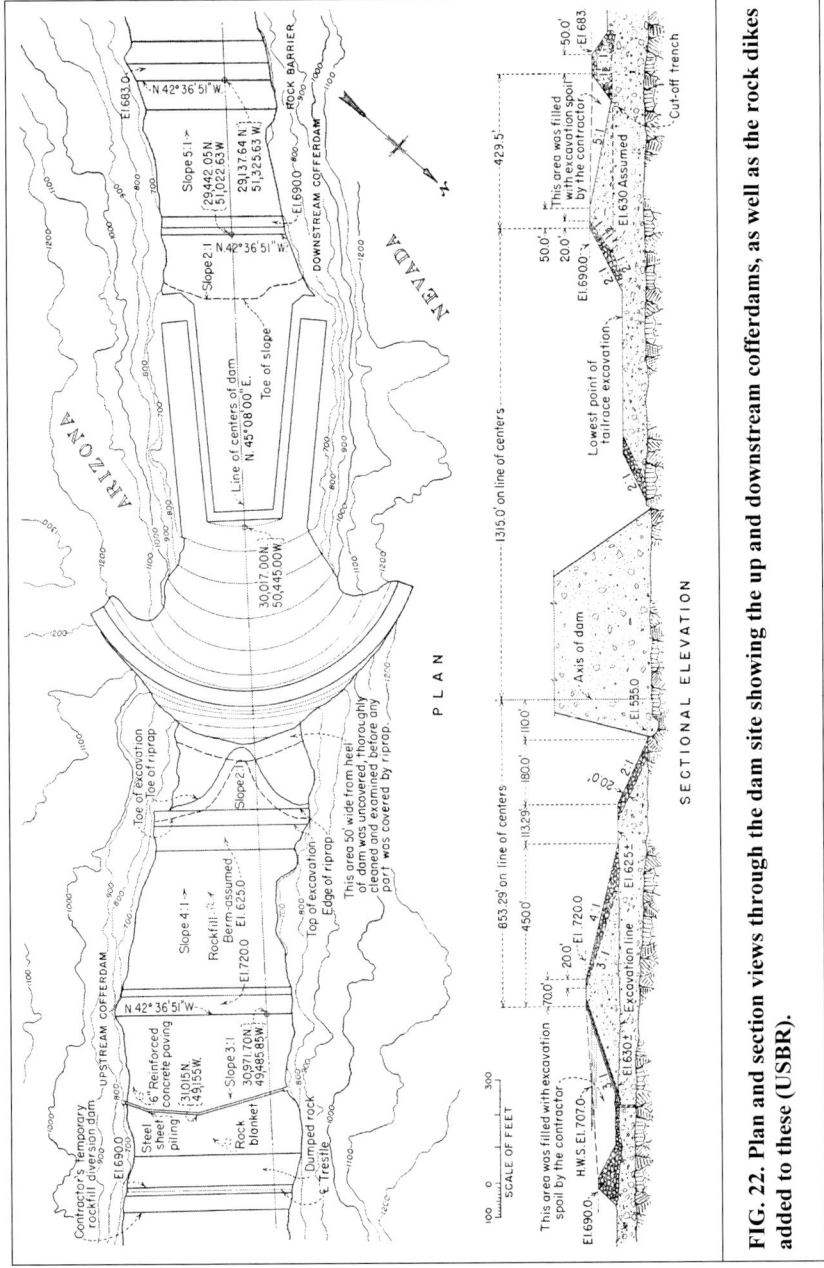

FIG. 22. Plan and section views through the dam site showing the up and downstream cofferdams, as well as the rock dikes added to these (USBR).

FIG. 23. Paving of the upstream face of the upstream cofferdam, as a protective layer and seepage barrier. The face was inclined at 3:1 (horizontal to vertical). Middle foreground shows the sheetpile cutoff wall being installed by a drag hammer attached to the tracked dragline. Note men for scale (USBR).

The Colorado River Board voiced some apprehension about the cofferdams because they were comprised of pervious channel gravels placed on channel gravels. Professor Mead worried that under 100 feet of pressure head during a flood, considerable seepage could be expected to percolate beneath and through these structures.

As an accommodation of this concern, Reclamation decided to require overexcavation of 250,000 yd^3 of channel fill to a depth of elevation 625 ft to provide a more stable and less pervious foundation for the cofferdam. This left 70 to 80 ft of channel gravel in-place beneath the embankment (Figure 22). Reclamation added a specification for installation of a sheetpile cutoff wall at the upstream toe of the cofferdam which extended 40 to 50 feet into the channel gravels, but they were unable to penetrate the boulders filling the inner gorge. They also required that the upstream face of the upstream cofferdam be covered with a 6-inch-thick reinforced-concrete mat (Figure 23), placed over a 3-foot-thick layer of tightly tamped rock, with sluiced fines brought in from Hemenway Wash (the only source of fine-grained material in the area).

To limit seepage along the abutments three rows of reinforced concrete "percolations stops" (cutoff walls) were constructed, which extended 30 feet into the embankment, along its axis. The upstream cofferdam had a crest elevation of 720 ft, about 30 feet above the crowns of the diversion tunnels. The maximum pool for the 25-yr flood (200,000 cfs) was assumed to reach elevation 707 ft, with a low water surface at 645 ft. This allowed 13 ft of freeboard beneath the crest of the upstream cofferdam. Another 77,000 yd³ of rock was used to construct a rock dike above the upstream cofferdam (Figure 22).

Construction of the upstream cofferdam began in September, 1932, two months before the river diversion began. A horseshoe-shaped rockfill dike protected the cofferdam on the Nevada side of the river. After the Arizona tunnels were completed, and the river was diverted through them, the remaining work on the Nevada diversion tunnels was completed much faster because all of the mining resources could be concentrated upon them. The upstream cofferdam contained 516,000 yd³ of re-worked channel fill (gravel and sand) and 157,000 yd³ of rockfill used on the facing shells (Figure 23), which came from the abutment excavations (Figure 18). During the month of December 1932 more than 400,000 cubic yards of material was placed in the upstream cofferdam, a record at the time.

Fill placement for the downstream cofferdam was delayed until the high-scaling of the canyon walls above the powerplant sites and outlet works was completed (Figure 9). The downstream cofferdam was built of rolled channel sands and gravels, with a crest elevation of 690 ft, 66 feet above the tailwater level and about 155 ft above the bedrock channel. The downstream cofferdam was originally designed for 230,000 yd³ of channel fill and 63,000 yd³ of rock shell. The rock shells were placed on the upstream sides of the downstream cofferdams to retard hydraulic sluicing or piping of low cohesion fill materials into the dam and powerhouse excavations, the "wet" side being downstream of the "dry" side of the embankments (Figure 23).

To lessen the backpressure against the downstream cofferdam that might occur under maximum spillage (~200,000 cfs) during construction, 98,000 yd³ of armor rock (> 18 inches in diameter) was placed to form an additional rock dike about 55 ft high, situated 350 feet downstream of the downstream cofferdam (shown on extreme right side of Figure 22). The area between this rock dike and the downstream cofferdam was eventually infilled with random rockfill, and the pay volume for the combination downstream cofferdam eventually reached 500,000 yd³. Both cofferdams were completed by March 1933, 13 months ahead of the government schedule.

Handling Flash Floods

The annual average peak flow of the Colorado River before construction of the dam was normally around 85,000 cfs during the month of June. One local flash flood did succeed in inundating the construction site on February 10-11, 1932, before any of the diversion tunnels were completed. These waters came from the Virgin River Basin in southwestern Utah, and reached a peak flow of something around 50,000 cfs (Figure 24). The flooding overwhelmed the rock dikes protecting the diversion tunnels, filling them with mud, and washed out the trestle bridge Six Companies had constructed at the dam site, which they quickly re-built (Figure 5). The damage necessitated full shutdown of the job to accommodate clean-up that lasted five days. Another flash flood occurred on August 31, 1932 when a thunder storm striking the lower Grand Canyon-Grand Wash Cliffs area brought 60,000 cfs of water down Black Canyon, overwhelming the tunnel cofferdam dikes and flooding all of the equipment working adjacent to the channel. The main cofferdams were completed by June 1st, 1933 just before the season's annual spring floods. The biggest flow the tunnels had to handle during construction occurred two weeks later, on June 16th 1933. That day 73,000 cfs flow was safely conveyed around the dam site through the diversion tunnels. The minimum flow of 1,000 cfs was recorded late the following summer, on August 26, 1934 and the reservoir began filling on February 1, 1935.

FIG. 24. The only flooding of the job site occurred in early February 1932, triggered by a local storm cell that hit the lower Grand Canyon area. This shows the flooded portals of Diversion Tunnels 3 and 4 on the Nevada side. Several wooden trestle bridges were washout out, further downstream (USBR).

STRUCTURAL CONSTRUCTION MILESTONES

Low-heat cement was not available in sufficient quantities to use on the entire dam mass, as more than 5,000,000 barrels of cement were needed for the job (USBR, 1947). Of these, only 400,000 barrels were Modified Low Heat type. The aggregate and feed water were chilled, but an additional cooling process was needed to remove 700 BTUs of heat per cubic yard of concrete because of seasonal impacts caused by the warm summer weather (Figure 25). Cooling pipes had only been used on a trial basis; at Merwin Dam on the Lewis River in Oregon (for which there was no reliable data) and in one small test section of Reclamation's Owyhee Dam, then under construction.

The plumbing details for the concrete cooling pipes were not worked out in advance of letting the contract for Owyhee Dam in 1929, which was the largest dam Reclamation had designed up to that point. It was serving as something of a prototype test bed for the various for many of the innovations being incorporated into the designs for Hoover Dam. There had been considerable delay in circulating the cooling water to the lowest portions of Owyhee Dam, where many of the dam's joints were 100 feet apart. In this area the internal temperature reached a maximum of 150 degrees F, causing the joints to open as much as 0.25 inch, which exceeded the capacity of the few embedded electrical-resistance joint-meters (their design range was 0.22 inch) and they were rendered inoperative. These gaps were subsequently grouted (Carlson, 1977).

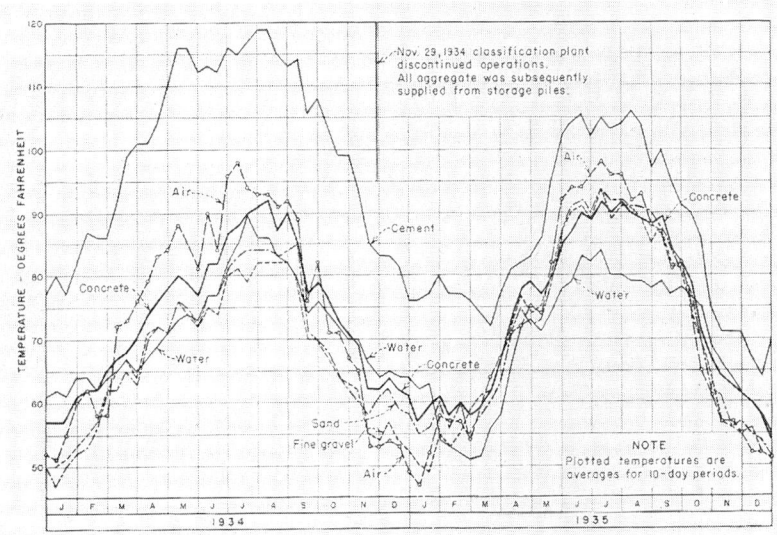

FIG. 25. Average temperatures of air, water, sand, gravel, and mass concrete as prepared at the High Level Mixing Plant throughout 1934-35. Note influence of warm summer temperatures (USBR, 1947).

By the time the contract for Hoover Dam was let in March 1931 Reclamation had opted to embed more than 582 miles of 1-inch diameter steel pipe in the concrete (Figure 26) and initially circulate river water through these, then shift to chilled water. The block temperatures were monitored, especially during the first 30 days. Chilled water came from a refrigeration plant that could produce up to 1,000 tons of ice every 24 hours. This water was chilled to between 35 and 40 degrees F, then pumped through the cooling pipes at a rate of 2,100 gallons per minute (gpm) with flow velocity of cooling water not less than 2 fps. The water temperature at exit was found to be between 42 and 65 degrees F, depending on the ambient temperature and time of year.

Cooling was completed in March 1935, allowing 125 years of curing to be completed in less than two years. A central slot 8 feet wide was left to provide access to the cooling pipes during construction (Figures 27 and 28).

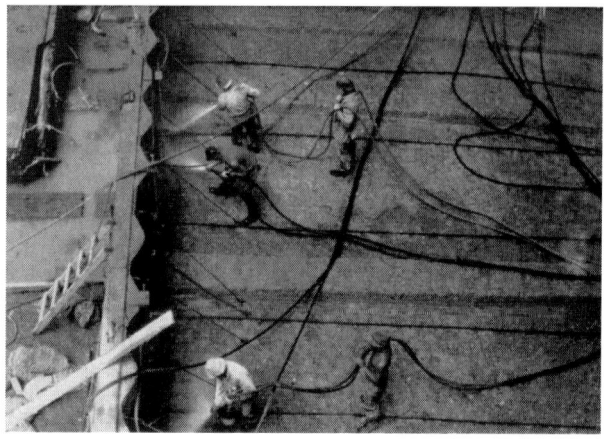

FIG. 26. Workers jetting off the laitance (scum) layer from freshly poured surface on one of the dam's internal blocks, in preparation for the next lift. These fine particles resulted from the tendency of free water to rise in response to aggregate separation, vibration, and agitation of the mix after placement (Troxell, Davis, and Kelly, 1968). Note one-inch-diameter cooling pipes spaced 5 ft apart (USBR, 1947).

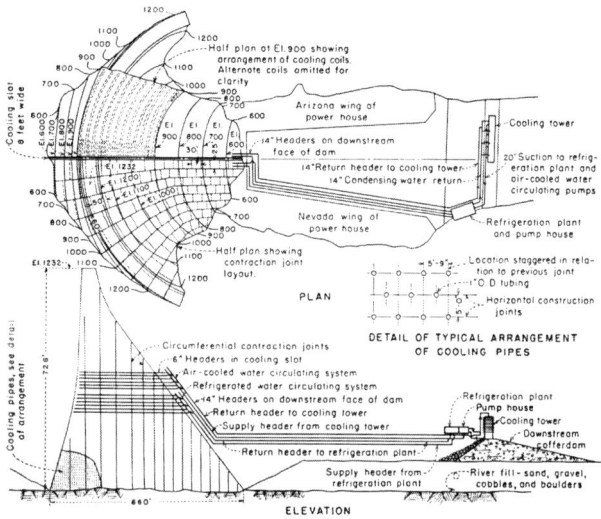

FIG. 27. Schematic of the cooling system installed in the dam as it was constructed. These pipes were subsequently grouted with cement prior to the dam's completion in June 1935 (USBR, 1941).

FIG. 28. 8-foot-wide "cooling slot" running vertically through the dam's axis afforded temporary access to the dam's cooling pipes until the concrete's heat of hydration was brought down to acceptable levels. This was grouted with a standard mix in 10 ft increments as the dam rose upward. (USBR)

Provision for Contraction Joints

In addition to using low-heat cement, two other measures were adopted to help alleviate problems with internal heating of the mass concrete during hydration. One was to cast the concrete in blocks small enough so that they would shrink as a monolithic block, and thereby avoid development of uncontrolled shrinkage cracks.

Dry mixes were specified to reduce shrinkage from moisture change. The dam was built in blocks or vertical columns varying in size from about 60 feet square at the upstream face of the dam to about 25 feet square at the downstream face, using steel forms. Adjacent columns were locked together by a system of vertical keys on the radial joints and horizontal keys on the circumferential joints (Figure 29). Lift heights in each block were limited to five feet in 72 hours, and 35 feet within 30 days. After the concrete was cooled, grout was forced into the spaces created between the columns by the contraction of the cooled concrete to form a monolithic (one piece) structure. Shrinkage was about 0.5%.

Water stops were employed near the up and downstream faces of block joints. Vertically serrated joints were used between blocks in the dam. These joints were grouted after the blocks had shrunk. Horizontally serrated joints used against the abutments. All joints between blocks were to be grouted in 100-foot lifts, after cooling occurred. All cooling pipes were grouted as well, after water circulation ceased.

FIG. 29. Interlocking and convergent nature of the dam's basal blocks can be appreciated in these views, taken during construction. Left view looking upstream, while that at right looks downstream. The blocks were poured intermittently to enhance heat dissipation into the air immediately after placement, caused by the concrete's heat of hydration (USBR).

Aggregate Processing

The aggregate quarries were located 12 miles upstream on the Arizona side of the river, on land owned by the government. The mined aggregate was transported by rail (Figure 30) to an aggregate plant capable of processing 20,000 tons of aggregate every 24 hours. This facility was located in Hemenway Wash, on the Nevada side of river (Figure 31). Six Companies employed an ammonia water and aggregate chilling plant. Aggregate mining ceased on November 29, 1934 and the remaining concrete was batched using existing stockpiles.

FIG. 30. Six Companies Inc. railroad train hauling aggregate from the terrace gravel deposits in Arizona across the Colorado River on a temporary timber trestle. This trestle survived all of the high flows until it was dismantled in the fall of 1934 (Six Companies Collection, USBR-LCR).

FIG. 31. Six Companies Gravel Classification Plant in Hemenway Wash, beneath what is now Boulder Bay of Lake Mead. The mined aggregate was sieved and stockpiled here for shipment to the concrete batch plants at the dam site. (USBR)

Low and High Level Mixing Plants

Six Companies constructed two concrete batch plants. The Low-Level Mixing Plant was located on an enormous rockfill pad placed along the Colorado River, using muck from the diversion tunnels (Figure 32). The plant was situated about 4,000 feet upstream of the dam at elevation 720 ft (same elevation as the crest of the upstream cofferdam), about 80 ft above the normal low flow level of the river, about 2,200 ft upstream of the massive diversion tunnel inlets (Yates, 1932). It was serviced by the low level railway line as well as a gravel road. This plant produced 300,000 yd^3 of concrete to line the four diversion tunnels and over half of the mass concrete placed in the main dam. It was equipped with four 4-yd^3 tilting mixers capable of producing 17 batches per hour, giving the plant a capacity of 280 yd^3 per hour. Some of these mixers and batchers were subsequently transferred to the High-Level Mixing Plant to increase its capacity when the low plant was shut down.

FIG. 52. Low-Level Mix Plant constructed by Six Companies about 4,000 ft upstream of the dam site, as seen in January 1932. Concrete was delivered to the dam site by rail, through tunnels 900 and 1,400 ft long on the Nevada side. This plant was shut down in November 1934 when the dam was 63% complete (USBR).

The High-Level Mixing Plant was assembled in a steep ravine about a 1,000 ft southwest of the dam's right abutment, at el. 1252 ft, 20 ft above the dam crest (Figures 33 and 34). The High-Level Mixing Plant produced almost half of the concrete used in the dam, as well as all of the concrete for the spillways, intake towers, and penstocks. It was placed in operation in March 1933, initially using just two batching and mixing units. The plant utilized conveyor-belt feeders with automatic batching and remote control of mixing using double screw feeders, which

FIG. 33. Classified aggregate being delivered to hoppers at the High-Level Mixing Plant by the Six Companies Inc. Railroad at an elevation of 1403 ft. The aggregate came from the classification plant in Hemenway Wash (USBR-LCR).

FIG. 34. High-Level Mix Plant constructed by Six Companies in a natural ravine about 700 ft downstream of the dam's right abutment crest. The lower silos were of 5,000 barrel capacity to store cement while the larger silos (at left) held aggregate. The plant's essential elements were 'stacked' vertically to take advantage of gravity in the material feed processing (USBR).

circumvented weighing the various batch components. Mix duration was controlled using consistency meters which operated by resistance feed-back. When both plants were operating they were capable of producing 24 yd^3 of concrete every 3.5 minutes, a record at the time (subsequently exceeded at Grand Coulee Dam). Two additional mixers were later brought in from Owyhee Dam and two more from the Low-Level Mixing Plant, when it shut down.

Most of the concrete was transferred to the dam in 8 yd^3 bottom dumping buckets carried on the overhead cableways, but smaller batches were also dispatched in mix trucks, which could reach all the locations on the job. When the dam reached about half of its design height (November 3, 1934) the Low-Level Mixing Plant was dismantled and shipped to the Parker Dam site downstream. From that time all concrete production shifted to the High-Level Mixing Plant (there was much less concrete volume in upper half of the dam).

CONCLUSIONS

The unprecedented size of Hoover Dam led to many innovations in construction, especially in scheduling and bidding of unit costs. Six Companies emerged as the model contracting organization for large complex jobs, leaving an indelible mark on heavy construction. Too large for a single contractor, the project required the formation of a joint venture of eight contractors, pooling their talents and resources, and sharing the risks. Six Companies also employed numerous innovations to avoid costly schedule delays, such as the use of construction access adits, to allow multiple headings to be advanced on underground excavations (at the contractor's expense), which forever shifted the manner by which underground jobs were designed or bid from thereafter. The complex geologic conditions were handled with what was state-of-the-art expertise at the time, with Reclamation hiring their first geologist. The project also set records for materials processing and handling, which proved of great value, and which figured prominently in the contractor being able to complete the project two years early.

ACKNOWLEDGMENTS

The author is most indebted to retired Reclamation engineer Richard Wiltshire, who appreciates civil engineering history and organized this symposium commemorating the 75th anniversary of Hoover Dam's completion. The author's interest in Hoover Dam was originally piqued by his initial viewing of the hour-long documentary film profiling the Boulder Dam project which he viewed in high school. A year later he visited the dams along the lower Colorado River and took an extended tour of Hoover Dam.

In 2009-10 the author was a Trent Dames Civil Engineering Heritage and Dibner Research Fellow at the Huntington Library in San Marino, California. This fellowship and the residency it afforded allowed the author to review of numerous serial publications, collections, archives, and ephemera that provide invaluable to understanding the engineering decisions and political pressures influencing those decisions, not just during construction (1931-35) but during the decade preceding and following the dam's completion. Huntington Archivists Dan Lewis and Bill Frank proved to be particularly valuable in ferreting out rare or obscure accounts from the Huntington's civil engineering and scientist manuscript collections, as well as rare maps and historic photos.

Between 1976 and 1988 the author conducted interviews with Roy W. Carlson (1900-1990), Milos Polivka (1917-1987), Jerome M. Raphael (1912-1989), and George E. Troxell (1896-1984), all professors in the civil engineering program at U.C. Berkeley. Carlson had been an ex-officio member of the Hoover Dam Concrete Research Board. Troxell, Carlson, Polivka, and Raphael were all protégées of Professor Raymond E. Davis (1885-1970), who supervised the concrete research for Hoover Dam during its final design and construction, between 1930-1935.

The author also wishes to thank Shelley Erwin of the Archives of the California Institute of Technology for researching the files of Professor F. Leslie Ransome (1868-1935), who served as the principal geologic consultant on the Boulder Canyon Project to Reclamation from 1922 to 1935, and alumni newsletters and records relating to Dr. Frank A. Nickell, who received all his degrees from Caltech (in civil engineering and geology) and became Reclamation's first engineering geologist in 1931. Special thanks is also due to Allen W. Hatheway, Jeffrey Keaton, Richard Proctor, and William K. Smith of the AEG Foundation, who provided documents from Frank Nickell's files.

Sincere appreciation for assistance is also rendered to the following people and their respective organizations: Reclamation historians Brit Allan Storey and Christine Pfaff, Dianne Powell of the Bureau of Reclamation Library; Reclamation archivists and staff Bonnie Wilson, Andy Pernick, Bill Garrity, and Karen Cowan of Reclamation's Lower Colorado Region (LCR) office in Boulder City; Boulder City Library Special Collections; Eric Bittner at the National Archives and Records Service Rocky Mountain Region depository at Denver Federal Center; Paul Atwood and Linda Vida at the University of California Water Resources Center Archives in Berkeley; Su Kim Chung of the University of Nevada-Las Vegas Libraries; civil engineering historian Edward L. Butts; journalist and author Michael Hiltzik; and a special thanks to the author's best friend Steve Tetreault, who accompanied the author of his first tour of the dam in February 1973. Over the past several decades Mr. Tetreault has provided invaluable support and logistical connections in support of the author's numerous visits to the University of Nevada-Las Vegas, Boulder City, and Hoover Dam.

REFERENCES

Clark, E. B. (1930). *William L. Siebert: The Army Engineer*. Dorrance & Co., Philadelphia.

Gerwick, B.C., Jr., and Woolery, J.C. (1983). *Construction and engineering marketing for major project service*s. John Wiley & Sons.

Hiltzik, M. (2010). *"Colossus: Hoover Dam and the making of the American Century."* Simon & Schuster.

House, P.K., Pearthree, P.A., and Perkins, M.D. (2008). Stratigraphic evidence for the role of lake-spillover in the birth of the lower Colorado River in southern Nevada and western Arizona, in Reheis, M.C., Hershler, R., and Miller, D.M., eds., Late

Cenozoic Drainage History of the Southwestern Great Basin and Lower Colorado River Region: Geologic and Biologic Perspectives: Geological Society of America *Special Paper 439*, p. 333-351.

Howard, K.A., and Bohannon, R.G. (2000). Lower Colorado River: Upper Cenozoic Deposits, Incision, and Evolution. Ch. 15 in R.A. Young and E.E. Spamer, eds., *Colorado River Origin and Evolution*. Grand Canyon Association. p. 101-106.

Howard, K.A., Lundstrom, S.C., Malmon, D.V., and Hook, S.J. (2008). Age, distribution, and formation of late Cenozoic paleovalleys of the lower Colorado River and their relation to river aggradation and degradation. In Reheis, M.C., Hershler, R., and Miller, D.M., eds., Late Cenozoic Drainage History of the Southwestern Great Basin and Lower Colorado River Region: Geologic and Biologic Perspectives: Geological Society of America *Special Paper 439*, pp 394-410.

LeTourneau, R.G. (1960). *Mover of Men and Mountains*. Prentice-Hall.

McAlpine, J., ed. (1996). *Paleoseismology*. Academic Press. 588 p.

Ransome, F.L. (1923). Geology of the Boulder Canyon and Black Canyon Dam Sites and Reservoir Sites on the Colorado River. Unpublished report to U.S. Bureau of Reclamation, Denver, April 1923, 22 p.

Ransome, F.L. (1931). *Report on the Geology of the Hoover Dam Site and Vicinity*. Unpublished report to U.S. Bureau of Reclamation, Denver, Nov. 30, 1931, 71 p., 22 pl.

Raphael, J.M. (1977). Personal communication.

Rocca, A.M. (2001). "*America's Master Dam Builder: The Engineering Genius of Frank T. Crowe.*" University Press.

Tassava, C.J. (2003). Multiples of Six: The Six Companies and West Coast Industrialization, 1930-1945. *Enterprise & Society*, Vol. 4, pp. 1-27.

Troxell, G.E., Davis, H.E., and Kelly, J.W. (1968). *Composition and properties of concrete*. McGraw-Hill, New York.

U.S. Bureau of Reclamation. (1941). General Features, Boulder Canyon Project Final Reports, Part IV – Design and Construction, Bulletin 1. Denver.

U.S. Bureau of Reclamation. (1947). Concrete Manufacture, Handling, and Control, Boulder Canyon Project Final Reports, Part IV – Design and Construction, Bulletin 1. Denver.

U.S. Bureau of Reclamation. (1950). *Geological Investigations*; Boulder Canyon Project Final Reports, Part III – Preparatory Examinations, Bulletin 1. Denver.

Weymouth, F.E. (1924). Colorado River Development. Senate Document No. 186, 70th Congress, 2nd Session, 231 p.

Wolf, D.E. (1996). *Big Dams and Other Dreams: The Six Companies Story*. University of Oklahoma Press.

Yates, J.P. (1932). Low-level Concrete Plant for Hoover Dam. *Western Construction News and Highways Builder*, v. 8:317-323.

Hoover Dam: Construction Milestones in Concrete Delivery and Placement, Steel Fabrication, and Job Site Safety

J. David Rogers[1], F. ASCE, D. GE, P.E., P.G.

[1]K.F. Hasslemann Chair in Geological Engineering, Missouri University of Science & Technology, Rolla, MO 65409; rogersda@mst.edu

ABSTRACT: The size of the Boulder Canyon Project necessitated a broad array of innovations in construction engineering and management which had enormous impacts on all of the large scale projects that followed it. Major milestones were achieved in mass concrete handling and placement across a large and sometimes treacherous job site; equally challenging problems with forming and pouring structural concrete elements for the project, such as the intake towers, penstocks, spillways, and outlet works; steel penstock fabrication and placement; and establishment of basic job site safety precautions that became increasingly common thereafter, including on-site medical care the provision of hard hats to all workers.

MILESTONES IN CONCRETE HANDLING AND PLACEMENT

Overlapping System of Cableways and Derricks

Six Companies General Superintendent for Hoover Dam was Frank Crowe. Crowe was renown in the dam construction industry for his ability to innovate, devising clever site-specific schemes for delivering critical components in a regular, timely manner, without creating undue delays. One of the innovations he was most remembered for was his novel employment of traveling cable hoists covering a dam site, which could be adjusted on any given day to cover expansive portions of the work under construction (Yates, 1933), shown in Figures 1 through 3.

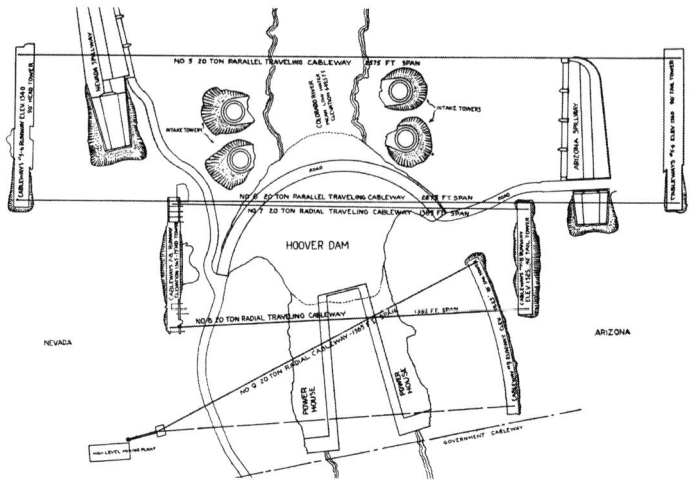

FIG. 1. Plan view of cross-canyon cableways designed for maximum flexibility to move buckets of concrete and construction materials to any spot on the dam, inlet towers, most of the spillways, outlet works, and powerhouses. Aerial delivery was not used on any of the diversion tunnels (Six Companies Collection, USBR-LCR).

This method of "overhead delivery" had been widely employed on steep mountains sides during the late 19th and early 20th Centuries using cableways to haul mining ore and timber, as well as grading earthen levees (Gillette, 1916; 1920). Overhead cableways were initially used in dam construction on some of Reclamation's earliest projects, such as Pathfinder Dam in Wyoming, built between 1905 and 1909, which Frank Crowe had observed early in his professional career with Reclamation. Cableways were also being used with great success by General Construction Company at Owyhee Dam (ENR, 1930a).

The system had innumerable advantages insofar that it avoided cumbersome transport across rough or uneven ground and avoided all manner of obstructions, provided the various cableways were aerially separated. Figures 1 through 5 show some of the essential components of the aerial cableway systems employed by Six Companies during construction of Hoover Dam. All but the 150-ton Government Cableway were dismantled after the powerhouse was completed in 1936.

FIG. 2. One of the 20-ton cableway towers mounted on rails, which allowed it and another similarly configured unit on the opposite canyon wall, to be moved 500 ft up or down the canyon, to facilitate precise placement of the 8-cubic-yard concrete buckets (USBR).

FIG. 3. Cableways 7 and 8 being set up on November 29, 1932 to carry 8-cubic-yard buckets of concrete from the high mix plant to the dam. These were equipped with electric drive cable hoists that allowed pinpoint precision in dropping the buckets on the downstream 2/3 of the dam (USBR).

FIG. 4. After the dam rose above the level of the rail line leading from the Low Level Mixing Plant this stiff-leg derrick positioned at elevation 930 ft to lift the concrete buckets from the rail cars to the blocks closest to the dam's upstream right abutment until the Low-Level Mixing Plant was shut down. (USBR)

FIG. 5. Another 20-ton capacity stiff-leg derrick in its precarious perch on the dam's right abutment to continue handling of concrete buckets coming from the Low-Level Mixing Plant after the dam rose 150 ft above the river bed. (USBR)

Innovations in Concrete Placement

The main dam contained 3.25 million yd^3 and Six Companies poured 4.36 million yd^3 of concrete in the entire project (including the diversion tunnels, spillways, intake towers, valve houses, and power plant). Most of the concrete placed using 8-cubic-yard cylindrical hopper buckets, each weighing approximately 20 tons loaded (16.2 tons of concrete and 3.6 tons for the buckets). The first concrete for the dam was placed on June 6, 1933 at approximately 135 feet below river level (Figure 6). As shown in Figure 7, by August 1932 concrete production zoomed up to 149,000 yd^3 per month, a record for that time (the previous record had been 50,000 yd^3/month at Owyhee Dam in June 1931). By October concrete placement exceeded 200,000 yd^3/month. In March 1934 concrete production reached a peak rate 262,000 yd^3, or about 1,100 buckets per day, basically one bucket every 78 seconds. The record one-day pour was 10,350 yd^3 on June 1, 1934. The last bucket of concrete for the main dam was poured on May 29, 1935.

Some of the essential elements of the concrete placement for the main dam structure are profiled in Figures 7 through 12, below. The most important aspects were the coordination that had to be maintained between adjacent blocks of the dam, and the working space that became increasingly congested as the dam rose and narrowed. The C-2 and C-5 cableways covered most of the two spillways, the intake towers, and the upstream portion of the dam, while the C-6 to C-7 cableways covered the abutments, which curled downstream. The pivoting C-8 cableway handled demands in the power plants, where a much lower volume of concrete was required for the reinforced concrete framework.

FIG. 6. Placing the first 8-cubic-yard bucket of concrete on the scrubbed bedrock surface of the inner channel on June 6, 1933. Note batter boards supporting forms. (USBR)

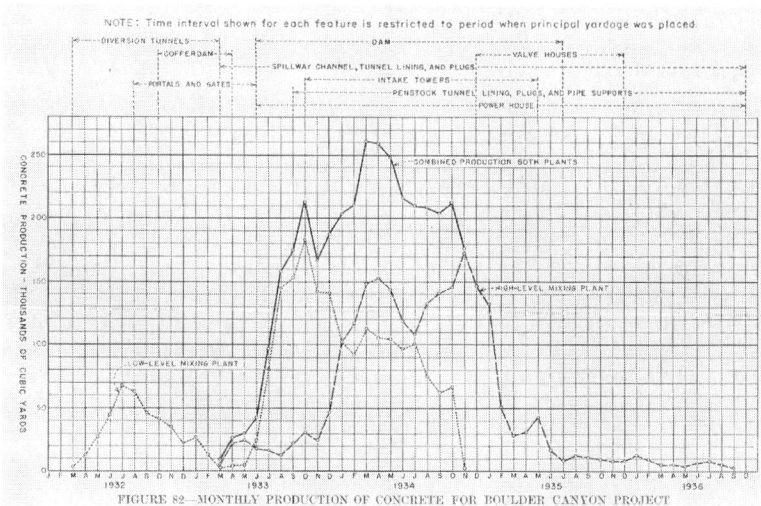

FIG. 7. Monthly production of concrete by the Low-Level and High-Level Mixing Plants at Hoover Dam (in thousands of cubic yards) versus the various project elements that were constructed (USBR, 1947).

FIG. 8. Concrete tampers flipping the release on the bucket containing 8 cubic yards of concrete. Stories about workmen being entombed in the dam's concrete were completely fallacious. When the bucket's contents were spread over a typical 50 x 50 ft pour block, it only amounted to one inch of concrete! (USBR)

FIG. 9. Workman using pneumatically powered concrete vibrator to remove some of the entrained air from wet concrete in one of the dam's blocks (USBR).

FIG. 10. Looking downstream on July 17, 1933, as the inner gorge was being filled with concrete. Note the 8 ft gap at center, to access the cooling pipes. Also note the excavation of the right abutment and temporary trestle which brought concrete buckets by rail from the Low-Level Mixing Plant to the dam site (USBR).

FIG. 11. Concrete placement was carried out around-the-clock, and the night shifts were a welcome respite during the summer. This time exposure shows the formwork for the powerhouses in the foreground with the dam in the background (USBR).

FIG. 12. Hoover Dam nearing completion, as seen along the downstream face, around New Years 1935. Note the vertical slot at midstream, which was not filled until after cooling water circulation was completed in mid-March 1935 (USBR).

Design Standards for Contraction Joints

One of the most foreboding aspects of the St. Francis Dam failure near Los Angeles in March 1928 was the existence of large shrinkage cracks that developed transverse to the dam's axis (Rogers, 1995). These cracks had been caulked with oakum across the dam's upstream face, which promoted the development of excessively high pore water pressures between the dam's blocks separated by these fractures. A significant technical debate erupted shortly thereafter wherein dam engineers argued about what values of hydraulic uplift beneath dams and inter-block pressures between adjacent blocks of the dam would be appropriate for design (Henny, 1928; Floris, 1928; Pearce, 1928a, 1928b; Jakobsen, 1928; and Hinds, 1929).

One aspect of mass concrete dam design everyone seemed to agree upon was to do everything in their power to avoid the development of uncontrolled transverse shrinkage cracks, and to seal all open fissures with cement grout to allay development of debilitating pore water pressure within the body of the dam (ENR, 1930b; Noetzli, 1930; Wiley, 1931; Henny, 1931; Houk, 1932; Weaver, 1932; and Terzaghi, 1934).

As a consequence, every conceivable measure was taken to control concrete shrinkage and employ an intricate system of criss-crossing expansion joints between pour blocks. The joints between adjacent blocks were equipped with interlocking shear keys, patented state-of-the-art water stops (Figure 13, and then grouted after sufficient curing (Figure 14) with an intricate system of grout pipes (Figure 15). A system of internal galleries with seepage collection gutters and measurement weirs was also built into the dam, with a much greater density of openings than had been originally envisioned when the project was approved in December 1928.

FIG. 13. Welder carefully brazing joints on a copper water stop placed in one of the dam's contraction joints. (USBR)

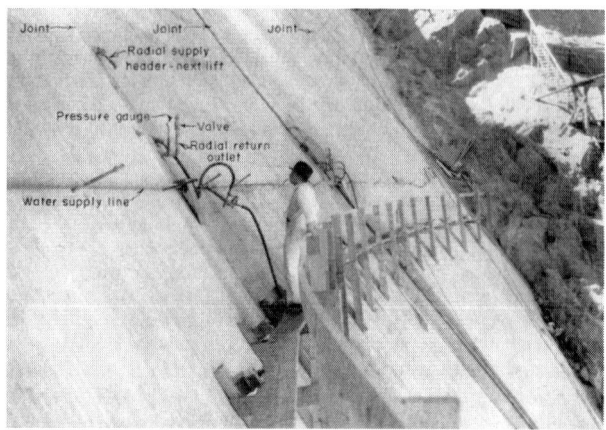

FIG. 14. Workman grouting one of the dam's radial joints from a scaffold walkway on the dam's downstream face (from USBR, 1941).

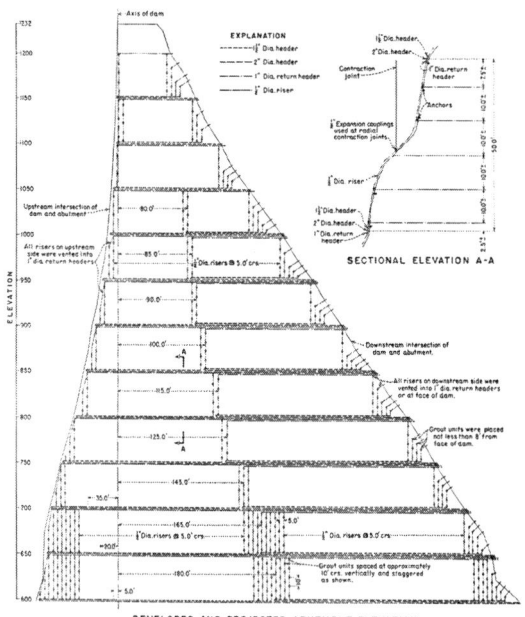

FIG. 15. Layout of grout pipes leading to contraction joints between the dam's monolithic blocks. Grout injection pipes were placed for every 30 to 50 ft^2 of joint surface area, using injection pressures between 100 and 300 psi (USBR, 1947).

MILESTONES IN PLACEMENT OF STRUCTURAL CONCRETE

Intake Towers

The four intake towers were constructed on cut benches upstream of the dam. The diameter of these towers is 82 ft at the base, 63.25 ft at the top, and 29.67 ft inside. Each tower is 395 ft high and extends above the crest of the dam (to accommodate periodic removal and cleaning of steel inlet screens). Each intake tower controls one-fourth of the water supply for the powerplant turbines. The water taken into each tower is funneled through two cylindrical gates, each 32 ft in diameter and 11 feet high. One gate is near the bottom and the other near the middle of each tower. The gates are protected by trash racks.

Their positions upstream of the main dam created some access problems that had to be overcome, as shown in Figures 16 and 17. The towers were constructed of high-strength reinforced concrete, using a 5-sack mix, traveling steel forms, and considerable timber falsework (Figure 18), because the towers taper upward.

FIG. 50. Intricate network of overhead cableways and suspended catwalks just upstream of the dam. These were used to carry workers and convey buckets of concrete to the spillways and intake towers. High strength structural concrete was used on the intake towers (Preston Collection Boulder City Library).

FIG. 17. Stiff-leg derricks were also erected on the Arizona abutment to place concrete to the Arizona intake towers. Taken on December 27, 1933 (USBR).

FIG. 18. Overview of the four intake towers approaching completion, as seen on March 26, 1935. The towers rise above the dam crest and are just under 400 ft high (USBR).

Penstocks and Outlet Works

Reservoir water is taken from the lake through 30-foot-diameter steel penstocks installed within 37- and 50-foot-diameter concrete-lined tunnels, shown in Figure 19. The upstream intake towers are connected to the inner diversion tunnels by 37-foot-diameter inclined tunnels. 37-foot-diameter tunnels also connect the downstream towers to the penstocks and outlet works.

The powerplant turbines are fed through four main penstocks, two on each side of the river (Figure 19). Wicket gates control water delivery to each turbine. The maximum pressure head is 590 ft and the minimum operating head is 420 ft. The average operating head assumed in design was between 510 and 530 ft. Water is fed to the Francis turbines by sixteen 13-foot-diameter steel penstocks (eight on each abutment) installed in 18-foot-diameter concrete-lined tunnels (Figures 20-23). The total length of these penstocks is about 5,800 feet. All of these figures were world records at the time of the dam's construction (USBR, 1938).

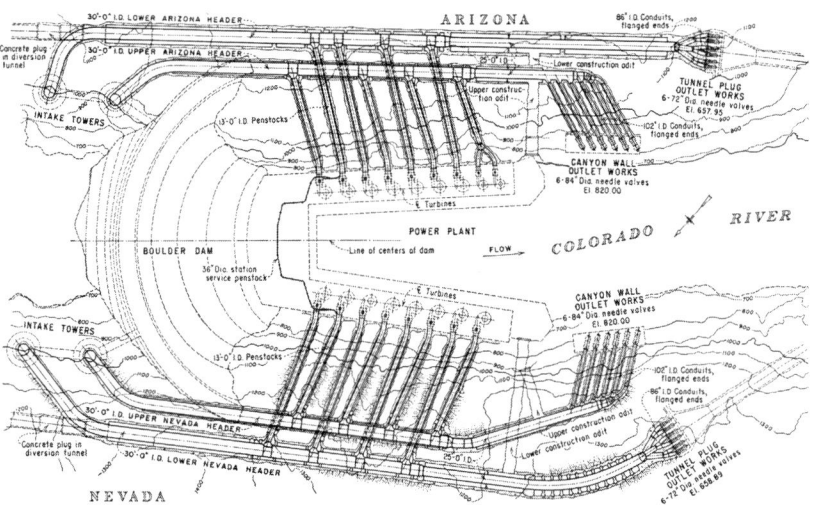

FIG. 19. Plan showing the network of tunnels and penstocks feeding into the dam's two power houses, canyon wall outlet works, and bypass tunnel outlet works. The intake tunnels lead to penstocks constructed within the four diversion tunnels. The headers leading into the powerhouses were 30 feet in diameter. These fed off the inboard and outboard 50-foot-diameter diversion tunnels (USBR).

FIG. 20. Single section of 30-foot-diameter lining for one of the steel penstock headers being dropped into one of the outboard 50-foot-diameter diversion tunnels, as viewed on Nov 16, 1934 (USBR).

FIG 21. Lining of one of the 14-foot finished diameter feeder tunnel leading to one of the Francis turbines in the power house, taken on Nov 16, 1934 (USBR).

FIG. 22. Oblique view of the eight 21-foot finished diameter penstock feeder tunnels connecting to the Arizona power house, under construction. The steel penstocks are 11 feet in diameter (USBR).

FIG. 23. Oblique view of the canyon wall valve house outlet works tunnels. The conduits were 7 ft in diameter and the tunnels 11 ft in diameter with a horseshoe shape. These tunnels were lined after the 102-inch-diameter outlet pipes were placed (Preston Collection-Boulder City Library).

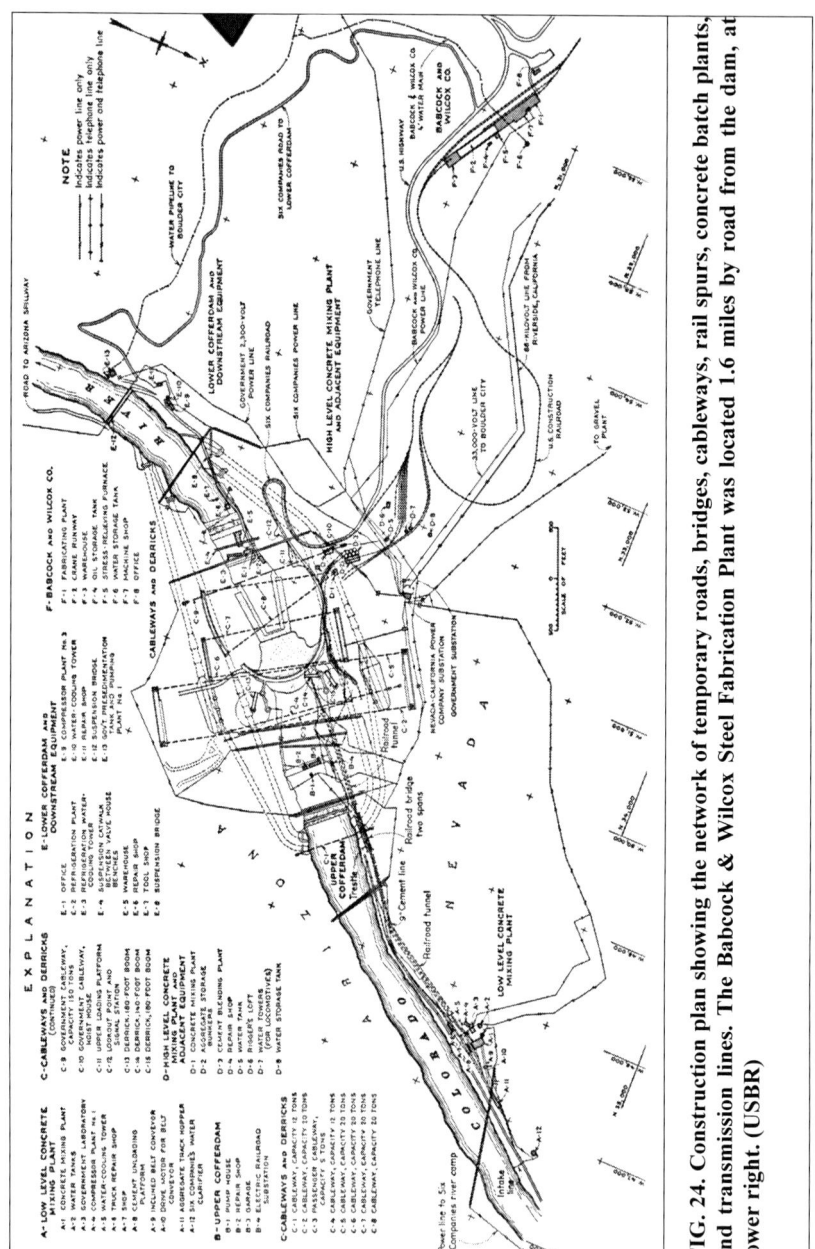

FIG. 24. Construction plan showing the network of temporary roads, bridges, cableways, rail spurs, concrete batch plants, and transmission lines. The Babcock & Wilcox Steel Fabrication Plant was located 1.6 miles by road from the dam, at lower right. (USBR)

Steel Fabrication Plant

Steel penstocks and the pressure feed lines serving the outlet works were fabricated at a specially constructed Steel Fabrication Plant constructed by Babcock & Wilcox approximately 1.6 miles southwest of the dam's right abutment, accessible by road (Figure 24). The fabrication plant cost $600,000 and was essential to the project

FIG. 25. Publicity still showing the scale of the 37-foot-diameter steel liners for the power house headers, in comparison to one of Six Companies steam locomotives (USBR).

FIG. 26. 11-foot-long sections of the 37-foot-diameter penstocks being bent and conjoined by arc-welding inside the Babcock & Wilcox Steel Fabrication Plant (Preston Collection-Boulder City Library).

FIG. 27. Caterpillar tractor pulling two 22-foot-long sections of the 13-foot-diameter steel penstocks to the dam site. All of the steel liners were conveyed to the dam site in this manner (USBR).

because the components were too large to transport by truck or by rail (Figures 25 and 26). The plant fabricated the required shapes from steel plate that varied between 5/8 inch and 2-3/4 inches in thickness. Penstock sections were fabricated in 11-foot-long sections, and adjoining sections were then attached to create 22-foot-long assemblies that could be transported the 1.6 miles to the dam site, as shown in Figure 27. 16,000 lineal feet of steel penstocks were be fabricated at this plant, more than had ever been fabricated for any other dam up to that time (USBR, 1938).

MILESTONES IN JOB SITE SAFETY

Ventilation Problems and Heat Prostration

The unprecedented scale of the Boulder Canyon Project and the hostile summer environment of Black Canyon led to a number of safety measures that were not anticipated prior to construction. During the first year of construction, housing, drinking water, and sanitation facilities were either lacking for many of the workers or were still under construction. In the summer of 1931 average daytime temperatures reached 120 degrees F at the dam site and heat exhaustion killed 14 of Six Companies workers.

In August 1931 the Industrial Workers of the World (IWW) trade union succeeded in convincing Six Companies' workers to strike. Six Companies' lawyers and local and federal law enforcement fought vigorously to break this and succeeding strikes, or threats thereof, without making formal concessions to the worker's demands (summarized in Stevens, 1988 and Hiltzik, 2010).

In total, 21,000 men worked on the dam over a four year period, with the workforce averaging 3,500 employees. At its zenith 5,218 men were on Six Companies payroll during June 1934. The labor troubles always stemmed from a job site safety problem of one kind or another, and Six Companies General Superintendent Frank Crowe was obliged to make whatever adjustments he could to keep the job moving and thereby appear concerned about the worker's safety (Hiltzik, 2010). These concessions included provision of potable water for the worker's dormitories, the hiring orderlies to carry ice water to workers on the job (Figure 28), transporting and feeding the workers using three shifts per day (Figures 29 and 30), establishing first aid stations with doctors and nurses closer on the job site instead of nine miles away at the Boulder City dispensary (Figure 31), and encouraging job site safety and security (Figure 32).

At the completion of the job the official death toll was 96 men, but this did not include any of the men who were fired or quit prior to dying. A common myth perpetuated for years thereafter was that the massive dam contained an unknown number of entombed bodies, which was wholly without merit (as shown in Figure 8). The only 'body' allegedly entombed in the dam resulted from a practical joke carried out during one of the graveyard shifts while one of the crews were pouring the concrete lining of one of the penstock feeder tunnels. They thought it would be humorous to embed a worker's hard hat, gloves, and boots within the tunnel lining, which were revealed when the forms were pulled off. Frank Crowe was not amused. He hunted down the perpetrators and promptly discharged them, to set an example that such nonsense would not be tolerated by the management (Stevens, 1988).

FIG. 28. Orderlies hired by Six Companies to carry ice water to its workers after the strike by IWW in August 1931 (USBR).

FIG. 29. 80 men being taken down the 10 by 12 ft skiff on the Nevada Abutment, heading to work on the 4 PM to Midnight swing shift. The skiff made the 565 ft trip in just over a minute (USBR).

FIG. 30. Some of the buses used to shuttle workers the nine miles between Boulder City and the dam site pose in front of the dam and power houses towards the end of the job, in July 1935 (Six Companies Collection, USBR-LCR).

FIG. 31. One of the medical dispensaries positioned on the job site after the deaths of more than a dozen men working in the diversion tunnels without forced-air ventilation. These were equipped with a doctor, nurse, ambulance, and driver (USBR).

FIG. 32. As the job progressed, increasing attention was paid to safety and security after accidents or carelessness had taken the lives of workers. These signs appeared after an accident involving a driver who failed to yield to one of the between-shift blasts in 1933 (USBR).

Hard-hats Issued to Every Worker

In the fall of 1931 some of the high scalers working on the cliffs began fashioning "hard boiled" work hats by overlapping two baseball caps with bills front and back, then dipping it to form a thin layer of tar and quenching it with cold water, to create a hard shell (Figure 33). Others used shellac applied to leather hats. Six Companies liked this idea so much, they eventually contracted for thousands of these so-called 'hard hats,' from a number of different suppliers, including the E.D. Bullard Company in Sausalito, California.

Six Companies stopped short of requiring the hats be worn, because of the oppressive heat, but issued them to each worker and suggested strongly that men wear them when working in exposed areas, where loose materials could fall on them (Pettitt, 1935). A large proportion of the men working in exposed areas began wearing the hard hats, as shown in Figure 34. Sadly, the most deadly of these falling projectiles were not loose rocks, but tools dropped by other workmen! Hand tools were so crucial to the work at hand that at one point on the job Frank Crowe actually ordered 144,000 crescent wrenches (Stevens, 1988).

FIG. 33. One of the high scalers at Hoover Dam wearing a home-made 'hard-boiled hat" fashioned by overlapping two baseball caps and dipping them in tar and chilling or brushing leather hats with shellac (USBR).

In 1915 the Bullard Company began designing protective hats made of leather for miners in California, based on the protective helmets then being used in the First World War. In 1917 the U.S. Navy asked Bullard to provide them with a protective cap for shipyard workers. Bullard responded with the first "hard-boiled hat," crafted from steamed canvas, glue, and black paint, which they patented in 1919 (because they employed a unique steam processing method to speed up production). Bullard employed headband liners similar to those used in steel 'Kelly helmets' used by American at that time. The Bullard "hard-boiled hats" used in the Navy shipyards were similar to the home-made hard-boiled hats fashioned by the high scalers before Six Companies began providing head wear.

FIG. 34. Six Companies worker wearing one of the mass-produced hard hats they issued free-of-charge to their workers, encouraging them to wear them wherever loose materials might fall upon them. They purchased five different kinds of hard hats, from different suppliers/manufacturers (USBR).

Bullard claims that the first "hard hat job" where workers were actually *required* to wear hard hats was in 1933, when work on the Golden Gate Bridge in San Francisco commenced. The project faced an unforeseen problem when the steel arriving by train from Pennsylvania began rusting rapidly in the foggy, salty air. To protect workers who were sandblasting the rusty steel Bullard designed a sand-blast respirator-helmet that covered worker's faces, provided a window for vision, and an air supply via a hose connected to an air compressor.

In 1938, Bullard designed and manufactured the first aluminum hard hat, which was considered very durable and reasonably lightweight for the time. This quickly became a favorite of forest-fire fighters, but utility workers soon learned that its one serious drawback was that aluminum is a great conductor of electricity. Bullard's three-ribbed heat resistant fiberglass hard hat was developed in the 1940s, and employed across the nation at various defense plants.

CONCLUSIONS

The unprecedented scale of Hoover Dam led to numerous innovations in construction management, especially with regards to critical path scheduling. All subsequent mass concrete dams more or less owe their existence to the advancements that were made during the construction of Hoover Dam. The innovations in mass concrete batching, handling, and placement left an indelible mark on the heavy construction industry, which had never witnessed such an efficient operation previously. Much of this was innovation grew out of the diverse and talented organization assembled by Six Companies, which was focused on problem solving to enhance the job's progress. This was impressive when considering that Reclamation required many more construction details requiring inspection and approval than any previous project they had supervised. Each new challenge was met with equally impressive and innovative technologies, including the most extensive system of overlapping cableways and standard gage rail spurs ever constructed, supplemented by multi-level rail trestles and an ever-changing array of stiff-leg derricks, which had to be managed 24-hours-per-day to avoid entanglements. The sheer scale of steel penstock elements necessitated unusual fabrication and method of transportation to the side-canyon tunnels. Although Six Companies did a poor job of ventilating their diversion tunnel excavations, they made numerous concessions during the course of the job to increase worker safety and morale, which became more of less standard for the heavy construction industry shortly thereafter.

ACKNOWLEDGMENTS

The author is most indebted to retired Reclamation engineer Richard Wiltshire, who appreciates civil engineering history and organized this symposium commemorating the 75th anniversary of Hoover Dam's completion. The author's interest in Hoover Dam was originally piqued by his initial viewing of the hour-long documentary film profiling the Boulder Dam Project which he viewed in high school. A year later he took his first extended tour of the dam.

In 2009-10 the author was a Trent Dames Civil Engineering Heritage Fellow and was a Dibner Research Fellow at the Huntington Library in San Marino, California. This fellowship and the residency it afforded allowed the author to review of numerous serial publications, collections, archives, and ephemera that provide invaluable to understanding the engineering decisions and political pressures influencing those decisions, not just during construction but during the decade preceding and following the dam's completion. Huntington Archivists Dan Lewis and Bill Frank proved to be particularly valuable in ferreting out rare or obscure accounts from the Huntington's civil engineering and scientist manuscript collections, as well as rare maps and historic photos.

Sincere appreciation for assistance is also rendered to the following people and their respective organizations: Reclamation historians Brit Allan Storey and Christine Pfaff, Dianne Powell of the Bureau of Reclamation Library; Reclamation archivists and staff Bonnie Wilson, Andy Pernick, Bill Garrity, and Karen Cowan of Reclamation's Lower Colorado Region (LCR) office in Boulder City; Boulder City Library Special

Collections; Eric Bittner at the National Archives and Records Service Rocky Mountain Region depository at Denver Federal Center; Paul Atwood and Linda Vida at the University of California Water Resources Center Archives in Berkeley; Su Kim Chung of the University of Nevada-Las Vegas Libraries; civil engineering historian Edward L. Butts; journalist and author Michael Hiltzik; and a special thanks to the author's best friend Steve Tetreault, who accompanied the author of his first tour of the dam in February 1973. Over the past several decades Mr. Tetreault has provided invaluable support and logistical connections in support of the author's numerous visits to the University of Nevada-Las Vegas, Boulder City, and Hoover Dam.

REFERENCES

Engineering News Record. (1930a). Heavy-Load Cableway Installation for Owyhee Dam. v. 105:62 (July 10, 1930).
Engineering News Record. (1930b). Western Engineers Giving Increased Study to Safety of Dams. v. 105:726 (Nov 6, 1930).
Floris, A. (1928). Uplift Pressure in Gravity Dams. *Western Construction News*, v. 3 (January 25).
Gillette, H.P. (1916). *Handbook of Rock Excavation, Methods and Cost*. McGraw-Hill, New York.
Gillette, H.P. (1920). *Earthwork and Its Cost: 3rd Ed*. McGraw-Hill, New York.
Henny, D.C. (1928). Important Lessons of Construction Taught by Failure of St. Francis Dam. *Hydraulic Engineering*, v. 4:731-33, 758.
Henny, D.C. (1931). Problems in Concrete Dam Design. *Engineering News Record*, v. 106:431-435 (March 12, 1931).
Hiltzik, M. (2010). "*Colossus: Hoover Dam and the making of the American Century.*" Simon & Schuster.
Hinds, Julian. (1929). Uplift Pressures Under Dams: Experiments by the United States Bureau of Reclamation. *Transactions* ASCE, v. 93, pp. 1527-1582.
Houk, I. E. (1932). Uplift Pressures in Masonry Dams. *Civil Engineering*, v. 2 (September).
Jakobsen, B.F. (1928). Letter to the Editor re: Uplift Under Gravity Dams by C.E. Pearce, *Western Construction News*, v. 3:789 (Dec 25, 1928).
Noetzli, F.A. (1930). Modified Types of Gravity Dams in Relation to Uplift. *Engineering News Record*, v. 105:884-886 (Dec 4, 1930).
Pearce, C.E. (1928a). Uplift Under Dams. *Western Construction News*, v. 3:569-571 (Sept 10, 1928).
Pearce, C.E. (1928b). Uplift Under Dams-Part II. *Western Construction News*, v. 3:662-664 (Oct 25, 1928).
Pettitt, G.A. (1935). *So Boulder Dam was built*. Press of Lederer, Street, and Zeuss, Berkeley.
Rogers, J.D. (1995). "A Man, A Dam and A Disaster: Mulholland and the St. Francis Dam." In D. B. Nunis, Jr., ed., "The St. Francis Dam Disaster Revisited." *Southern California Quarterly*, Vol. LXXVII, n. 1-2, pp. 1-110.
Stevens, J.E. (1988). *Hoover Dam: An American Adventure*. University of Oklahoma Press, Norman.

Terzaghi, K. (1934). Discussion of Stability of Straight Concrete Dams. *Transactions* ASCE, v. 99:1107.
U.S. Bureau of Reclamation. (1938). *Dams and Control Works, 2nd Edition*. U.S. Gov't Printing Office, Washington, DC, 269 p.
U.S. Bureau of Reclamation. (1941). General Features, Boulder Canyon Project Final Reports, Part IV – Design and Construction, Bulletin 1. Denver.
U.S. Bureau of Reclamation. (1947). Concrete Manufacture, Handling, and Control, Boulder Canyon Project Final Reports, Part IV – Design and Construction, Bulletin 1. Denver.
Weaver, Warren. (1932). Uplift Pressures on Dams. *Journal of Math and Physics*, n. V. (June).
Wiley, A.J. (1931). Past Experience with High Dams and Outlook for the Future. *Transactions* ASCE, v. 95: 130-138.
Yates, J. P. (1933). Cableways Place Hoover Dam Concrete. *Western Construction News and Highways Builder*, v. 9:377-381.

Hoover Dam: Operational Milestones, Lessons Learned, and Strategic Import

J. David Rogers[1], F. ASCE, D. GE, P.E., P.G.

[1]K.F. Hasslemann Chair in Geological Engineering, Missouri University of Science & Technology, Rolla, MO 65409; rogersda@mst.edu

ABSTRACT: Hoover Dam was a monumental accomplishment for its era which set new standards for post-construction performance evaluations. Reclamation based many of their decisions on surveys of previous high dams to compare their physical, geologic, and hydrologic features with those proposed at Hoover Dam. Perhaps the greatest triumph was the hydroelectric power generation, which repaid the cost of the project with interest over a term of 50 years, which quickly became the economic model for similar high dam projects, world-wide. In the aftermath of the behemoth project, the project did suffer a number of unexpected failings, principally, the failure of the grout curtain to function as intended (requiring retrofitting between 1936 and 1948) and severe cavitation of the spillway elbows in 1941 and 1983 (requiring installation of air ducts in 1983-84). Along the way Hoover Dam was the first dam to be singled out as a terrorist target (by German agents) and provided electricity for the world's largest magnesium production facility from 1942 onward.

OPERATIONAL MILESTONES

First Hydroelectric Power Generation by Reclamation

One of the most controversial aspects of the Boulder Canyon Project was the eventual proposal for the dam to generate electricity and thereby pay for itself over a term of 50 years. It was the first hydropower project undertaken by a federal agency that gained congressional approval, essentially placing the U.S. Department of the Interior in competition with private energy providers (who vociferously opposed the project for this reason). Hydroelectric power plants had been constructed by private interests here and there across the United States since the turn of the 20th Century (the small hydroelectric powerplant at Roosevelt Dam, an early Reclamation project, was never operated by Reclamation). No hydroelectric generation scheme on the scale of what was proposed at Hoover Dam had ever been undertaken by Reclamation.

The Hoover power plant was split into two separate power houses on the Arizona and Nevada banks of the Colorado River, just downstream of the dam. Many critics felt there would be undue transmission loss caused by turbulence and spray of bypassed flows (those flows which are not run through the power houses), so the design was altered to move the valve houses, outlet works, and spillway aprons downstream of the power houses (Wilbur and Mead, 1933).

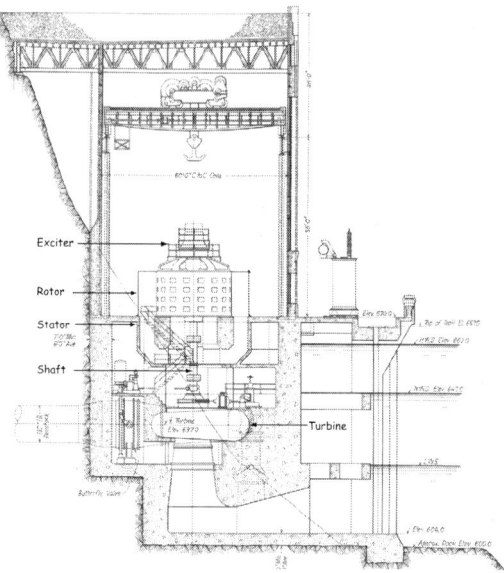

FIG. 1. Elevation views of the Francis turbines originally used at Hoover Dam, with a shaft height of 70 ft, fed by an 11 ft inside diameter penstock (modified from Wilbur and Mead, 1933).

Each power house is 650 feet long with 10 acres of floor space to accommodate eight Francis turbines and two smaller Pelton water wheel turbines, used for internal power production. Reclamation purchased turbines from every manufacturer in the United States as a purposeful economic incentive during the Great Depression. The generating units were comprised of an exciter, rotor, stator, and shaft, as shown in Figure 1.

The exciter is itself a small generator that makes electricity, which is sent to the rotor, charging it with a magnetic field. The rotor is a series of electromagnets, also called poles. The rotor is connected to the shaft, so that the rotor rotates when the shaft rotates. The stator is a coil of copper wire. It is stationary. The shaft connects the exciter and the rotor to the turbine. Water strikes the turbine causing it to spin.

The government installed all of the generators at Hoover Dam, after Six Companies "delivered" the completed powerhouses in November 1935 (Figure 2). The first generator to go into operation was N-2, shown in Figure 3. It began operating on October 26, 1936. The second generator to go into operation was N-4 on November

14, 1936, and Generator N-1 began producing electricity on December 28, 1936. Generators N-3 and A-8 begin operating on March 22 and August 16, 1937.

FIG. 2. The last major structures to be constructed at the dam were the two power houses, shown here, looking upstream on June 24, 1935, as the powerhouse neared completion. Note the depth of the forebay (Bechtel Collection, USBR-LCR).

By late summer 1938 storage in Lake Mead reached 24 million acre-feet, and the new reservoir stretched 110 miles upstream, into the lower Grand Canyon. Generators N-5 and N-6 came online on June 26th and August 31st, respectively. By 1939 storage in Lake Mead reached 25 million acre-feet, or more than 8 trillion gallons. Generators A-7 and A-6 began operations June 19th and September 12th, respectively. With an installed capacity of 704,800 kilowatts (kW), Hoover Powerplant was the largest hydroelectric facility in the world, a distinction held until surpassed by the Grand Coulee Dam Powerplant in 1949.

In 1940 Hoover Dam produced 3 billion kilowatt-hours of electricity. On October 9, 1941, generator A-1 was placed into operation, bringing to ten the number of units in service. Generator A-2 began operations in July 1942, followed by unit A-5 in January 1943, and unit N-7 in November 1944. In October 1946, a ceremony was held at the dam commemorating ten years of commercial power production. In 1952, units A-3, A-4, and A-9 were placed into service, and during 1952 and 1953, a record of 6,400,000 kilowatt-hours (kW-hr) was generated. The final generating unit was placed into service in 1961, when unit N-8 was placed on-line, bringing the capacity of the powerplant to 1,334,800 kW.

FIG. 3. Gantry crane in the Nevada Powerhouse lowering a 500-ton rotor of Generator N-2, one of the General Electric 82,500 kilovolt-ampere generating turbines, and the first to brought online at Hoover Dam, on October 26, 1936 (USBR).

In the early 1980s, Reclamation began updating the power units at Hoover Dam, and by 1990, ten of the 82,500-kW units had been upgraded to 130,000-kW, and two to 127,000-kW. The remaining 82,500-kW units have been upgraded to 130,000-kW with the 40,000-kW unit upgraded to 61,500-kW and the 50,000-kW unit to 68,500-kW. The upgrade was completed between 1986 and 1993. Today there are thirteen 130,000 kilowatt, two 127,000 kilowatt, one 61,500 kilowatt, and one 68,500 kilowatt generators. All machines are operated at 60 cycles. There are also two 2,400 kilowatt station-service units driven by Pelton water wheels. These provide electrical energy for the powerhouse and dam.

The average annual net generation for Hoover Powerplant for operating years 1947 through 1994 was about 4 billion kilowatt-hours. The maximum annual net generation at Hoover Powerplant was 10,348,020,500 kW-hr in 1984, while the minimum annual net generation since 1940 was 2,648,224,700 kW-hr in 1956.

In the original power contracts negotiated during the dam's construction in 1931-35, the principal consumers negotiating long-term contracts for Hoover Dam's hydroelectric power were: Arizona - 18.9527 %; Nevada - 23.3706 %; Metropolitan Water District of Southern California - 28.5393 %; City of Los Angeles 15.4229 %; and Southern California Edison Co., 5.5377%.

Repayment of Project Cost

The cost of construction completed and in service by 1937 was repaid by the sale of electricity on May 31, 1987. All other costs will be repaid within 50 years of the date of installation or as established by Congress. Repayment of the $25 million construction cost allocated to flood control was subsequently deferred beyond 1987, and further action will be subject to Congressional direction. Arizona and Nevada each receive $300,000 annually, paid from revenues, and $500,000 annually is set aside from revenues for further irrigation and power development of the Colorado River Basin.

Spillway Cavitation in 1941 and 1983

One of the unprecedented challenges at Hoover Dam was the sheer scale of the spillway tunnels. Reclamation recognized that the greatest potential for problems existed at the transition elbows, shown in Figure 4. At these transitions the diversion tunnel was plugged with concrete and a curved transition to the inclined section dropping 500 vertical feet from the side channel spillway troughs had to be formed with smooth, high-strength concrete. When the lining of these transitions was completed in March 1934 it was discovered that there was a 3-inch misalignment working the spillway's inclined section from top-and-bottom simultaneously (McClellan, 1950). No one was overly concerned and the slight imperfection was gleaned over with an application of neat grout.

Water impoundment behind Hoover Dam began on February 1, 1935, four months before the last concrete was placed on the dam. The dam was dedicated on September 30th and Reclamation Commissioner Elwood Mead died a few months later, on January 28, 1936. The new reservoir was christened Lake Mead in his honor shortly thereafter, in April 1936. The reservoir continued to fill and by early June 1941 had reached an elevation of 1205 feet, flush with the crest of the side channel spillways.

FIG. 4. Elbow transition between the inclined spillway feeder and diversion tunnel No. 3, just after tunnel was plugged with concrete to create a smooth hydraulic transition. Velocities reached 175 fps in this elbow, hastening cavitation (USBR).

FIG. 5. Water from Lake Mead discharging over the four drum gates of the Nevada spillway during spillway tests in August and October 1941 (USBR).

The drum gates were raised and the reservoir continued to fill, reaching 1220.45 feet by July 30, 1941, within 11.55 feet of the dam crest. On August 6th the drum gates were gradually lowered and several months of spillways testing ensued, which continued through early October (Figure 5). Relatively modest flows, never exceeding 13,000 cfs, were passed through both spillways for four months (Keener, 1943). Even with the modest flows, velocities at the elbows reached 175 fps and cavitation damage ensued on both spillways. The cavitation was most severe on the Arizona spillway elbow, where a hole 112 ft long, 35 ft wide, and up to 36 ft deep was eroded into the high strength reinforced concrete on the Arizona spillway (Figure 6).

Reclamation chose the new Prepakt method to repair the spillway damage during the winter of 1941-42 (Keener, 1943; Chadwick, 1947). Despite this disappointing performance, Reclamation remained unconvinced that it was ascribable to anything but "roughness" and "irregularities" in the concrete lining. Their repair directed the contactor to "remove surface irregularities" (McClellan, 1950). These repairs were not completed until 1943 because of wartime shortages of reinforcing steel.

FIG. 6. Cavitation damage observed in December 1941 in the elbow of the Arizona spillway after modest discharges passed through the spillway between August and October 1941. These were a great surprise to the Bureau of Reclamation (from McClellan, 1950).

In the spring of 1945 Reclamation let a contract to correct problems with the outboard spillway-diversion tunnel on the Nevada side and make improvements to the river channel (Arlt, 1954). These problems had been revealed during the four months of spillage in the fall of 1941, when the outflow swept considerable debris downstream, clogging the outlet area, downstream of the inboard outlet tunnel. The river channel was also deepened immediately downstream of the dam to increase the net energy head on the turbines, by lowering the tailwater. This work was deferred until 1945 because of wartime materials priorities (Figure 7).

That same year (1945) Reclamation began investigating the possibility of installing aeration devices in the spillway tunnels, but concluded that the air introduced into the water dispersed too rapidly to prevent cavitation along the tunnel invert (Frizell and Mefford, 1991). For the next half century Reclamation held a minority view amongst dam engineers that cavitation was ascribable to imperfections in the smoothness of concrete linings, not simply by velocities exceeding 100 to 114 ft/sec, which was advocated by a number of people, but not widely until many years later (Vennard, 1947; Russell and Sheehan, 1974; Cooke, 1979; Chanson, 1989; and Kenn, 1992).

FIG. 7. Crews from Guy F. Atkinson Construction Co. extending the Nevada spillway tunnel outlet on April 26, 1945, as part of a $2 million improvement project that was delayed due to wartime steel shortages. The work shown here improved the hydraulic transition between the outboard Nevada tunnel with the river channel, to correct damage that occurred during the 1941 spillway tests (USBR).

The roughness assumption was finally proven false to everyone's satisfaction during spillage at Glen Canyon and Hoover Dams in the summer of 1983 (Figure 8). That year the Upper Colorado River Basin experienced an abnormally late mountain snow accumulation, followed by an accelerated snowpack ablation, which triggered early seasonal flooding. Many Reclamation reservoirs found themselves without sufficient flood storage to handle the unanticipated inflows (Vandivere and Vorster, 1984).

The tunnel spillways at Glen Canyon and Hoover Dams were run through most of that summer, with Hoover spilling between July 2nd and September 6, 1983. The left spillway at Glen Canyon Dam passed up to 32,000 cfs, but neither of those at Hoover Dam ever exceeded 14,000 cfs (about 4.5 ft above the raised drum gates). Despite these relatively low flows, both dams experienced the same style of cavitation damage that had previously afflicted Hoover Dam in 1941.

Reclamation had been studying aeration ducts for the previous 15 years because of concerns about cavitation damage at Yellowtail Dam in 1967. After the 1983 floods Reclamation undertook a comprehensive program to retrofit their high dams to alleviate cavitation, using aeration slots. These slots were added to the Hoover Dam spillways in 1985-86 (Pugh and Rhone, 1988).

FIG. 8. In July and August 1983 the Nevada and Arizona side channel spillways discharged excess storage for the second time in the dam's history (USBR-LCR).

Problems with a Leaky Grout Curtain and Uplift

During the exploration of Black Canyon prior to 1931, 22 exploratory borings were advanced in the Colorado River channel beneath the proposed dam, along four lines across the channel. The primary focus of the exploratory borings was to ascertain the depth and character of the channel fill and the profile of the bedrock. One deep boring was drilled to a depth of 545 feet below low water level, to ascertain whether the andesite breccias continued to great depth. Assessments of the dam abutments were limited to shallow six-inch cores, which were extracted for unconfined compression tests. Reclamation also excavated exploratory adits into both abutments, to explore foundation conditions.

The principal rocks identified during these surveys were latite flow breccias, dam breccia, and basalt dikes which perturbed both abutments. The latite breccia was characterized by locally intense fractures, especially along faults and shear zones. These inclined faults also crisscrossed one another (Ransome, 1923; 1931).

Some hot springs were also noted at the upstream base of the right abutment, at river level (McKay, 1981). Crude percolation tests were employed, using a gravity-feed reservoir that fed to drill holes through sealed pipes. These could not replicate pore pressures induced by 500 to 800 feet of head. The geology of the dam base was mapped after excavation of the channel gravels. Ransome (1931) noted dozens of faults and adjacent zones of intense shearing.

Original Program of Foundation Grouting

A specially designed joint was employed between the dam concrete and the rock abutments. These joints were not grouted prior to the reservoir filling because the designers believed the dam would deflect downstream under full reservoir load. A conventional grout curtain was installed beneath the dam's upstream axis. This included a single line of holes 100 to 125 feet deep, about 14 to 21% of the dam height (Figure 9).

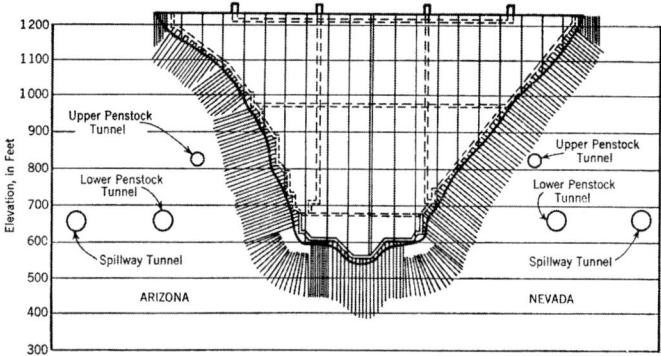

FIG. 9. Elevation view showing the design depths of the original grouting program, performed in 1932-33. These holes extended between 14 to 21% of the dam height (from Simonds, 1953).

These depths were based on a survey of existing dams with grout curtains that Reclamation made prior to construction (Simonds, 1953). Foundation grouting was carried out during construction in 1932-33 along a single line of grout holes. This included grouting of some of the principal faults on both abutments (USBR, 1950). On the Nevada abutment, between elevations 840 and 940 (Figure 10-left), several grout holes penetrated two minor faults, and four holes had to be abandoned, because of excessive grout take and leaks (Simonds, 1953).

When the reservoir reached 1100 ft elevation in 1937, the faults daylighted in the right abutment, and water began entering the fault zone. At this time the abutment drains in the Nevada side began discharging cool water. Warm water from the natural hot springs was collected along the right abutment drainage gallery near elevation 555, emanating from several "shattered zones" (Figure 10-left). It turned out that original grouting of this area was ineffective due to premature set of the cement grout, caused by the elevated water/ground temperatures.

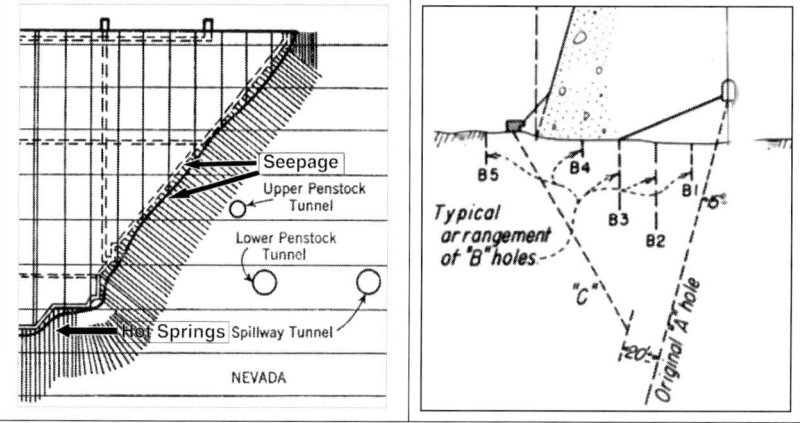

FIG. 10. Left diagram shows the original grout holes on the Nevada abutment, the zone of seepage between elevations 840 and 940 ft, and the hot springs near elevation 555 (annotated from Simonds, 1953). Right figure shows the network of grout holes drilled beneath the dam's upstream heel (McClellan, 1950).

A profile view of the original grout curtain is shown in Figure 10-right. The layout included four rows of shallow B-holes, drilled 30 to 50 ft deep and spaced 20 ft apart. These were considered to be "dental work." The C-holes were drilled on an incline from outside the upstream heel of the dam on 10-foot spacings, to a maximum depth of 100 feet. The C-holes were grouted with pressures of up to 900 psi prior to drilling of the A-holes, which were inclined upstream, from the lower drainage gallery (Figure 10-right). The A-holes forming the curtain were 150 feet deep, on five-foot centers, inclined 15 degrees upstream. 191 A-holes were drilled on the Nevada side, but 33 ended up being abandoned, due to loss of circulation. 202 A-holes were advanced on the Arizona side. Of these, 21 of the holes had to be abandoned.

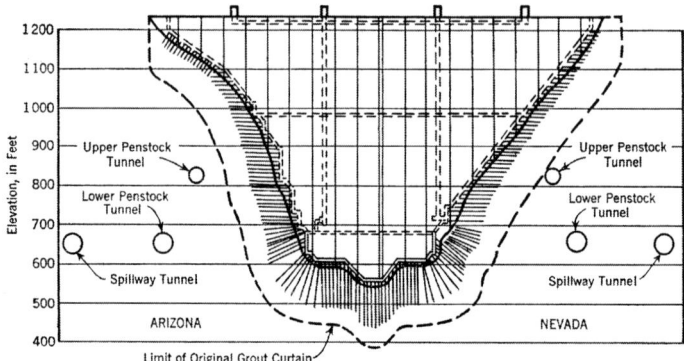

FIG. 11. Elevation view of the uplift relief (drainage) curtain holes, advanced 100 ft from the lower foundation galleries. Dashed line shows limit of the original grout curtain, upstream of the relief wells (Simonds, 1953).

A line of vertical drain holes 100 feet deep was drilled just downstream of the grout curtain. Figure 11 presents the profile of dam centerline showing the pattern of original uplift relief (drainage) holes, which extended a maximum of 100 feet. Note ratio of dam height to depth of the uplift relief holes.

Uplift Problems

By the second year of operation (June 1937), abnormally high uplift pressures began developing beneath the right center of the dam (Figure 12). The inflowing abutment seepage began overwhelming the lower galleries, pouring out of the canyon wall above the Nevada Powerhouse. In addition to these unforeseen levels of seepage, alkaline water seeping into the lower penstock header tunnel began accelerating corrosion of the steel penstock. Hot alkaline water also began seeping through the concrete liner of the inboard 56-foot-diameter Nevada diversion tunnel and spilling onto the 30-foot-diameter steel penstock feeder, causing accelerated corrosion of the steel penstock. These seepage problems were mitigated by additional grouting around the 56-foot-diameter diversion tunnel.

A number of through-going faults were also exposed in the Nevada spillway excavations. The Nevada spillway shaft experienced significant seepage after the reservoir filled, along brecciated zones adjacent to these faults. An extensive program of post-construction grouting was carried out during the 1940s to extend a grout curtain beneath the Nevada spillway and intake towers. This program succeeded in mitigating the seepage problems that arose in 1937 (Simonds, 1953). Excessive seepage also manifested itself along two fault strands through the right abutment when the reservoir reached elevation 1100 feet, 132 feet below crest (Figure 13).

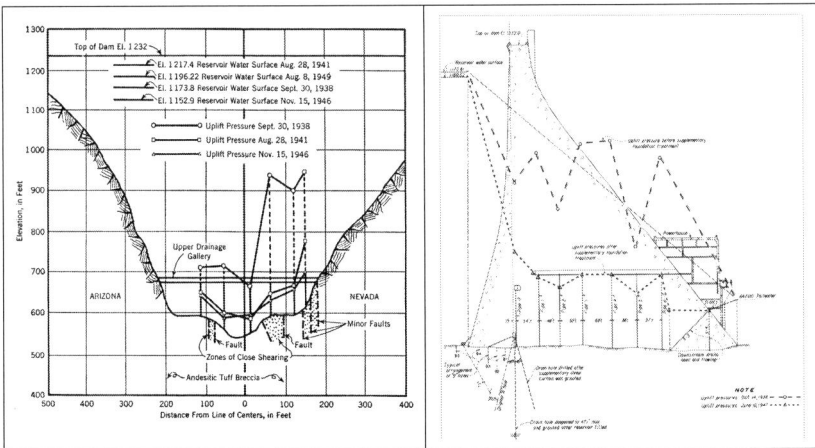

FIG. 12. Left - Uplift pressure gradients along centerline of upper drainage gallery. Note increased pressures on Nevada side, above the fault zones (Simonds, 1953). Right – Profile image of uplift pressures measured before and after the post-construction grouting, between 1939 and 1947 (McClellan, 1950).

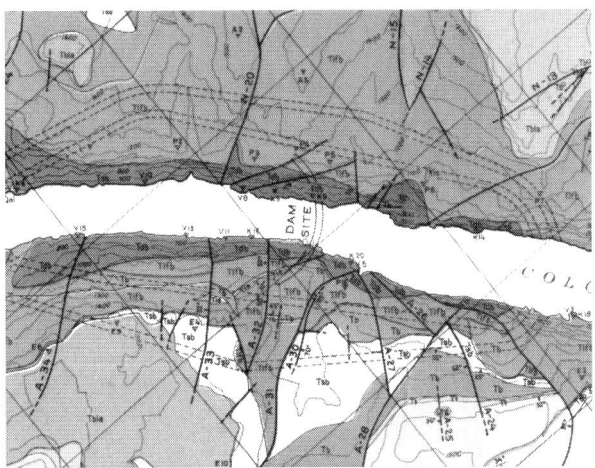

FIG. 13. Portion of the geologic map of the dam site prepared by Ransome in 1931 and reprinted in USBR (1950). The two diagonal faults cutting the dam's right (upper side in this view) abutment (on the Nevada side) preferentially directed seepage through brecciated zones when the reservoir rose to within 130 ft of the dam crest.

Post-construction Foundation Evaluations

The reservoir uplift reached its maximum levels in September 1938. At that juncture the decision was made to drill a series of BX size cores in the foundation beneath the dam. The drilling revealed that the grout curtain was much too shallow on the faulted abutments, because six zones of intensely sheared rock were feeding water into the foundation and a series of crisscrossing manganese gouge seams were perching the underseepage, causing abnormally high pore pressures to develop (Figure 14).

The dam's grout curtain was extended deepened extensively between 1938 and 1947 (Figure 15). The grout holes were extended to depths of 300 feet beneath the dam's foundation, then pumped under pressure of full reservoir head. These were drilled from the dam's system of internal inspection galleries, shown in Figure 16. A schematic view of the deepened grout curtain, as it appeared in 1947 is shown in Figure 17.

During the 12-year supplemental drilling program, 410,000 linear feet of grout and drainage holes were drilled, and 422,000 cubic feet of grout were injected under pressure. This remedial program cost an additional $3.86 million (Simonds, 1953). Uplift pressures were significantly reduced, as shown on the right side of Figure 12.

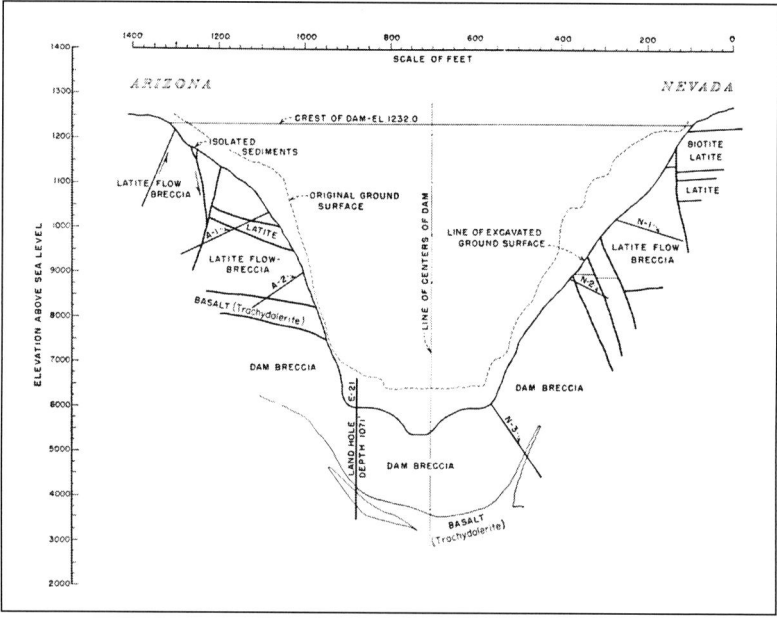

FIG. 14. System of block faults identified during construction. Note absence of data beneath the dam (USBR, 1950).

FIG. 15. Cramped working spaces typified the 9- year program of extending the grout curtain, between 1938 and 1947 (USBR).

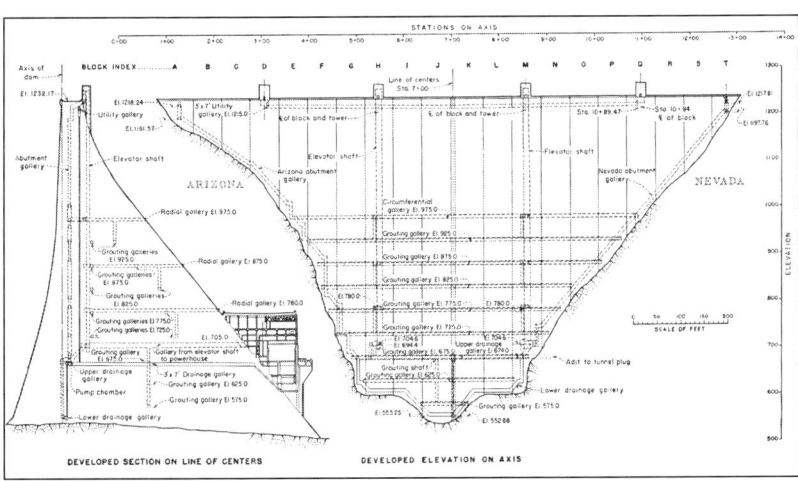

FIG. 16. Distribution of internal galleries in Hoover Dam. Note lower drainage gallery (arrow), from which the new, deeper grout curtain and drainage holes were drilled (USBR).

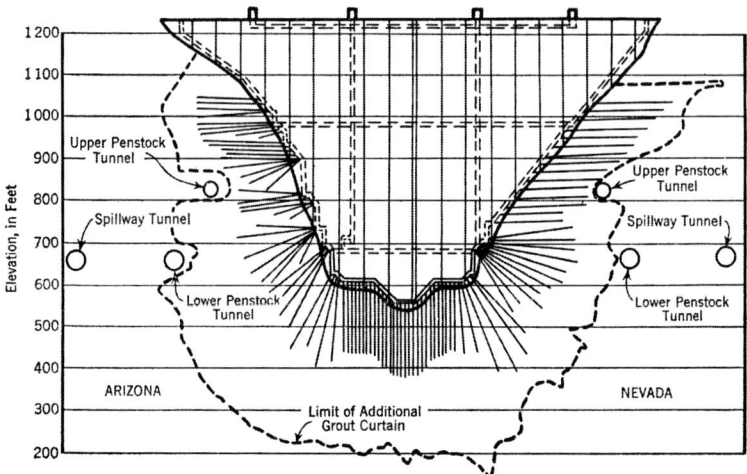

FIG. 17. Profile of dam centerline showing deepened grout curtain, extended between 1938 and 1947 (from Simonds, 1953).

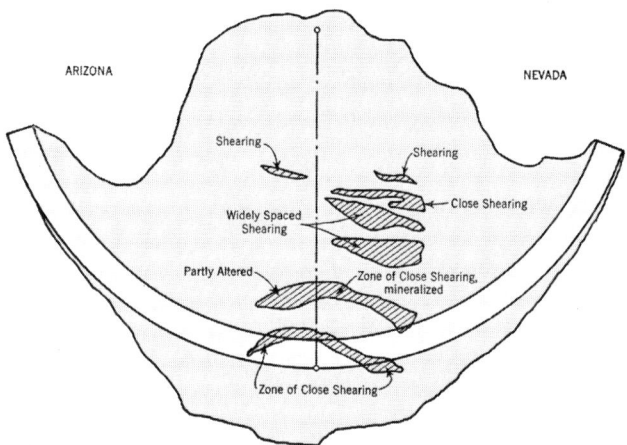

FIG. 18. Manganese-rich gouge zones discovered in the dam foundation, along faults and shear zones (modified from Simonds, 1953).

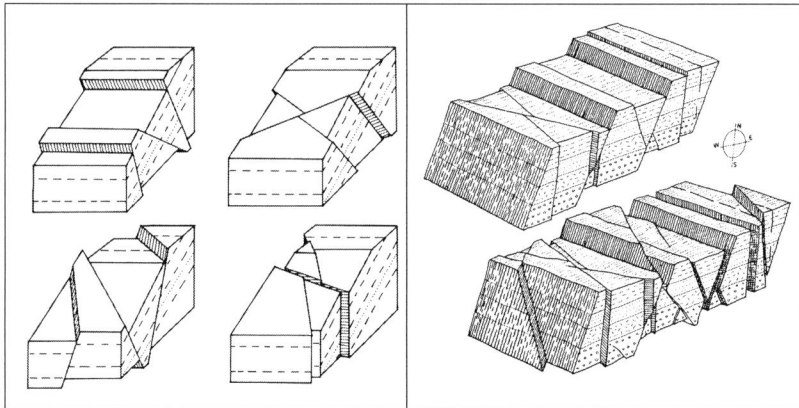

FIG. 19. Left diagram - Four basic types of conjugate fault sets exposed at Hoover Dam, relative to tilt of flow foliation. Upper left shows early normal faults; upper right is early strike-slip faults; lower left late normal faults; and lower right is late strike-slip faults. Right diagram - Block diagrams illustrating tectonic evolution of the dam site. Upper right diagram shows the main tilting stage, typified by NE-SW extension; while the lower diagram shows the principal post-tilt stage, typified by WNW-ESE extension (from Angelier et al., 1985).

Reasons for the Grout Curtain Failure

The failure of the Hoover Dam grout curtain was ascribable to some manganese-rich gouge zones deep in the dam foundation, which developed by chemical weathering along faults and shear zones that perturbed the volcanic strata, as shown in Figure 18. The failure to recognize these features demonstrated that mapping surficial geology in detail doesn't help anyone unless that effort is accompanied by a cogent understanding of how the geologic conditions might impact the proposed structure. The Bureau of Reclamation's Board of Consulting Engineers did not include an engineering geologist, so there was no technical oversight of the geological information collected in the field during construction. If the design assumptions about foundation conditions had been shown to be invalid, some sort of action should have been taken.

Black Canyon turned out to be an area that is geologically complex, pervasively sheared by more than 500 mapped faults. Detailed structural geologic assessments were carried out in the vicinity of Hoover Dam and Black Canyon in the early 1980s by Angelier et al. (1985). This work revealed that two stages of extensional tectonism occurred at the dam site, summarized in Figure 19. Extensional tectonism can be expected to perturb a brittle ground mass, creating numerous shears, faults and brecciated zones. Hydrothermal activity can cause mineralization and infilling of voids with secondary products not visible from the ground surface.

There was insufficient exploration and characterization of the foundation materials beneath and adjacent to the dam; especially the faults, shears, and breccia zones. The grouting program was not sufficiently deep or redundant to provide an adequate seepage cutoff. The cost of the supplemental grouting program was $1.84 million, about 2.37% of the cost of the dam (Simonds, 1953).

STRATEGIC IMPORT OF THE DAM

Attempted Sabotage and Security Precautions during World War II

On November 30, 1939, just three months after the Second World War began, but two years before America's entry into the conflict, the U.S. Embassy in Mexico City received a tip of a plot by German agents in their Mexico City embassy to bomb the intake towers at Hoover Dam. The attack was intended to paralyze the American aviation industry, which was showing signs of aiding Great Britain (Pfaff, 2003). Two German agents were already living in Las Vegas, one of whom was an "explosives expert." One of these agents allegedly made a dozen visits to the dam to scope the site out, and a "German national" was observed by a National Park Service ranger taking dozens of detailed photos of Hoover Dam in early October.

The German plan envisioned the pair posing as fishermen and renting boat at Boulder Bay Marina, making their way to the intake towers after dusk and planting the explosives inside the towers (no details were ever revealed about how they would pierce the stainless steel screens on the gates of the towers). A second prong of this attack was to have included setting similar charges off at the Mead Substation near Boulder City, where the two high-voltage transmission lines bifurcate.

The State Department promptly notified the Bureau of Reclamation in Washington, D.C., but asked that they not reveal any details of the plan, lest the news leak out to the public. Reclamation banned all private boats from Black Canyon, and a few days later placed restrictions upon employees and visitors to the dam (Pfaff, 2003). On December 9th Reclamation requested the aid of the FBI to assess the security of the dam against sabotage. A steel mesh net supported by cross-canyon cables was installed 300 ft upstream of the intake towers to prevent boats from passing and floodlights were set up to illuminate the lake upstream of the dam.

In January 1940 the FBI issued a report listing 38 security precautions that could be implemented to increase physical security of the dam, including additional security training for the dam's rangers. 149 men subsequently received this specialized training from the FBI. Rumors of plots to sabotage the dam spread quickly because of all the precautionary measures being taken. In February 1940 information of yet another alleged plot was passed onto Reclamation, this one involving sabotage of electrical power generation stations in southern California. The German agents slated to carry out these attacks were supposedly coming from Havana, Cuba.

In June 1940 the Department of the Interior asked the War Department to furnish armed guards to protect and patrol the potentially vital features of Hoover Dam and power plant, as well as the switchyard just outside at Boulder City. The War Department denied the request, but Nevada Senator Patrick McCarran introduced legislation to establish an Army post at Boulder City to protect vital federal property. In July 1940 a Reclamation warehouse at Parker Dam was burned down in a suspicious fire.

In response to these political pressures, in December 1940 the Army announced that it would establish a cantonment with 800 soldiers near Boulder City to train military policemen (Moehring, 1986). This new facility was christened 'Camp Siebert' (Figure 123), named after Corps of Engineers General William L. Siebert, who had chaired the Colorado River Board between 1928-32 (it was later renamed Camp Williston when a larger Camp Siebert was established in the late general's home state of Alabama). In July 1941 the Army agreed to allow some of these military police soldiers to patrol the transmission switchyards.

Reclamation continued allowing tours of the dam until Pearl Harbor was attacked on December 7, 1941. That day the dam was closed to the public at 5:30 PM and remained closed during the duration of World War II. The dam's lights were shut off and roadblocks were established on either side of dam along Hwy 93 that evening. From that point onward the Army agreed to provide convoy escort for vehicles crossing over the dam on U.S. Hwy. 93 between Boulder City and Kingman, Arizona (Figure 20), until these soldiers were transferred to overseas assignments.

FIG. 20. Camp Siebert was constructed on the southern side of Boulder City in the late summer of 1941 to train military policemen. The Army agreed to help patrol transmission switching yards and provide convoy security across the dam until the soldiers were deployed. The camp was later renamed Camp Williston and reconfigured to become Boulder City High School after the war (USBR).

FIG. 21. Army escort for convoy of civilian vehicles cued up along U.S. Highway 93, waiting to cross Hoover Dam during the Second World War (USBR).

The dam did not reopen to the public until August 1945. On December 8th Reclamation formally requested Army assistance in protecting Hoover, Grand Coulee, and Parker Dams, which produced half of all the electrical power used in Pacific Coast wartime production plants (Parker also provided pumping power for the Colorado River Aqueduct, vital to southern California). Security precautions then shifted to defense against aerial attack, and a no-fly zone was established around the dam, although there was no means to enforce the restriction (General George Patton flew over the dam at a height of just 500 ft on July 6th, 1942, while he was commanding desert maneuvers a few hundred miles to the south).

During the summer of 1942 Reclamation and the Army began exploring various means of protecting Hoover, Parker, Grand Coulee, and Shasta Dams from aerial attack, including smoke generators and a number of elaborate camouflage schemes (Shasta Dam was actually under constructed throughout the war's duration, being completed in 1945). At that time Army Engineer battalions included dedicated 'camouflage companies,' trained and equipped solely for that purpose.

In an April 1943 letter to the Chief of Engineers, Colonel W.J. Matteson provided a review of the plan to camouflage Hoover Dam by Colorado artist Allen True, as well as outlining additional or alternative measures that might be employed to protect Hoover, Parker, Grand Coulee, and Shasta Dams (Matteson, 1943). True's scheme was classified 'secret' and involved the erection of a cable-supported dummy dam upstream of the actual dam, as shown in Figure 22. The ruse would have employed a network of highly reflective features attached to welded wire mesh draped over heavy supporting cables, attached to the steep walls of Black Canyon. This scheme was designed to look like concrete and rock. Similar techniques were employed by the

Germans and the British to camouflage high value targets, such as factories, airfields, locks, bridges, highways, etc. (Stanley, 1998).

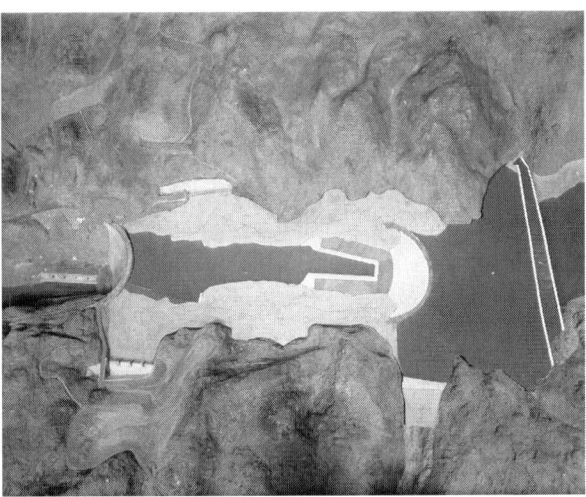

FIG. 22. Painted plaster model of a camouflage scheme believed to have been designed by Allen True, the Colorado artist who had previously designed the western and Native American themes depicted in Hoover Dam's terrazzo floors. His design included a 'dummy dam' and false torpedo nets about a quarter mile upstream of the actual dam, both which would be lit at night. This model was informally preserved as a curiosity at the dam for many years thereafter (USBR).

An alternative plan for a three-quarter scale wire frame 'decoy dam' downstream of the actual dam was also conceived, but never implemented. Matteson felt that it would be far less expensive to employ smoke screens to screen the dam and power house, which "would render precision bombing impossible." He noted that the smoke screens could be deployed in just 15 minutes, provided they were set-up and manned. He also exercised considerable doubt about the difficulty of employing similar camouflage schemes at the other three dams because they were not situated in narrow steep canyons like Hoover Dam. Rumors still abound about a World War II-era 'floating decoy dam' with lights having been deployed, but these are not true.

In September 1943 the Army informed Reclamation that they were re-positioning their military police soldiers in preparation for overseas deployment, and that they would no longer be available to help guard the dam anymore or assist in convoying vehicles. Reclamation appealed to help from the FBI, but was rebuffed (Pfaff, 2003). In January 1944 an Army intelligence officer prepared a report which found some of the allegations about lax security to be without merit, but others to ring true. The report concluded that since no security incidents had occurred, it would be difficult to justify additional military resources being detailed to the dam. In 1946 normal public visitation resumed and the security force dwindled to seven watchmen and the 29 rangers that performed a myriad of collateral duties, in addition to security.

World's Largest Magnesium Production Facility

In the late summer of 1940 the German Luftwaffe began hammering the British Isles, initially focusing their efforts on bombing airfields, aircraft factories, and munitions plants. In January 1941 the United States entered into a massive Lend-Lease Agreement with Great Britain. The U.S. agreed to supply the British with massive amounts of magnesium, because of its value to aircraft production. Magnesium's low specific weight (about 2/3 that of aluminum) was most advantageous when blended with aluminum and zinc. Magnesium and phosphorus were also requested by the British for munitions.

The Las Vegas Valley quickly emerged as a likely site to construct the world's largest magnesium processing facility because of the vast magnesium deposits in southern Nevada and the proximity to unlimited quantities of fresh water and electrical power that could be drawn from Lake Mead and Hoover Dam, essential to the two-step electrolytic processes to process magnesium ore. The government moved quickly, selecting a site on the alluvial fan between Las Vegas and Boulder City and constructing the Basic Magnesium Industries (BMI) Plant, beginning in June 1941 (six months before the attack on Pearl Harbor). The government's plans included the initial construction of 1,000 homes for workers and their families, in what was christened the "Basic Townsite" (it was renamed Henderson, Nevada after the war).

The BMI Plant received top government priority for materials and draft deferments for working in critical defense industries. These advantages allowed it to begin operations in November 1941, just five months after the first land was purchased. The plant continued expanding throughout 1942-43, triggering feverish building activity to keep up with the housing demand. In 1942 the Defense Housing Corporation met the accelerated demand by constructing their first 300-unit "Victory Village" complex, which included schools, a recreation center, apartments, and dormitories for single workers (Moehring, 1986). Despite these accommodations, there was a chronic shortage of defense workers, in large measure due to competition from aircraft plants, steel mills, and shipyards in nearby southern California, where the weather was more favorable and the environment more "family-friendly" (no casinos).

BMI met their manpower shortfall by importing African American workers from Mississippi, who were undaunted by southern Nevada's dry heat. This was during the era of segregation, so the Defense Housing Corporation scrambled to build a separate 324-unit Victory Village which they christened "Carver Village" (for George Washington Carver) on the western side of Las Vegas, which was pretty desolate (Moehring, 1986). By mid-1943 this sprawling complex would include ten separate plants employing 5,000 workers (Figure 23). By war's end BMI would employ almost 10,000 workers.

FIG. 23. The sprawling Basic Magnesium Industries Plant and the community of Henderson were built during the war to supply this strategic metal for the aircraft industry in southern California (Nevada Historical Society).

CONCLUSIONS

Hoover Dam was Reclamations first hydroelectric generating project of any consequence, one of the world's largest hydroelectric power plants. Its detractors did not envision a need for so much electricity in the sparsely populated American Southwest. The last generating unit was not brought online until 1961, 26 years after the dam's completion. The most important aspect of Reclamation entering the utility marketplace was the ability of this revenue to repay the cost of increasingly large and ambitious reclamation projects that almost no one envisioned for the American West prior to the construction of Hoover Dam, which was viewed more-or-less as something of a one-time aberration, made possible by the political intrigue of southern California. The Reclamation projects that followed in quick succession, across the American West, bear testimony to an era of massive reclamation and public works construction undertaken by the federal government which proved to be of inestimable value during the Second World War (such as the Basic Magnesium Inc plant) and in the post-war southwestern expansion, which continues to present.

A much-less publicized aspect of Hoover Dam were the operational failings that occurred, from which much valuable engineering information was gleaned, which impacted dam engineering world-wide. The failure of the dam's grout curtain led to excessive uplift pressures beneath the right side of the dam which were of unprecedented scale. These were ascribable to complex faulting and geologic structure, associated with Tertiary age volcanic units located in a tectonically-active portion of the Basin and Range Province. The geologic reconnaissance which presaged the dam's construction was of a fundamental nature, with very little energy expended to drill into the abutments and ascertain the nature of the volcanostratigraphy or block faulting. That the site was complexly faulted there was no question, and it would be difficult to envision the siting of a mass concrete dam at the same site today, given the state-of-the-practice in dam safety considerations (there are 26 faults less than 4 Ma age cutting the dam's foundation). Engineering geology was in its embryonic years when Hoover Dam was designed and constructed, and very little attention was paid to evaluating stratigraphy and structure at any significant depth beneath or adjacent to the dam's foundation. Considerations of block kinematics were altogether nonexistent until the 1970s, after the failure mode triggering the 1959 Malpasset Dam failure became more understood and appreciated (Londe, 1973; 1987).

The cavitation damage experienced in the spillway elbows at Hoover Dam in 1941 and again in 1983 were perhaps Reclamation's greatest failing. In the 1930s there were very few high dams (> 200 ft) with any appreciable record of operation where velocities > 150 ft/sec were reached. At that time many engineers noted that smooth finished concrete tended to exhibit less tendency towards cavitation than rougher surfaces (Vennard, 1947). Reclamation engineers attributed the 1941 cavitation damage at Hoover Dam to surveying imprecision between the diversion tunnels and the inclined shafts coming from the abutment spillways, which caused a misalignment of ½ inch (Kenner, 1943). Their 1942-43 repairs focused exclusively on replacing the damaged elbows with a thoroughly smooth transition, bereft of any "alignment bumps" (Warnock, 1947). This same approach was used on all subsequent Reclamation high dams built in the 1960s, including Glen Canyon and Yellowtail.

During this same interim (1945-83) many dam engineers took exception with Reclamation's views, assuaging that cavitation was most ascribable to implosion of entrained air at velocities between 114 and 150 ft/sec (Raphael, 1977; Cooke, 1979; Vanoni, 1990). This view was being born out in the massive post-war projects being built elsewhere around the globe, which had suffered grievous cavitation damage until air ducts were installed (Russell and Sheehan, 1974). Reclamation studied these largely foreign projects (Many of the consultants on those projects were former Reclamation engineers) and developed contingency plans for retrofitting their highest dams with air ducts, but the funds were not forthcoming (Pugh and Rhone, 1988).

When the 1983 spring flows necessitated modest levels of spillage at Glen Canyon and Hoover Dam, all of the spillway elbows suffered cavitation damage that required emergency repairs. Reclamation's repairs included the insertion of air ducts, bringing Reclamation in line with general practice, world-wide (Frizzell and Medford, 1991; Kenn, 1992).

ACKNOWLEDGMENTS

The author is most indebted to retired Reclamation engineer Richard Wiltshire, who appreciates civil engineering history and organized this symposium commemorating the 75th anniversary of Hoover Dam's completion. The author's interest in Hoover Dam was originally piqued by his initial viewing of the hour-long documentary film profiling the Boulder Canyon Project which he viewed in high school. A year later he visited the dams along the lower Colorado River and took an extended tour of Hoover Dam.

In 2008 the author served as one of the speakers participated in a Conference on The Fate and Future of the Colorado River, sponsored by the Huntington-USC Institute on California & the West and the Water Education Foundation at the Huntington Library. This led to many valuable contacts which led to a myriad of sources for information and photographs relating the Boulder Canyon Project and subsequent water resources development in the Lower Colorado River Basin.

In 2009-10 the author received a Trent Dames Civil Engineering Heritage Fellowship and was a Dibner Research Fellow at the Huntington Library in San Marino, California. This fellowship and the residency it afforded allowed the author to review of numerous serial publications, collections, archives, and ephemera that provide invaluable to understanding the engineering decisions and political pressures influencing those decisions, not just during construction (1931-35) but during the decade preceding and following the dam's completion. Huntington Archivists Dan Lewis and Bill Frank proved to be particularly valuable in ferreting out rare or obscure accounts from the Huntington's civil engineering and scientist manuscript collections, as well as rare map and historic photos.

Between 1976 and 1988 the author conducted interviews with Professor Jerome M. Raphael (1912-1989) in the civil engineering program at U.C. Berkeley and Professors Vito Vanoni (1904-99) and George Housner (1910-2008) at Caltech, regarding the historical aspects of hydraulic uplift that developed beneath Hoover Dam and the various controversies surrounding cavitation, which was the one technical area Recalamtion engineers found themselves most at-odds with consulting engineers throughout the post-war era (after 1945). Other key interviews included J. Barry Cooke (1915-2005) in 1979, 1983, and 1984 regarding the evolution of understanding what causes cavitation on high dams and methods to mitigate it. Cooke had been a vociferous critic of Reclamation's long-held views about surface roughness preventing cavitation.

The author also wishes to thank Shelley Erwin of the Archives of the California Institute of Technology for researching the files of Professor F. Leslie Ransome (1868-1935), who served as the principal geologic consultant on the Boulder Canyon Project to Reclamation from 1922 to 1935, and alumni newsletters and records relating to Dr. Frank A. Nickell, who received all his degrees from Caltech (in civil engineering and geology) who become Reclamation's first engineering geologist in 1931. Special thanks is also due to Allen W. Hatheway, Jeffrey Keaton, Richard Proctor, and William K. Smith of the AEG Foundation, who provided original documents from Frank A. Nickell, former Chief Geologist of the Bureau of Reclamation, who in addition to being the resident geologist, had supervised studies of reservoir triggered seismicity and crustal deflection at Lake Mead from 1935 till 1949.

Sincere appreciation for assistance is also rendered to the following people and their respective organizations: Reclamation GIS specialist Steven Belew; Dr. Brit Allan Storey and Christine Pfaff, historians with the U.S. Bureau of Reclamation at the Denver office, Dianne Powell of the Bureau of Reclamation Library in their Denver office; Reclamation archivists and staff Bonnie Wilson, Andy Pernick, Bill Garrity, and Karen Cowan of Reclamation's Lower Colorado Region (LCR) office in Boulder

City; Boulder City Library Special Collections; Eric Bittner at the National Archives and Records Service Rocky Mountain Region depository at Denver Federal Center; Paul Atwood and Linda Vida at the University of California Water Resources Center Archives in Berkeley; Su Kim Chung of the University of Nevada-Las Vegas Libraries; Lee Brumbaugh of the Nevada Historical Society; civil engineering historian Edward L. Butts; and a special thanks to the author's best friend Steve Tetreault, who accompanied the author of his first tour of the dam in February 1973, who moved to Las Vegas in 1986. Over the last 24 years Mr. Tetreault has provided invaluable support and logistical connections in support of the author's numerous visits to the University of Nevada-Las Vegas, Boulder City, and Hoover Dam.

REFERENCES

Angelier, J., Colletta, B., and Anderson, R.E. (1985). Neogene paleostress changes in the Basin and Range: A case study at Hoover Dam, Nevada-Arizona. *GSA Bulletin*, v. 96:3, p. 347-361.

Arlt, A.W. (1954). Hoover Dam, Boulder Canyon Project, Arizona-California-Neva, in *Dams and Control Works, Third Ed.*, U.S. Bureau of Reclamation, Denver, pp. 33-46.

Chadwick, W.L. (1947). Hydraulic Structure Maintenance Using Pneumatically Placed Mortar. *Journal American Concrete Institute*, Vol. 43:17 (February).

Chanson, H. (1989). Study of air entrainment and aeration devices. *Journal of Hydraulic Research*, Vol. 27, No. 3, pp. 301-320.

Cooke, J. Barry. (1979). Personal communication.

Frizell, K.H., and Mefford, B.W. (1991). Designing Spillways to Prevent Cavitation Damage. *Concrete International*, v.13:5, pp. 58-64.

Keener, K.B. (1943). Erosion causes invert break in Boulder Dam Spillway Tunnel. *Engineering News Record*, Vol. 131 (November 18), p. 762.

Kenn, M.J. (1992). Discussion of Reduction of cavitation on spillways induced by air entrainment, by J.A. Kells and C.D. Smith. *Canadian Journal of Civil Engineering*, vol. 19, p. 924-925.

Londe, P. (1973). Analysis of Stability of Rock Slopes. *Quarterly Journal Engineering Geology*, v. 6: 93-127.

Londe, P. (1987). The Malpasset Dam failure. *Engineering Geology*, v. 24:295-329.

Matteson, W.J. (1943). *Comments on "A Study of Camouflage for Boulder Dam and Power Plant."* From Assistant Executive Officer, The Engineer Board, Fort Belvoir, to Chief of Engineers, U.S. Army, April 6, 1943, National Archives & Records Service, College Park, MD. File 618.33 (Boulder Dam) in Record Group 77 (Records of the Office of the Chief of Engineers.

McClellan, L.N. (1950). *Deterioration of Large Dam Structures*. Unpublished manuscript. U.S. Bureau of Reclamation, Denver, 12 p., 10 pl.

McKay, W. A. (1981). Hydrogeochemical inventory and analysis of thermal springs in the Black Canyon-Hoover Dam area, Nevada and Arizona. *Transactions*, Geothermal Resources Council, v. 5, p. 185-187.

Moehring, E.P. (1986). Las Vegas and the Second World War. Nevada Historical Society *Quarterly*, v. 29 (Spring), pp. 1-14.

Pfaff, C. (2003). Safeguarding Hoover Dam during World War II. *Prologue*, v. 35:2, pp. 10-21.

Pugh, C.A., and Rhone, T.J. (1988). Cavitation in Bureau of Reclamation Spillways. *Proceedings* of the International Symposium on Hydraulics for High Dams, Beijing, 8 p.

Ransome, F.L. (1923). Geology of the Boulder Canyon and Black Canyon Dam Sites and Reservoir Sites on the Colorado River. Unpublished report to U.S. Bureau of Reclamation, Denver, April 1923, 22 p.

Ransome, F.L. (1931). *Report on the Geology of the Hoover Dam Site and Vicinity.* Unpublished report to U.S. Bureau of Reclamation, Denver, Nov. 30, 1931, 71 p., 22 pl.

Raphael, J.M. (1977). Personal communication.

Russell, S.O., and Sheehan, G.J. (1974). Effect of Entrained Air on Cavitation Damage. *Canadian Journal of Civil Engineering*, Vol. 1:1, pp. 97-107.

Simonds, A.W. (1953). Final Foundation Treatment at Hoover Dam. *Transactions* ASCE, v. 118:78-112.

Stanley, R.M. (1998). *To Fool a Glass Eye: Camouflage versus Photoreconnaissance in World War II.* Smithsonian Institution Press, Washington, D.C.

Vandivere, W.B., and Vorster, P. (1984). Hydrology analysis of the Colorado River floods of 1983. *GeoJournal*, Vol. 9:4 (Dec), pp. 343-350.

Vanoni, Vito. (1990). Personal communication.

Vennard, J.K. (1947). Nature of Cavitation, in Cavitation of Hydraulic Structures: A Symposium. *Transactions* ASCE, Vol.112, pp. 2-15.

Wilbur, R.L., and Mead, E. (1933). The Construction of Hoover Dam; Preliminary Investigations, Design, of Dam, and Progress of Construction. U.S. Dept. of Interior, Gov't Printing Office, Wash., D.C.

Hoover Dam: Scientific Studies, Name Controversy, Tourist Attraction, and Contributions to Engineering

J. David Rogers[1], F. ASCE, D. GE, P.E., P.G.

[1]K.F. Hasslemann Chair in Geological Engineering, Missouri University of Science & Technology, Rolla, MO 65409; rogersda@mst.edu

ABSTRACT: Hoover Dam was a monumental accomplishment for its era which set new standards for post-construction performance evaluations. Many landmark studies were undertaken as part of the Boulder Canyon project which shaped the future of dam building. Some of these included: comprehensive surveys of reservoir sedimentation, which continue to the present; the discovery of turbidity currents operating in Lake Mead; the nature of nutrient-rich sediment contained in these density currents; cooperative studies of crustal deflection beneath the weight of Lake Mead; and reservoir-triggered seismicity. These studies were of great import in evaluating the impacts of large dams and reservoirs, world-wide, and have led to a much better understanding of reservoir siltation than previously existed. Hoover Dam was also the first dam to be fitted with strong motion accelerometers and Lake Mead the first reservoir to have an array of seismographs to evaluate the impacts of reservoir triggered seismicity. Along the way, there has been considerable confusion about the name of the dam, which was changed in 1931, 1933, and 1947. Most of the project documents are filed or referred to by the several names employed by Reclamation between 1928-1947. The article concludes with a brief description of the Boulder Canyon Project Reports, which have been translated into many different languages and distributed world-wide. This is followed by a summary of the unprecedented influence Hoover Dam has exerted on dam and reservoir construction, not only in the United States, but also abroad.

MONITORING MILESTONES

First Crustal Deflection Studies

During the design of Hoover Dam, it was recognized that the tremendous weight of the dam and lake, more than 41,000,000,000 tons, might have a localized effect on the Earth's crust. Bell (1942) estimated that the Colorado River deposited 232 million tons of sediment, or 875,000 tons per week, in just six years, between February 1935 and

June 1941. Estimates made prior to construction indicated that there could be as much as three feet of deformation due to the weight, assuming a granitic continental mass lying on basalt, lying upon a somewhat denser crustal layer (Westergaard and Adkins, 1934).

Three series of precise leveling surveys were carried out between 1935 and 1950 to measure the actual movement of the Earth's crust (Figure 1). These measurements were carried out over a triangulation net of 711 miles, running the calculations to one-third higher accuracy than had ever been carried out previously on permanent benchmarks across the United States (Raphael, 1954). The leveling surveys used Cane Springs, north of Moapa, as the reference datum point. The three leveling surveys were carried out in 1935-36 (zero reservoir condition), 1940-41 (reservoir pool at el. 1221.5 ft), and 1949-50 (reservoir pool at el.1174 ft), allowing the crust to adjust to the reservoir load. These revealed up to seven inches of settlement of the Earth's crust in the fifteen years following completion of the dam and 8-1/2 years after the reservoir pool reached its normal operating level (Figure 2).

FIG. 1. Triangulation surveying of established monuments around Lake Mead began in 1935, to ascertain the extent of crustal deformation caused by the weight of the rising reservoir (USBR).

These precise leveling surveys also noted that the rate of sinking of the crust was increasing during the first 15 years, Hoover Dam having dropped about 2 inches in 1940, increasing to 5 inches by 1950. Since the reservoir weight did not increase after 1941, this suggested that plastic flow or creep (continuing strain under sustained loading) was playing some sort of role in the movements. By extrapolating the leveling data Raphael (1954) concluded that the ultimate maximum settlement would reach a value on the order of 10 inches. He also assuaged that the somewhat patchy pattern of settlement suggested some sort of regional warping of the crust, towards the south. This suspicion was confirmed by subsequent work (Smith, et al., 1960; Angelier et al., 1985).

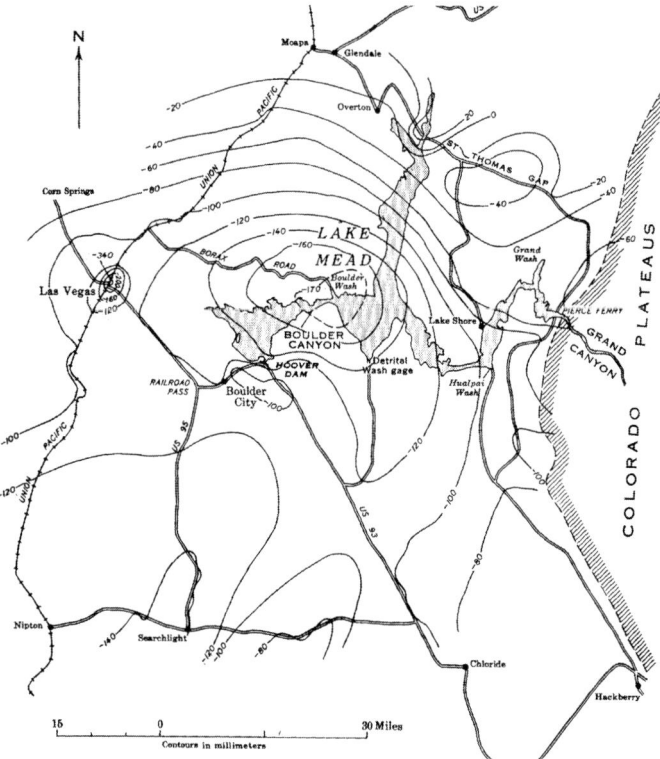

FIG. 2. Ground settlement contours, based on measurements around Lake Mead between 1935 and 1950 (Longwell, 1960).

First Evaluations of Reservoir Triggered Seismicity

A few weeks after Lake Mead reached its peak elevation of 1025 ft in the summer of 1936, a number of earthquakes began occurring, which garnered some publicity. The first seismograph had been installed at Boulder City in 1935, but the paucity of other nearby instruments prevented any meaningful triangulation to determine the precise epicenters and focal depths. The following year earthquakes began occurring in the eastern end of the reservoir, where the greatest volume of sediment was beginning to accumulate. Their interest piqued, seismologists gathered for the annual meeting of the Seismological Society of America in 1938 drafted a resolution asking the Bureau of Reclamation to install a seismic array around Lake Mead to record any possible relationship between reservoir filing and seismic activity. This suggestion met with positive approval of Reclamation's Board of Consulting Engineers, which included

Professor W.F. Durand at Stanford, a colleague of Professor Bailey Willis, one of the originators of the resolution.

In response to this inquiry, Reclamation contracted with the U.S. Coast & Geodetic Survey and the National Park Service to undertake a cooperative seismological investigation of Lake Mead and vicinity. In February 1938 two additional seismograph stations were established at Overton and Pierce Ferry (Figure 3). With the station already established at Boulder City, the seismographs were arranged in a rough equilateral triangle, about 50 miles apart, allowing the first study of reservoir triggered seismicity (Berkey and Nickell, 1939). In addition, that same year three strong motion accelerographs were installed in vicinity of the dam; one on the dam, one in the oil house, and another in the Nevada intake tower (Raphael, 1954).

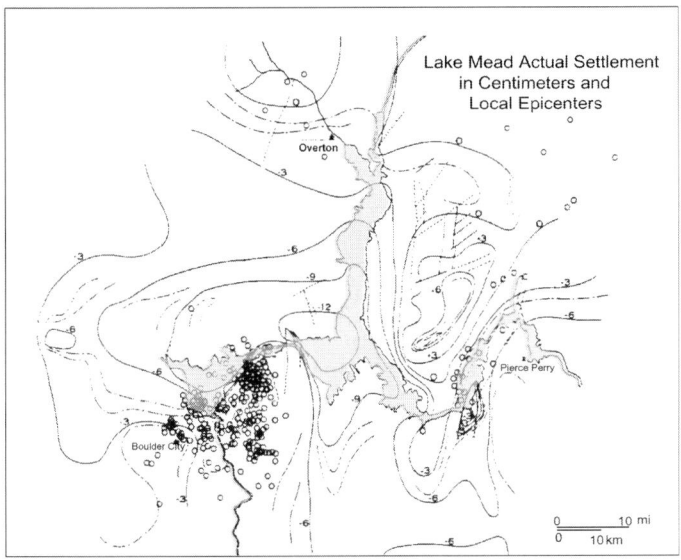

FIG. 3. Crustal settlement in vicinity of Lake Mead, showing the locations of the three seismographs and earthquake epicenters between 1937 and 1947 (modified from Carder and Small, 1948). This was the first documented study of reservoir triggered seismicity.

Recorded earthquakes from these seismographs were triangulated to locate the epicenters of earthquakes. During the first ten years of operation, more than 6,000 minor tremors were recorded in the vicinity of Lake Mead where no tremors had been recorded for the fifteen years prior to construction of the dam (Carder and Small, 1948). Lake Mead reached an average mean lake level of ~1174 ft in 1938 and the seismic activity shafted back to the deepest portion of the lake, closer to the dam. Most of the felt earthquakes varied in magnitude between 3.0 and 5.0. The strongest of these early events was a Magnitude (M) 5.0 quake that occurred on May 4th, 1939, setting off the accelerographs at the dam. This quake was felt as far away as Parker Dam, 200 miles to the south. This earthquake was associated with a swarm of 509 quakes recorded in May 1939, which garnered considerable interest (Berkey, and Nickell, 1939; Carder, 1945).

The vast majority of earthquakes recorded during the first 15 years were clustered within 10 miles of Hoover Dam (Figure 3), in an area of intermediate subsidence (having dropped just 3 inches over 15 years). Scientists sought to unravel any obvious pattern between earthquake activity, the rate of reservoir filing, or the total reservoir load. Longwell (1936) had mapped several potentially active faults through Boulder and Black Canyons, as well as other parts of the reservoir and its margins.

Throughout the 1940s earthquake activity in vicinity of Lake Mead seemed to correlate somewhat with the annual high water stands in 1940-45, then with the low water stands between 1946 and 1952 (Carder and Small, 1948). A sizable quake occurred in 1958 after rapid filling of the reservoir during the previous year, while another Magnitude 4 quake occurred after a rapid filling sequence in 1962-63. This was followed by a series of Magnitude 3.4 to 3.9 quakes when the reservoir dropped again, between 1963 and 1965. No correlation was ever found between the measured subsidence and the pattern of earthquakes.

Since 1965 only four Magnitude 3.7 to 3.9 earthquakes have occurred, despite repeated cycling of the reservoir. Post-1966 records suggest that seismic activity in vicinity of Lake Mead is no greater than that of the surrounding area (Rogers and Lee, 1976). There continues to be abundant microseismic activity ($M < 4$), especially when the reservoir pool cycles more than 35 ft. Subsequent work by a host of scientists working on Basin & Range tectonics has revealed that the Black Hills and Frenchman Mountain faults are seismically active, expressing northwest-directed tectonic extension. A swarm of microearthquakes were recorded along the Black Hill fault in 1972-73. Small magnitude quakes and fresh scarps along the Mead Slope fault several miles east of the dam suggest that it is tectonically active (O'Connell and Ake, 1995).

First Comprehensive Studies of Reservoir Sediment Accumulation

The river's name Rio Colorado is Spanish for "colored" or "colored red," so-named because of its distinctive reddish-brown color during the late summer months, when it is choked with red mud and silt from the highly erodible San Juan River and Little Colorado River watersheds. In December 1928 the Colorado River Board estimated the silt load of the Colorado River was likely between 80,000 and 137,000 acre-feet per year (ENR, 1928). For purposes of planning they recommended that silt accumulation during the 50-year repayment period be estimated to be about 3,000,000 ac-ft. Before infilling of Lake Mead, Chester Longwell detailed studies of the reservoir area (Longwell, 1936). These geological and topographic data provided a basis for evaluation of sedimentation processes after the reservoir filled. At the time the project was approved (December 1928) the maximum reservoir capacity of 30.5 million acre-feet was made before the project was constructed, which was subsequently adjusted to 30.25 million ac-ft (Smith et al., 1960).

A fascinating aspect of Hoover Dam is that the lower half of the dam only retains 1% of the reservoir's original design volume (Figure 4), based on the rating curve. When the dam was designed some allowance was made for sediment accumulation in the deepest portions of the reservoir floor, below the sill elevations of the dam's cylinder gates, at the base of the four intake towers, at elevation 895 ft. When the reservoir began filling on February 1, 1935 the estimated sediment storage below this elevation was 3,223,000 ac-ft (Smith et al., 1960). Accumulation of sediment above this elevation served to reduce the useable storage space of the reservoir, and thereby impacted the regulation of river flow.

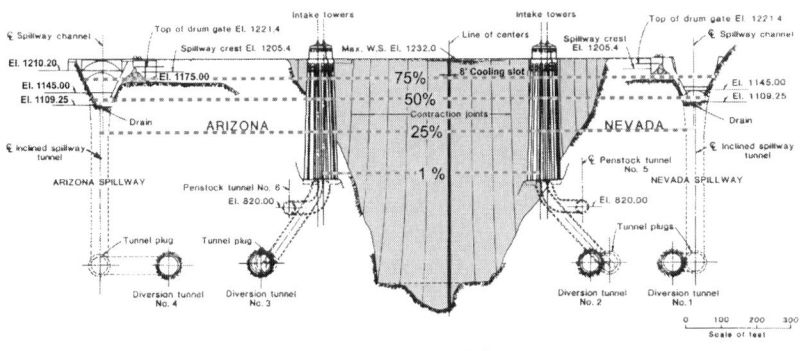

FIG. 4. Upstream elevation view of Hoover Dam, with lines delineating 1%, 25%, 50%, 75%, and 100% levels of the design reservoir storage, based on the original stage curve. The "dead storage" is that zone lying below el. 895, at the base of the intake towers. (USBR)

At the spillway design crest elevation Lake Mead extends 120 miles upstream of the dam, into Lower Grand Canyon, with a maximum surface area of 160,000 acres and 550 miles of contiguous shoreline. The reservoir occupies Boulder, Virgin, Temple, and Gregg Basins, which are separated by narrow canyons, where the Colorado River crosses resistant ridges (Figure 5). These basins were up to 650 ft deep when the dam was constructed and up to five miles wide. The reservoir includes several arms in tributaries that have been inundated, the longest of which is the Overton Arm, along the lower reaches of the Virgin River.

The largest tributaries are the Colorado River, Virgin River, Moapa River, Muddy Creak, and Las Vegas Wash. Other tributary valleys are normally dry and only contribute runoff during brief periods of storm activity (Twichell et al., 2003; Smith, et al., 1960).

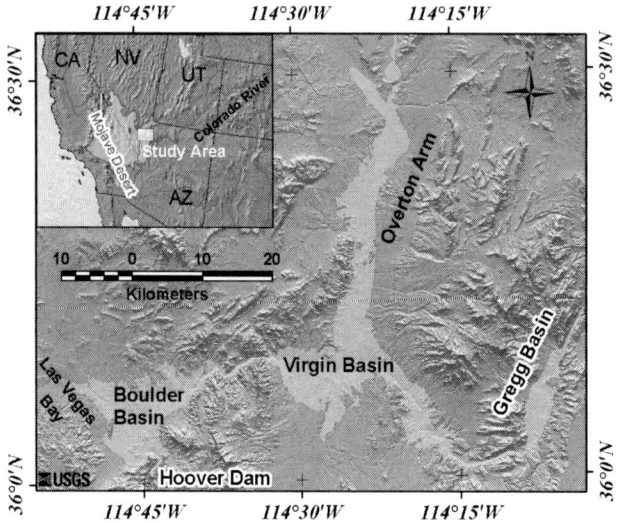

FIG. 5. Lake Mead extends 120 miles upstream of Hoover Dam. It is divisible into five distinct basins, separated by narrow bedrock canyons. The Pierce Basin is upstream, off the right side of the map (from Twichell et al., 2003).

Detailed studies of sediment accumulation were undertaken at Lake Mead as soon as the dam's diversion tunnels were shut down on February 1, 1935 (Figure 6). Despite warnings from several prominent engineers that sediment accumulation could endanger the economic models used to justify megaprojects like Hoover Dam (Stevens, 1946), the storage capacity of Lake Mead was only reduced about 5% through sediment accumulation during the first 14 years of operation (1935-49).

The post-war sediment studies included intensive surveys of the Lower Grand Canyon and Pierce Basin, where sediment accumulation reached a maximum thickness of 270 ft by 1948 (Smith et al., 1960). These studies revealed that the Colorado River delivered a daily average of 400,000 tons per day of sediment into Lake Mead. The post-war studies also revealed that the reservoir actually stored about 12% more water than predicted by reservoir stage plots because reservoir water in "bank storage," that moisture which seeps into pervious beds and banks of the reservoir (Horton, 1933). In addition, the sediment compacts under its own weight, as shown in Figure 7.

By combining these two unforeseen dividends, the total storage capacity in 1948 was actually increased by 13%, to 35 million ac-ft, over that predicted by the 1935 reservoir stage curve (Gould, 1960). But, the lake's sediments are deposited in two principal deltas, one along the Colorado River channel, and a much smaller delta within the inundated Virgin River channel, beneath the Overton Arm. The Virgin

River only supplies about 10% of the accumulated sediment, so the Overton Arm will likely never fill completely with sediment.

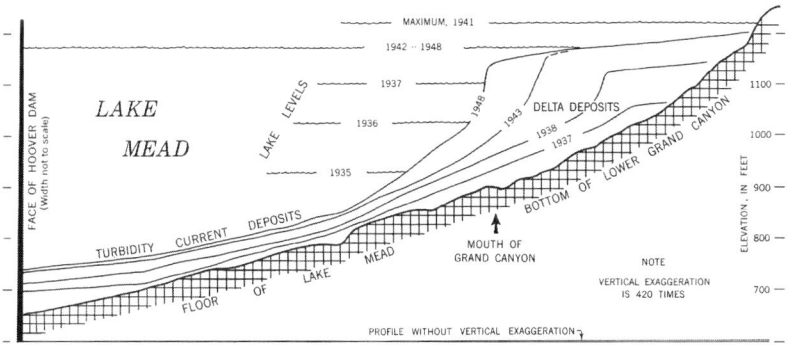

FIG. 6. Vertically exaggerated profile of the Colorado River channel in Lake Mead, showing the advancing sediment delta surveyed between 1935 and 1948. In the first 13 years 105 ft of sediment had already accumulated against the dam (modified from USGS PP 295, 1960).

FIG. 7. Left – Geologist sampling recently deposited silt beds in Pierce Basin of Lake Mead in March 1939 (USBR). Right – Sediment water content versus specific weight for silt and clay, plotted as a function of depth, showing the sediment compaction data for Lake Mead (from USGS-PP295).

The sediment brought down by the Colorado River averaged about 45% sand and 55% silt and clay, with very little bed load. Practically all of the silt and clay was deposited by turbidity flows into the lowest parts of the reservoir (described below). The slope of the advancing delta has consistently been observed to dip sharply for a distance of about 1-1/2 miles, while the slope of the accumulated sediment diminishes with increasing distance downstream, as shown in Figure 8.

The volume of suspended sediment measured at Bright Angel Creek in the Grand Canyon (after the station was established in 1923) compared favorably with that deposited in the reservoir over the first 14 years, being within 2%. This unexpected loss of sediment was a source of considerable consternation and controversy, leading to pioneering studies within the Grand Canyon in mid-1965, soon after the gates at Glen Canyon Dam were closed (Leopold, 1969).

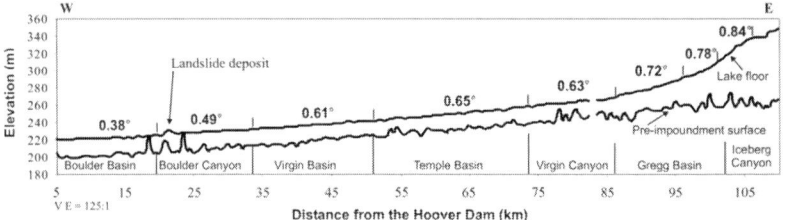

FIG. 8. Sediment accumulation along thalweg of Colorado River through Lake Mead, between 1935 and 2002 (from Twichell et al., 2003). Note diminishing gradient with distance from the river's delta, which by 2000, had advanced to the mouth of Iceberg Canyon, 65 miles upstream of Hoover Dam.

The scientific teams reported that if sediment transport rates continued more or less at the levels recorded in Grand Canyon between 1926 and 1950, it would take slightly more than 400 years for before Lake Mead filled with sediment, noting that the rate of accumulation after 1935 was actually about 20% below the 15-yr average. Construction of sizable upstream reservoirs at Glen Canyon (1963), Flaming Gorge (1964), and Fontenelle (1965) served to intercept much of the sediment along the upper Colorado and Green Rivers, reducing the sediment load impacting Lake Mead by over 90%.

Between 1935 and 1963, an average of 91,000 acre-feet of sediment was deposited in Lake Mead each year (Figure 9). Before construction of the Glen Canyon Dam (370 miles upstream), the Colorado River transported about 500,000 tons of silt and sediment per day through the Grand Canyon into Lake Mead. The peak flow rate of the Colorado before construction of the dam would normally have been around 85,000 cfs for the month of June. By examining river sediments, scientists determined that on a number of occasions over the past 4,000 years, the river reached peak flows of over 250,000 cfs (Swain, 2008). The peak flows routed through the Grand Canyon after construction of Glen Canyon Dam are normally between 12,000 and 30,000 cfs, depending on the severity of the snowpacks in source areas (1983 and 1984 were exceptions, see Vandivere and Vorster, 1984). Since the gates at Glen Canyon Dam were closed in March 1963 the rate of sedimentation in Lake Mead has dropped by about 91.5% (from an average rate of 93,300 ac-ft/yr to just 7,900 ac-ft/yr), as shown in Figure 9.

The sediment studies in the early years within Lake Mead electrified everyone working in sedimentation because so much new information was gleaned from the ongoing measurements (Fry, 1950), which were among the first to utilize Sonar profiling for bathymetry as well as sediment profiling of the accumulated sediments, which revealed sediment "drapes" over the pre-existing topography.

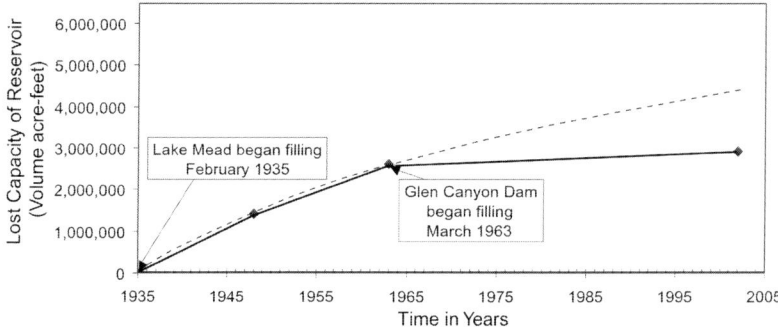

FIG. 9. Loss of storage capacity in Lake Mead due to sediment accumulation between 1935 and 2002, based on bathymetry surveys in 1947-48 and 2002-03. The apparent flattening of the curve between 1935 and 1963 is ascribable to compaction of the silt and clay fraction. Dashed line shows pre-Glen Canyon Dam prediction, accounting for compaction of the accumulated sediment.

Discovery of Turbidity Currents

The bypass gates at Hoover Dam were initially closed to initiate reservoir storage on February 1, 1935, about 18 months ahead of schedule. The last diversion tunnel was closed on May 1, 1936. During that 15-month period there were at least four sequences of turbid underflow being passed through the diversion tunnels, which drew water from the lowest elevations of the rising lake. The first turbid outflows were noted in March and April 1935, when turbidity tests at Willow Beach, 10 miles downstream of the dam verified a sudden shift to more turbid outflow.

Lake Mead crested at elevation 928.5 ft in 1935, about 422 ft above the lowest point of its foundation. In late September a noticeably turbid flow passed the Bright Angel Canyon stream gage in Grand Canyon, 265 miles upstream of Hoover Dam. About 45 hours later this turbid mixture disappeared into the clear waters of Lake Mead over an abrupt line, like that shown in Figure 120-left. The inflow at the time was about 9,360 cfs. Six days later a turbid outflow suddenly issued from the dam, which was passing 9,900 cfs. The increased turbidity and dissolved solids suites were again noted at Willow Beach. Grover and Howard (1938) showed that the percentages of sulphate recorded at Bright Angel Canyon and Willow Beach followed one another with convincing regularity. These 'sediment streams' continued downstream and were similarly noted at Topock. Bell (1942) estimated that the flow moved through Lake Mead with an average flow velocity of about 0.86 ft/sec over a distance of about 87 miles. That subsurface flows would be carried all of the way to the dam in such a short span of time, so soon after the lake partially filled, was a startling development.

A similar pattern of turbid sediment flow through Lake Mead was noted during 1936, when the lake rose to el. 1026 ft. Up through May 1, 1936 there were 46 days of turbid discharge being passed by the dam, carrying approximately 8.4 million tons of sediment, the great majority of which was fine silt and clay (90% finer than 20 microns). During this time the reservoir was between 70 and 90 miles long and contained 4 to 5 million ac-ft of water. The four density current events recognized in 1935-36 passed about 6 million tons of silt and clay through the dam (Howard, 1960). After May 1st, 1936 the lowest elevation from which water could be released from Lake Mead was the base of the intake towers at el. 895 ft, about 270 ft above the original channel.

FIG. 10. Left - Turbid flow of the Colorado River being subducted as underflow at the precipice of the delta being formed in Pierce Basin in 1948, 77 miles upstream of Hoover Dam. Enormous "islands" of driftwood accumulated at the demarcation between the fluids of contrasting density. Right – Divers engaged in sampling sediments in the floor of Lake Mead in the 1940s (both USBR).

Around this same time the National Bureau of Standards had been making preliminary studies of density currents, comparing field data with physical and analytical models, using data provided by Reclamation and the U.S. Geological Survey (USGS). The Soil Conservation Service made a series of laboratory studies at the California Institute of Technology, which seemed to replicate what was being observed at Lake Mead (Bell, 1942). The National Academy of Sciences decided to convene a special panel of the National Research Council to investigate the density flow phenomenon in the spring of 1937. This committee was comprised of individuals selected from the National Research Council's Division of Geology and Geography and christened the Interdivisional Committee of the National Research Council on Density Currents, headed by Herbert N. Eaton of the National Bureau of Standards, which initially convened in June 1937. Two subcommittees were then appointed to study density current phenomena at Elephant Butte Reservoir and at Lake Mead (Reclamation began monitoring sediment accumulation at Elephant Butte in 1931, 16 years after completion). The Subcommittee on Density Currents in Lake Mead was chaired by Carl P. Vetter, an engineer with the Bureau of Reclamation. The results of these studies were compiled in three volumes containing 900 pages of data, which were released between 1940 and 1949 (National Research Council, 1949).

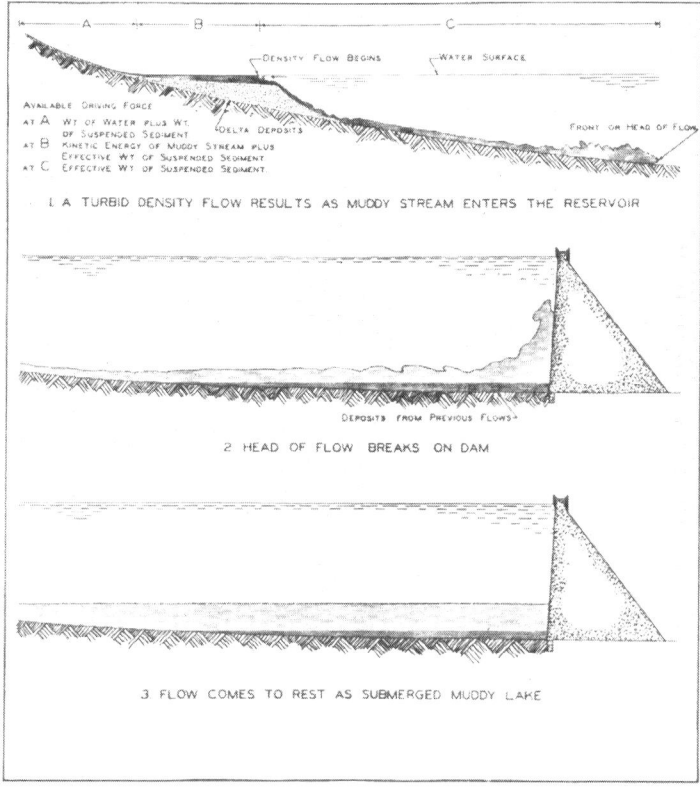

FIG. 11. Diagrammatic representation of a subaqueous density flow in a reservoir, showing the advancing delta in the upper part of the lake and the manner by which the fine grained debris piles up against the dam (from Bell, 1942).

C. S. Howard of the USGS and T.C. Mead of Reclamation began collecting samples and taking measurements of the Lake Mead sediments in May 1937 (Figure 10-right). They initiated a program of taking samples each month at five locations between the Virgin River and Hemenway Wash, as well as from behind Hoover Dam and at Pierce Ferry. Other locations were sampled less frequently (two or three times per year). They soon learned that the density currents only occurred when conditions were favorable to their development at certain times of the year, when flash floods spilled down the major tributaries in the Grand Canyon, increasing sediment concentrations markedly. These usually occur in the late spring and/or late summer, driven by thunder storms. Howard (1960) reported that 21 such events were documented between 1935 and 19 50, and only one of these, in 1941, persisted for more than a week (the 1941 sequence lasted almost five months, while that in 1983 persisted for three months).

The Subcommittee on Lake Mead discovered large accumulations of fine silt and clay were being deposited in the old river channel, at times with densities as low as 1.0008 times that of water. This fine debris was basically moving as a "submerged stream" within the lake, infilling all of the lowest spots in the reservoir, extending all of the way to the dam (Figure 11), where it reached a maximum thickness from piling up against this obstruction (Bell, 1942). The ability of these fine materials to be transported over such a great distance through broad basins and sinuous bedrock narrows on an initial hydraulic gradient of about five ft/mile was of great interest to the engineering and scientific communities, evidenced by the fact that Grover and Howard's 1938 article generated 57 pages of technical discussions from 20 contributors (Grover and Howard, 1938; Vanoni, 1990).

This widespread interest led to the development of the multi-agency team led by Carl Vetter, Chief of the Office of River Control for the Bureau of Reclamation, which made comprehensive evaluations of reservoir sedimentation in Lake Mead in 1947-48, summarized in Smith et al. (1960).

Unexpected Discovery of Warm Sediment

Another intriguing aspect of the density currents was the temperature of the muddy fine-grained sediment, which was noted by Grover and Howard (1938) in the recent sediments deposited in Virgin Canyon, Boulder Canyon, and Black Canyon in 1937. In addition, the temperature sensors embedded in the dam's concrete detected anomalously high temperatures towards the upstream heel of the newly completed dam (Carlson, 1977). The elevated temperature of the oozy low-density sediment was unanticipated, and it quickly drew considerable attention. Page (1938) thought it ascribable to the hot springs along the river channel just upstream of the dam, but this was discounted by Grover and Howard (1938) because all of the fine-grained sediment deposited by density currents was giving off measurable heat (68 to 70° F), while the deep lake water (greater than 100 ft deep) varied from 52 to 55° F.

Sampling soon revealed that the organic silt was hosting more than one million bacteria per gram, comparable to the bacteria count in raw sewage! Near the surface of this mud the bacteria concentration soared to 10 million bacteria per gram, producing methane (Sisler, 1960). Reservoir water just 12 inches above the mud contained only

100 bacteria per gram. These were much higher concentrations of bacteria than had been previously encountered in either marine or lacustrine clays. The Lake Mead samples were also unique insofar that the high bacterial populations were distributed uniformly throughout the deposited layer, which was of very low relative density (Sisler, 1960).

Laboratory tests suggested that the activity of microflora (bacteria and other microorganisms) caused heat to be generated in the nutrient-rich silt and clay. This activity also depletes the hydrogen ion concentration of the entrained water, which shifts the pH from 7.25 to 10, hastening a 22-percent volume reduction, resulting in accelerated compaction of the colloid clay particles (Sisler, 1960).

The heat against the upstream heel of the dam was insufficient to cause uncontrolled tensile cracks because the concrete's curing cycle had been accelerated through the use of cooling pipes and use of low heat cement (Carlson, 1977).

Page (1938) showed remarkable insight in noting that the density currents were predominately fine grained mixtures of colloidal clay, which contained "a high percentage of salts." The role of salts in promoting dispersion of clay was subsequently recognized, four decades later (Sherrard at al., 1976). Most of the red shales exposed in the San Juan, Little Colorado, and Lower Colorado Basins are of dispersive character, which promotes their suspension (marked turbidity), increasing the density of the water-sediment mixture. This increased density serves to increase susceptibility to debris flows and density currents (Rogers, 1985).

Continued Monitoring of Sediment Accumulation

Reclamation has continued monitoring sediment accumulation in Lake Mead, even though the rate of sedimentation has dropped dramatically since the completion of Glen Canyon Dam in 1964 (Figure 9). A comprehensive study of the lake floor was undertaken between 1999 and 2002 by the U.S. Geological Survey, in cooperation with the Lake Mead/Mohave Research Institute, and the University of Nevada-Las Vegas (UNLV) (Twichell, 1999; 2001; 2003). In 1999 the Boulder Basin portion of the lake was surveyed. In 2000 surveys were carried out in the northwestern portion of Las Vegas Bay, and in 2001 the eastern part of the lake bed was mapped. In 2002 UNLV researchers evaluated cores of the lake sediments to ground-truth of the results of the geophysical surveys, conducted from boats. Evaluation of sediment accumulation and the distribution of sediment and any associated pollutants were the principal objectives of this study.

The reservoir floor was remotely mapped using sidescan-sonar and high-resolution seismic-reflection profiling. High resolution seismic reflection profiling is derived from recording impulse signals reflected from the floor and subsurface interfaces. It allows a graphic "picture" of the floor geology along the profile of observation, like that shown in Figure 12. The sidescan-sonar also records the acoustic energy scattered on the lake bottom and the resulting digital images provide considerable detail about the structure of the sediments and the underlying strata.

The most recent sedimentation studies show that sediments are concentrated in the deepest parts of the lake along the pre-impoundment tributary valleys. These sediments have accumulated as a "continuous cover" along the Colorado River trough, from the eastern part of Lake Mead all the way to Hoover Dam. The thickest sediment through Lake Mead has accumulated in the Lower Granite Gorge, where the channel is most narrow.

The sediment also reaches thicknesses of 225+ ft in the delta it has deposited in the Pierce Basin and Grand Bay (Twichell, 2003). The sediment thickness decreases to 50 to 80 ft in the central part of the reservoir, and gradually increases to 100 ft at Hoover Dam. The maximum sediment thickness in Boulder Basin at the west end of the lake is 148 ft. In the Overton Arm, occupying the original Virgin River valley, there is only 3 to 13 ft of sediment because Muddy Creek and the Virgin River have relatively low mean annual discharge (Twichell, 2003).

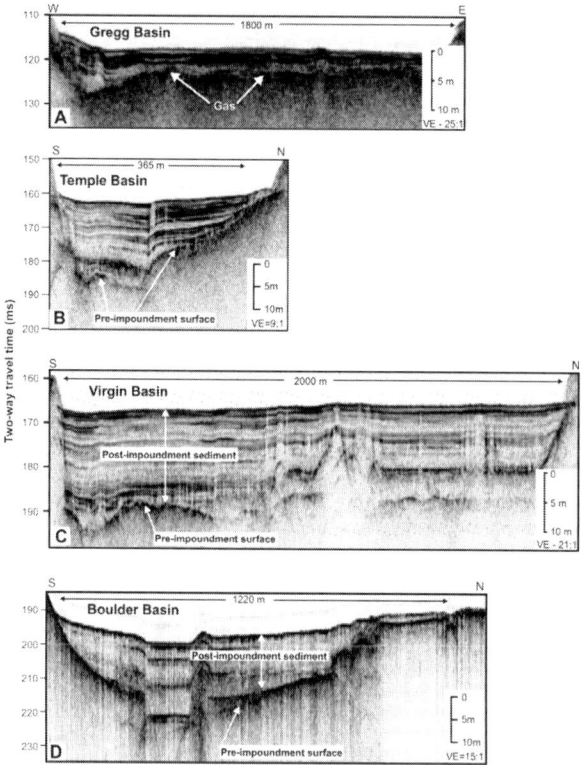

FIG. 12. High-resolution seismic reflection profiling used to measure sediment accumulation across Gregg, Temple, Virgin, and Boulder Basins of Lake Mead between 1999 and 2001. Gas-saturated sediments tend to attenuate the acoustic signal, as seen in profile A. Note high definition of original lake bottom (from Twichell, 2003).

These most recent analyses suggest that the Colorado River contributes about 98% of the sediment in Lake Mead. The geometry of the sediment lenses suggests its accumulation is due to the density flows that run from the mouth of the Colorado River to the Hoover Dam (Twichell, 2003).

The high-resolution seismic surveys also internal layering of the sediment deposited in the western basins, which appears to be sand covered by mud. The bulk of recent deposition in the eastern portion of the reservoir was found to be coarse sand. This shift from sand deposition may be explained by the reduced sediment loads and mollified flows since the construction of Glen Canyon Dam in 1964 (Twichell et al., 2002).

THE NAME CONTROVERSY

Boulder Canyon Dam

In 1921 Congress authorized detailed studies of the Lower Colorado River Basin, with particular emphasis on Boulder and Black Canyons, below Grand Canyon. The Fall-Davis Report released in February 1922 concluded that a great dam could be constructed at either Boulder or Black Canyons. This was the seminal document that led to the Boulder Canyon Project Act, introduced in the first session of the 67th Congress in April 1922.

The other Colorado Basin states reacted warily to California's zealous proposition of an appropriation of unprecedented magnitude that would funnel the great majority of the Colorado River water into southern California, where it might easily be swallowed up in perpetuity under the common law doctrine known as "prior rights."

The Coolidge Administration sought to head-off an impasse between the basin states by convening a commission to draft a Colorado River Compact, which would seek to establish an equitable distribution of water between Wyoming, Colorado, Utah, New Mexico, Arizona, Nevada, and California. The commission was organized formally in January 1922, with an appointed commissioner from each state and a chairman appointed by the President. Calvin Coolidge chose his Secretary of Commerce Herbert Hoover, who had technical training as a mining engineer and had distinguished himself in a myriad of public service roles, most notably, as head of European relief efforts for President Woodrow Wilson after the First World War (25 years later he was appointed to a similar position by President Truman after the Second World War).

Public hearings began in March 1922 and negotiations wound along till November 24th, when the Colorado River Compact was agreed upon at Bishops Lodge in Santa Fe, New Mexico. It was the most notable water agreement in American history up to that time and Hoover gained a considerable degree of notoriety, for his encouragement of equitable development of the nation's natural resources. Many newspapers portrayed him in a favorable light, noting his high level of education and apparent incorruptibility because he was independently wealthy.

Boulder Canyon Project

The name 'Boulder Canyon Project' was chosen by Reclamation engineers because they assumed the great dam would be built near the head of Boulder Canyon, where the granite outcrops create an extremely narrow channel, perfect for a dam. Most dam engineers believed granite to be about as stable of a foundation for a dam as could be found, anywhere. The Boulder Canyon Project Act was introduced twice each year, to the 67th, 68th, 69th, and 70th Congresses.

FIG 13. Witnesses to the signing of the Boulder Canyon Project Act on December 21, 1928, from left: Reclamation Commissioner Elwood Mead, Congressman Phil Swing of Imperial Valley, President Calvin Coolidge, Senator Hiram Johnson of California, Congressman Addison T. Smith (Chair of the House Committee on Irrigation & Reclamation), and W. B. Matthews, counsel to the newly formed Metropolitan Water District, who subsequently purchased long term contracts for 28.5% of all the power generated by Hoover Dam. President-elect Herbert Hoover was not present because he did not endorse the maximum height dam at Black Canyon, believing it to be unnecessary. (USBR)

The Act finally succeeded in gaining approval in late 1928, passing through the Senate on December 14th and the House on December 18th, with approval by President Coolidge on December 21st (Figure 13). By that juncture Herbert Hoover was the President-elect, but he refrained from partaking in any of the publicity attached to the Act's passage because he had stopped short of endorsing a dam of record height, maintaining that major dam of somewhat less proportions would be adequate to the tasks at hand.

Hoover's support for a lower dam did not endear him with the project's boosters in southern California, although he always enjoyed the support of the Los Angeles Times (Hiltzik, 2010). The Bureau of Reclamation saw the mighty dam as an instrument of unspoken manifest destiny, which would allow them to transform the west from barren wasteland into a 'perennial breadbasket' by irrigating the Palo Verde, Yuma, Imperial, and Coachella Valleys, whose crops could supply the nation with green vegetables year round (Lyons, 1947).

Black versus Boulder Canyons

Prior to the Act's passage it was pretty clear that there were many advantages in siting the dam in Black Canyon as opposed to Boulder Canyon. These were summarized by Professor W.F. Durand in a report to Congress titled "Development of the Lower Colorado River" (Emerson et al., 1928, pp. 394-95), dated January 9, 1928. After listing nine advantages posed by the Black Canyon site, Durand summarized the consensus view on why the Boulder Canyon name should be retained:

"In order to avoid confusion with the name it should be noted that the name Boulder Canyon dam and reservoir was first given to a proposed dam to be located in Boulder Canyon and to the reservoir formed thereby. Subsequent investigation gave evidence that a site some 20 miles lower by the river, and located in Black Canyon, might be more advantageous in certain respects. The reservoir, in either case, would flood approximately the same territory.

Actually, therefore, it is one project with a choice of two dam sites and in order to emphasize this viewpoint it has seemed desirable to retain the original name for the general project. When, however, it is desired to distinguish the Black Canyon site and development in a specific manner it may be designated as the Boulder (Black) Canyon project or development."

By the time the fourth version of the Boulder Canyon Act with all the amendments recommended by the Colorado River Board was put forth for congressional approval in late 1928, the decision had long since been made that the dam would be constructed in Black Canyon, not Boulder Canyon. As stated above, no one in Washington, DC seemed to think that it would serve any good purpose to change the bill's title to the 'Black Canyon Project Act,' which had a rather sinister ring to it.

Naming it Hoover Dam (1931)

Republican President-elect Herbert Hoover took the oath of office on March 4, 1929. On May 27th Democratic Congressman Ed Taylor of Colorado introduced a bill to name the proposed dam in Black Canyon after the new president, but this effort failed. Through the following 16 months various legal opinions were issued from the Department of Justice concerning water and power contracts being negotiated by the government, and the dam was referred to as 'Boulder Dam' or the 'Boulder Canyon Dam.' Even the order Interior Secretary Ray Lyman Wilbur issued to Reclamation Commissioner Elwood Mead on July 7, 1930 authorizing construction referred to the 'Boulder Canyon Project' and mentioned 'Boulder Dam' four times.

On September 17, 1930 Ray Lyman Wilbur went to Nevada to commemorate the beginning of the project. In his dedication speech, he announced that the dam would from that point on be officially known as Hoover Dam, in honor of the dominant role

Hoover played in effecting original Colorado River Compact and in encouraging congressional approval of the Boulder Canyon Project.

After Wilbur's announcement at the 'Silver Spike Ceremony' on September 17th all references to the dam's name were changed. The new water and power contracts were changed from 'Boulder' to 'Hoover Dam' as they were amended and published. During Congressional hearings for the dam's initial appropriations on December 12, 1930, Congressman Ed Taylor, by now the ranking Democratic member of the Interior Department Subcommittee on Appropriations, made a florid speech justifying his committee's official naming of the project's kingpin structure as 'Hoover Dam' in the appropriations bill before them, stating: *"There is another feature of this... bill under consideration that I feel ought not to be passed over in silence. I refer to the three words in the second line, "The Hoover Dam."*

This appears to have been the first time that 'Hoover Dam' appeared in any bill or official act of Congress. Taylor justified the committee's decision:

> *"Members of the committee felt this decision was simply following precedents that had previously been applied in the naming of the Roosevelt Dam during Theodore Roosevelt's administration, Wilson Dam during Woodrow Wilson's administration, and the Coolidge Dam during Calvin Coolidge's administration, so President Hoover was justly entitled to the same distinction, so we unanimously and very gladly wrote into this action those words... so that the dam is now officially named by both the Secretary of the Interior and by Congress."*

The appropriation passed on February 14, 1931, and in the next four appropriation acts passed by Congress in 1932-33, the structure was referred to as 'Hoover Dam.' For the remainder of Hoover's administration all official references to the dam, as well as tourist and other promotional material issued during this period called it Hoover Dam.

After his first year in office, Hoover's popularity as president waned as the Great Depression put more and more people out-of-work. Hoover was soundly defeated in his bid for re-election by Franklin Roosevelt on November 8, 1932, just six days before the Colorado River was initially diverted through the project's bypass tunnels. Hoover made his only trip to the dam site by making a slight detour on his way back to Washington, D.C. from California, shortly after losing his re-election bid. He arrived and departed from the dam site in the cloak of darkness on the evening of November 12th-13th. When he visited one of the mess halls in Boulder City workers booed him. He then paid a visit to the dam site to view the massive diversion tunnels (Figure 14), making a brief speech, but never referring to the dam by name:

> *"This is not the first time I have visited the site of this great dam. And it gives me extraordinary pleasure to see the great dream I have long held taking form in stone and cement... This dam is the greatest engineering work of its character ever attempted at the hand of man... The waters of this great river, instead of being wasted in the sea, will now be brought into use by man".*

FIG. 14. The presidential entourage accompanying Herbert Hoover's midnight visit to the dam site on November 12-13, 1932, four days after he was defeated by Franklin Roosevelt. Some of the notables included: Walker R. Young (far left), Reclamation Chief Engineer Raymond F. Walter (third from left), Mrs. Lou Hoover and the President (center), Mrs. Wilbur, Interior Secretary Ray Lyman Wilbur, Reclamation Commissioner Elwood Mead, and Frank Crowe of Six Companies. It would be another seven months until the first concrete was poured for the dam. (USBR)

Re-naming it Boulder Dam (1933)

Franklin Roosevelt took office as president on March 4, 1933, and he named Harold Ickes as his Interior Secretary. Roosevelt had initially offered the Interior Secretary position to California Senator Hiram Johnson, a Republican and co-sponsor of the Boulder Canyon Project Act. Johnson turned him down, but suggested Ickes because he was an ex-Republican. Even as a Republican Ickes had never supported Herbert Hoover, only those candidates who opposed him. Enjoying considerable consensus support, Ickes went onto serve as Interior Secretary for an astonishing 13 years, the longest tenure of any cabinet secretary in American history.

On May 8, 1933 Ickes sent a telegram to Reclamation Commissioner Elwood Mead informing him that the name of the new dam would thereafter be 'Boulder Dam.' He assuaged that this was the "original name," but that was actually "Boulder Canyon Dam" (used in Reclamation designs dating back to 1920). No one doubted that this decision was politically motivated, but nobody in Washington barked too loudly about it because Hoover's approval rating was at an all-time low.

Boulder Dam was dedicated by President Franklin Roosevelt on September 30, 1935. Among the dignitaries in attendance was Elwood Mead, Commissioner of Reclamation; Harold Ickes, Secretary of Interior; John Savage, Raymond Walter, and Walker Young, of the Bureau of Reclamation; and the Governors of California, Utah, Arizona and Wyoming. Representing the Six Companies was Harry Morrison of

Morrison-Knudsen; Steve and Kenneth Bechtel of the Bechtel Corporation; and Frank T. Crowe, Six Companies General Superintendent. President Roosevelt's remarks were carried by radio stations and appeared in numerous newsreels. Speaking to thousands of onlookers, the President referred to the dam as "*an engineering victory of the first order - another great achievement of American resourcefulness, skill and determination.*"

When his turn came to speak at the lectern, Reclamation Commissioner Dr. Elwood Mead referred to the structure as 'Hoover Dam' instead of Boulder Dam, infuriating Interior Secretary Ickes. But, his miscue drew little attention from anyone else. Mead was by that time 77 years old. He had been reclamation commissioner for almost 12 years and his retirement seemed imminent. Mead died a few months later in Washington, D.C., on January 26, 1936, 10 days after his 78th birthday. A grateful nation moved quickly to memorialize his contributions. On February 6th, a motion was made in Congress to name Boulder Reservoir after him, christening it as 'Lake Mead,' an integral part of the new Boulder Dam Recreation Area that was to be administered by the National Park Service, beginning May 1, 1936 (the name was changed to the Lake Mead National Recreation Area in 1964).

FIG. 15. National Park Service brochure covers before (left) and that which appeared for years following (right) the name change in April 1947, when the name of the surrounding area also changed, from Boulder Dam Recreational Area to the Lake Mead Recreation Area (becoming the first national recreation area in 1964). As the memory of the Boulder Canyon Project faded, so did confusion about the name (USBR-LCR files).

Re-naming it Hoover Dam (1947)

In April 1947 the Republican-controlled 80th Congress officially changed the name back to Hoover Dam, creating more confusion, which gradually subsided over the years (Figure 15). In part, this bi-partisan change of heart came about after Herbert Hoover served admirably as President Truman's Chief of the European Relief Commission (the same post he was appointed to following the First World War, by President Wilson). By this time, most people realized that Hoover was not solely responsible for the Great Depression.

ENGINEERING FEAT BECOMES TOURIST ATTRACTION

During Construction (1931-36)

Hoover Dam became a tourist attraction almost from the day construction commenced in the sweltering heat of May 1931, as newsreels in theaters throughout America heralded its construction, which rivaled that of the Panama Canal a generation previous. During the first few years there were insufficient accommodations in Boulder City or Las Vegas to even house project workers, let alone tourists. But, that soon changed.

Throughout 1933 and 1934 numerous hotels sprang up, mostly in Las Vegas, because it lay along the Union Pacific Railroad and U.S. Highway 91, which ran between Los Angeles and Salt Lake City. Boulder City even constructed a hotel for guests in 1933. By 1934 Las Vegas was rapidly emerging as a serious tourist destination, especially during the cooler winter months. It offered a frontier Wild West ambiance with gambling, speak-easys (before Prohibition ended), marriage chapels, even tourist flights over the Grand Canyon. It also sat a few miles from the record-setting dam that was taking shape in nearby Black Canyon (Figures 16 and 17).

FIG. 16. "Gawkers" gather at the temporary overview at the terminus of the government highway about 3/8 mile downstream from the dam's Nevada abutment, as seen in February 1935 (USBR).

FIG. 17. Once the dam was dedicated and opened to the public the National Park Service had to cope with traffic congestion and finding suitable parking areas. This 1938 view shows the first parking lot, just upstream of the voluminous Arizona Spillway intake. Another higher lot, from which this photo was taken, was added a short while later. Visitors walked to the dam and took the public elevator down to the power house (USBR).

FIG. 18. Union Pacific's M10000 Streamliner excursion train stopped at the steel fabrication plant, on March 9, 1934. The streamlined Art Deco style of the high-tech diesel-electric engines seemed futuristic in an age of steam locomotives (USBR).

In 1934 alone a staggering 266,436 visitors secured entry to the federal reservation to view the massive dam under construction, being heralded as one of the 'Seven Man-

Made Wonders of the World' (a title that was officially conferred in 1955). The dam is still listed as one of the 'Seven Man-Made Wonders of the United States' and among the 'Seven Wonders of the Industrialized World' (Cadbury, 2004).

In 1930 the government paid for a branch rail line from a junction on Union Pacific's Los Angeles & Salt Lake rail line, from a point just south of Las Vegas to Boulder Junction, at the Boulder City town site. The first passenger train on this new line arrived on April 25, 1931, just as construction of the dam was starting.

During the dam's last two years of construction the Union Pacific Railroad began running special excursion trains to the dam site weekly throughout the cooler months. On March 9, 1934 they staged a publicity photo op by running their ultra-modern diesel electric streamliner train the M10000 to the site, driving the train through a 30-foot-diameter section of pipe next to Babcock & Wilcox's fabrication plant, memorialized in all the government's subsequent project films (Figures 18 and 19).

In October 20, 1934 ten trainloads of Shriners descended upon the dam site, using Six Companies' lower spur line to convene a candlelight ceremonial service on the upstream cofferdam. Such special privileges were likely secured through Frank Crowe or one of the other Six Companies' officers who were Shriners.

FIG. 19. Another view of the M10000 Streamliner visit in 1934, on the wooden trestle of Six Companies' low-level spur, at the Nevada abutment on the dam's upstream face (USBR).

The Government line and Six Companies' low spur rail line leading to the Colorado River and around Cape Horn were abandoned after the dam's completion and Lake Mead began filling. The last government train took a generator to the dam's Nevada abutment in 1967 and this line was removed a few years later, after being used in some Hollywood films. The Union Pacific continued to actively use its spur line as far as Henderson, and still does today. In 1985, the railroad donated the Henderson-Boulder City segment to the Nevada State Railroad Museum. This segment remains intact, although the grade crossing near Railroad Pass has been paved over. Once restorative work is completed the museum hopes to start operating excursions over the old line.

Post Construction (after 1936)

The first Visitor's Center was constructed by Reclamation in 1936, and a one-story concrete exhibit house was opened to the public in October 1946 (Arlt, 1954). The number of visitors to Hoover Dam and Lake Mead has steadily increased since visitor services began in 1937.

During World War II the Las Vegas Valley and adjoining area underwent rapid growth, driven by the establishment of the Army Air Corps Aerial Gunnery School at Las Vegas Army Airfield (renamed Nellis Air Force Base in 1951) and Indian Springs Airport, which handled 6,000 students at a time for 6-week training courses; the military police school at Camp Siebert in Boulder City; and the construction of the Basic Magnesium Industries Plant between Las Vegas and Boulder City. In 1940 the population of the Las Vegas area had been 8,400. By mid-1942 these figures had swelled to over 30,000 people (Moehring, 1986).

FIG. 20. The 'snackateria" on the Nevada abutment of Hoover Dam as it appeared in 1961. More and more people were venturing to Las Vegas by automobile from southern California, increasing the demand for services at the dam (USBR-LCR).

Prior to the emergence of Las Vegas as a tourist destination after the Second World War few people appreciated that the Las Vegas area would eventually draw so many permanent residents. The low property cost and common employment of central conditioning eventually succeeded in attracting many vacationers and retirees.

In a strange twist of fate, it would be Las Vegas, situated next to the largest man-made lake in the world, that would have to "make do" with an annual water allotment of just 300,000 ac-ft and 4% of all surplus water, which the Nevada representative had agreed to back in 1922, when the Colorado River Compact was signed.

By 1951, 2,000,000 people per year were visiting the Lake Mead Recreation Area (of which, almost 400,000 toured the dam). In 1953, 448,081 people toured Hoover Dam, and by the end of 1958, over 7,000,000 people had toured the dam and powerplant since it was opened to the public-at-large. In 1959, a new annual record of 472,639 visitors was set, and in 1962, the record was raised to 500,000 visitors.

During the 1960s Las Vegas began billing itself as "The Entertainment Capital of the World" as increasing numbers of talented entertainers performed live in floor shows and a number of the biggest headliners settled there, performing in specially-constructed theaters. As Las Vegas grew, so did visitation to the dam (Figure 20). By 1967, the number of yearly visitors exceeded 600,000, and in 1968, the 12,000,000th visitor toured the facilities.

The 15,000,000th visitor was recorded in 1972, and in 1983, on the eve of the dam's fiftieth anniversary, the 23,000,000th person visited Hoover Dam. In the late 1980s Las Vegas began a rapid expansion driven by lower house prices and attractive climate for people retiring from California, and a conscious shift aimed at attracting families as an annual vacation destination. By 2008 it has become the 28th most populous city in the United States with a population of 558,383. The estimated population of the Las Vegas metro area was 1,865,746 in 2008. In 1989 the Bureau of Reclamation began construction of a new Visitors Center and a multi-level parking structure at Hoover Dam. The new visitor facilities vastly improved visitor safety, interpretive capability, and visitor capacity at the dam.

Traffic congestion continued to worsen through the 1980s and 1990s. Semi-trailer commercial truck traffic was banned from U.S. Hwy 93 over the dam after the 9/11 attacks in 2001. In 2004 construction began on a prestressed concrete arch bypass bridge over Black Canyon a mile downstream of the dam, which will eliminate almost a mile of highway and allow vehicles to cross Black Canyon in five minutes instead of 17 minutes (absent any traffic). The Hoover Dam Bypass Bridge for U.S. Hwy 93 is scheduled to be opened in late 2010.

CONCLUSIONS

Boulder Canyon Project Final Reports

The scale of the Boulder Canyon Project was so massive that it gave rise to an unprecedented volume of scientific research and engineering analyses, which was of inestimable value to the civil engineering community, which was so voluminous it could not be summarized in short articles within traditional engineering journals. This information was of enormous interest to those nations contemplating water resources development in the post-World War II era, as well as the financial institutions funding such mega-projects, such as The World Bank.

The Bureau of Reclamation envisioned publishing a series of 34 bound volumes summarizing the technical aspects of planning, design, construction, and operation of the Boulder Canyon Project. Over the years, they revised this effort downward and eventually published 21 of the 32 volumes, between 1939 and 1950. These were collectively known as the "Boulder Canyon Project Final Reports," and they were sold at 1940s prices of $1 to $1.50 each for blue soft-bound and $1.50 to $3 for dark blue cloth bound volumes. They were sold to the general public by over-the-counter sales or by mail order from Reclamation offices in Washington, D.C. and Denver (and Boulder City from 1939-46). They were purchased by many engineering students between 1945 and 1965, by engineers working in water resources engineering, and most of the docents at Hoover Dam Visitor's Center. Many of the volumes remained available until the stocks were eventually exhausted in the late 1980s.

The Boulder Canyon Project Final Reports were originally subdivided into seven broad categories, Parts I thru VII are summarized below:

1) Part I – Introductory reports, consisted of three volumes that provided an overview of the entire project, titled: General Description of the Project; Hoover Dam and Water Contracts and Related Data. The third volume was titled Legal and Financial Problems. Scheduled for release after all the other technical volumes, no titles from Part 1 were ever released.

2) Part II – Hydrology reports, was to have consisted of two volumes titled: Stream Flow and Project Operation; and Utilization of Water. No titles from Part II were ever published.

3) Part III – Preparatory Examinations were summarized in a single volume titled: Geological Investigations (this was the last report to be released, which included several color plates)

4) Part IV – Design and Construction reports was the most expansive of the seven categories, consisting of 10 separate volumes, titled: General Features; Boulder Dam; Diversion, Spillway, and Outlet Structures; Concrete Manufacturing, Handling, and Control; Penstocks and Outlet Pipes; and Imperial Dam and Desilting Works. Proposed volumes titled Hydraulic Valves and Gates; Power Plant Structures and Handling Facilities; Power Plant Generating Equipment; and All-American Canal and Canal Structures were never published.

5) Part V – Technical Investigations was divided into seven major categories, six of which were eventually published: Trial Load Method of Analyzing Arch Dams; Slab Analogy Experiments; Model Tests of Boulder Dam; Stress Studies for Boulder Dam; Penstock Analysis and Stiffener Design; Model Tests of Arch and Cantilever Elements. A proposed volume on Research Measurements at Dam was never published.

6) Part VI – Hydraulic Investigations were summarized in four of the five volumes originally proposed. Those published included: Model Studies of Spillways; Model Studies of Penstocks and Outlet Works; Studies of Crests for Overfall Dams; Model Studies of Imperial Dam and Desilting Works, All-

American Canal Structures (the studies on Imperial Dam and the All-American Canal were combined in to a single volume).
7) Part VII - Cement and Concrete Investigations were summarized in four of six proposed volumes, as follows: Thermal Properties of Concrete (largely covering the various tests and measurements carried out at Owyhee Dam and incorporated into Hoover Dam); Investigations of Portland Cements; Cooling of Concrete Dams; and Mass Concrete Investigations. The proposed volumes on Contraction Joint Grouting and Volume Changes in Mass Concrete were never released.

The first volume that appeared was from Part V - Stress Studies for Boulder Dam, which was released in 1939. The last one to be published was from Part III – Geological Investigations, released in late 1950. The remaining volumes were released during the 1940s. As described above, the name Hoover Dam was not re-established until April 1947, so the reports were always referred to as the "Boulder Canyon Project Final Reports." Many of the more important volumes were translated into other languages, including French, Italian, Portuguese, Russian, and Mandarin Chinese.

When the author first visited the Three Gorges Project office near Wuhan, China in 1989, he was amazed to see the entire 21-volume set of the Boulder Canyon Project reports on the senior Chinese engineer's office book shelf! Most of the senior Chinese engineers (educated before the Cultural Revolution in the mid-1960s) had been educated in the Soviet Union. They were familiar with these volumes because they were the model texts for mass concrete dam design engineers used around the world for the half century following Hoover Dam's completion.

Great Engineering Feat of the 20th Century

In 1955, the American Society of Civil Engineers selected Hoover Dam as one of the Seven Modern Civil Engineering Wonders of the United States. In 1985, the Society named the dam as a National Historic Civil Engineering Landmark. Also in 1985, in recognition if the dam's contribution to the history of the southwest, it was designated as a *National Historic Landmark* by the Department of Interior. In 2000 the American Society of Civil Engineers christened Hoover Dam as a 'Monument of the Millennium.' For years engineers and politicians have touted the benefits of Hoover Dam to society. These benefits include the following:

1) Storing a two year supply of average flow of the Colorado River, which can be released as needed;
2) Over 15 million people use water taken from the Colorado River, including cities in Los Angeles, Orange and San Diego Counties, and Phoenix and Tucson;
3) Water from the Colorado River is diverted to irrigate 750,000 acres in California and Arizona, as well as 470,000 acres in Mexico;
4) The Imperial and Coachella Valleys have become the 'salad bowls' of the southwest, providing the entire United States with lettuce, carrots and other crops during cool winter months, cash crop valued at over $1 billion annually;

5) The Hoover powerplant produces 4 billion kilowatt hours of clean non-polluting electrical energy each year, providing power for 1.3 million people in Nevada, Arizona, and California;
6) The project generates $1 billion in economic benefits to the American Southwest each year;
7) Hoover Dam attracts more than 700,000 visitors each year; and
8) The Lake Mead National Recreation Area encompasses the largest man-made lake in North America and recreation activities generate about 9 million visitors per year.

Unprecedented Influence on Dam and Reservoir Construction

Even before it was completed, Hoover Dam was recognized as one of the greatest engineering feats of the 20th Century. For more than two decades after its completion, Hoover Dam stood as the tallest dam in the world. Today, more than twenty dams are higher, but all owe their existence to the advancements that were made during the design and construction of Hoover Dam. The unprecedented size of the dam led to studies in almost every aspect of dam design and construction, including concrete composition and cooling, stress analysis, hydraulic design, and hydraulic and structural modeling. Even the form of the contracting organization was a new development that changed the face of the construction world. Too large for a single contractor, the project required the formation of a joint venture of six contractors, pooling their talents and resources and sharing the risks. This organizational structure became a model for future large construction projects.

The economic success of Hoover Dam had an enormous influence on the engineering profession all over the world, providing a model for funding high dams through the sale of hydro-electricity over typical terms of 50 years. Many countries were influenced to devise similar schemes to pay for mega dam projects. These included the Indians at Bhakra, the Egyptians at Aswan, the Pakistanis at Tarbella, the Brazilians at Guri, the Taiwanese at Chinman, the Hondurans at El Cajon, and the Chinese at Three Gorges. Everyone sought to emulate the concept for providing a master operating unit controlling a significant river, using hydroelectric generation to pay for the project while providing flood control and irrigation releases as the justification. It was probably over-emulated, especially in less favorable locations, like Aswan.

No one can deny the enormity of Hoover Dam's impact, not only on dam engineering world-wide, but in the industrial development of the American West, where the hydroelectric power proved essential to critical wartime industries (twinning of aluminum and magnesium, and production of plutonium, for many years thereafter), and the construction of water resources infrastructure that has been the pivotal element sustaining agricultural and urban development in arid and semi-arid climes of the western United States, where land values depend, more than anything else, on the availability of water.

ACKNOWLEDGMENTS

The author is most indebted to retired Reclamation engineer Richard Wiltshire, who appreciates civil engineering history and organized this symposium commemorating the 75th anniversary of Hoover Dam's completion. The author's interest in Hoover Dam was originally piqued by his initial viewing of the hour-long documentary film profiling the Boulder Dam project which he viewed in high school. A year later he visited the dams along the lower Colorado River and took an extended tour of Hoover Dam.

In 2008 the author served as one of the speakers participated in a Conference on The Fate and Future of the Colorado River, sponsored by the Huntington-USC Institute on California & the West and the Water Education Foundation at the Huntington Library. This led to many valuable contacts which led to a myriad of sources for information and photographs relating the Boulder Canyon Project and subsequent water resources development in the Lower Colorado River Basin.

In 2009-10 the author was a Trent Dames Civil Engineering Heritage and Dibner Research Fellow at the Huntington Library in San Marino, California. This residency allowed the review of numerous serial publications, collections, archives, and ephemera that provide invaluable to understanding the engineering decisions and political pressures influencing those decisions, not just during construction (1931-35) but during the decade preceding and following the dam's completion. Huntington Archivists Dan Lewis and Bill Frank proved to be particularly valuable in ferreting out rare or obscure accounts from the Huntington's civil engineering and scientist manuscript collections, as well as rare map and historic photos. The author is also indebted to engineering geologist James Shuttleworth, a volunteer in the Huntington's manuscripts department, well versed in the history of the lower Colorado River, who provided innumerable suggestions of inestimable value to the project.

Between 1976 and 1988 the author conducted interviews with Roy W. Carlson (1900-1990) and Jerome M. Raphael (1912-1989), professors in the civil engineering program at U.C. Berkeley. Jerry Raphael had worked for the Bureau of Reclamation from 1938 to 1953 and had served as head of Reclamation's Structural Behavior Group between 1949 and 1953, when he supervised the evaluations of crustal deformation at Lake Mead and reservoir-triggered seismicity.

In 1990 interviews were also conducted in Pasadena with Caltech Professor Vito Vanoni (1904-99), regarding some of the hydraulics research issues that emanated from Hoover Dam and Lake Mead, which were examined by Caltech faculty, staff, associates, and alumni.

Special thanks is also due to Allen W. Hatheway, Jeffrey Keaton, Richard Proctor, and William K. Smith of the AEG Foundation, who provided original documents from Frank A. Nickell, former Chief Geologist of the Bureau of Reclamation, who in addition to being the resident geologist, had supervised studies of reservoir triggered seismicity and crustal deflection at Lake Mead from 1935 till 1949.

Sincere appreciation for assistance is also rendered to the following people and their respective organizations: Reclamation GIS specialist Steven Belew; Dr. Brit Allan Storey and Christine Pfaff, historians with the U.S. Bureau of Reclamation at the Denver office, Dianne Powell of the Bureau of Reclamation Library in their Denver office; Reclamation archivists and staff Bonnie Wilson, Andy Pernick, Bill Garrity, and Karen Cowan of Reclamation's Lower Colorado Region (LCR) office in Boulder City; Boulder City Library Special Collections; Eric Bittner at the National Archives and Records Service Rocky Mountain Region depository at Denver Federal Center; Paul Atwood and Linda Vida at the University of California Water Resources Center Archives in Berkeley; Su Kim Chung of the University of Nevada-Las Vegas Libraries; civil engineering historian Edward L. Butts; journalist and author Michael Hiltzik; and a special thanks to the author's best friend Steve Tetreault, who accompanied the author of his first tour of the dam in February 1973, who moved to Las Vegas in 1986. Over the last 24 years Mr. Tetreault has provided invaluable support and logistical connections in support of the author's numerous visits to the University of Nevada-Las Vegas, Boulder City, and Hoover Dam.

REFERENCES

Arlt, A.W. (1954). Hoover Dam, Boulder Canyon Project, Arizona-California-Neva, in *Dams and Control Works, Third Ed.*, U.S. Bureau of Reclamation, Denver, pp. 33-46.

Angelier, J., Colletta, B., and Anderson, R.E. (1985). Neogene paleostress changes in the Basin and Range: A case study at Hoover Dam, Nevada-Arizona. *GSA Bulletin*, v. 96:3, p. 347-361.

Bell, H.S. (1942). Density Currents as Agents for Transporting Sediments. *Journal of Geology*, v. 50:5, pp. 512-547.

Berkey, C.P., and Nickell, F.A. (1939). *Report on Seismic Activity at Boulder Dam.* Report to the Chief Engineer, U.S. Bureau of Reclamation, Denver, Aug 24, 1939, 61 p.

Cadbury, D. (2004). *Seven Wonders of the Industrialized World*. Harper-Perennial, London.

Carder, D.S. (1945). Seismic investigations in the Boulder Dam area, 1940-44, and the influence of reservoir loading on local earthquake activity. *Seismological Society America Bulletin*, v. 35:175-192.

Carder, D.S., and Small, J.B. (1948). Level divergences, seismic activity, and reservoir loading in the Lake Mead area, Nevada and Arizona. *AGU Transactions*, v. 29:767-771.

Carlson, R.W. (1977). Personal communication.

Emerson, F. C., W.F. Durand, J.G. Scrugham, and J. R. Garfield. (1928). Development of the lower Colorado River: reports by special advisors to the Secretary of the Interior. U.S. Bureau of Reclamation, Government Printing Office, Wash., D.C., pp. 365-435.

Engineering News Record. (1928). Report of Boulder Dam. v. 101:887-889 (Dec 13, 1928).

Fry, A.S. (1950), Sedimentation in Reservoirs, Ch. 20, in Trask, P. D., Ed., *Applied Sedimentation*, John Wiley & Sons, New York, pp. 347-363.

Horton, R.E. (1933). Natural Stream Channel-storage. *Transactions American Geophysical Union*, v. 14, pp. 446-460.

Howard, C.S. (1960). Character of the Inflowing Water. Part 4-K., in W. O. Smith, C.P. Vetter, G. B. Cummings, and others, Comprehensive Survey of Sedimentation

in Lake Mead, 1948-49. USGS *Professional Paper 295*, pp. 103-113.
Gould, H.R. (1960). Sedimentation in Relation to Reservoir Utilization, Part 4-S., in W. O. Smith, C.P. Vetter, G. B. Cummings, and others, Comprehensive Survey of Sedimentation in Lake Mead, 1948-49. USGS *Professional Paper 295*, pp. 215-230.
Grover, N.C., and Howard, C.S. (1938). The Passage of Turbid Water through Lake Mead. *Transactions* ASCE. v. 103, pp. 720-790.
Hiltzik, M. (2010). *"Colossus: Hoover Dam and the making of the American Century."* Simon & Schuster.
Howard, C.S. (1960). Character of the inflowing water. Part 4-K., in W. O. Smith, C.P. Vetter, G. B. Cummings, and others, Comprehensive Survey of Sedimentation in Lake Mead, 1948-49. USGS *Professional Paper 295*, pp. 103-114.
Leopold, L.B. (1969). The Rapids and the Pools-Grand Canyon. The Colorado River Region and John Wesley Powell: USGS *Professional Paper 669*, pp. 131-145.
Longwell, C.R. (1936). Geology of the Boulder Reservoir Floor, Arizona-Nevada. *GSA Bulletin*, v. 47:1393-1476.
Longwell, C.R. (1960). Interpretation of the Leveling Data, Part 2-F., in W. O. Smith, C.P. Vetter, G. B. Cummings, and others, Comprehensive Survey of Sedimentation in Lake Mead, 1948-49. USGS *Professional Paper 295*, pp. 33-38.
Lyons, B. (1947). *They subdued the desert: the story of irrigation as told to Barrow Lyons by the men who apply water, till the land and feed their flocks and herds.* U.S. Bureau of Reclamation, Washington, DC.
Moehring, E.P. (1986). Las Vegas and the Second World War. Nevada Historical Society *Quarterly*, v. 29 (Spring), pp. 1-14.
O'Connell, D.R., and Ake, J.P. (1995). Ground motion analysis for Hoover Dam, Boulder Canyon Project, U.S. Bureau of Reclamation *Seismotectonic Report 94-1*, Denver, 114 p.
National Research Council Interdivisional Committee on Density Currents, Subcommittee on Lake Mead. (1949). Lake Mead density current investigations 1937-40, v.1, 2; 1941-46, v.3 1946-49, U.S. Bureau of Reclamation, Washington, 904 p.
Page, J.C. (1938). Discussion of the Passage of Turbid Water through Lake Mead. *Transactions* ASCE, v. 103:745-747.
Raphael, J.M. (1954). Crustal Disturbances in the Lake Mead Area. U.S. Bureau of Reclamation *Engineering Monograph No. 21*.
Raphael, J.M. (1977). Personal communication.
Rogers, A.M. and Lee, W.H.K. (1976). Seismic study of earthquakes in the Lake Mead, Nevada-Arizona region. *Bulletin Seismological Society America*, v. 66:5, pp. 1657-1681.
Rogers, J.D. (1985). Impacts of Climatic and Environmental Change on Slope Morphology in the Colorado Plateau: *Penrose Specialty Conference on Geomorphic and Stratigraphic Indicators of Climatic Change in Arid and Semi-arid Environments,* Geological Society of America, Lake Havasu, Arizona.
Sherrard, J.L., Dunnigan, L.P., and Decker, R.S. (1976). Identification and Nature of Dispersive Soil. Journal of Geotechnical Engineering, ASCE, v.102:GT4, pp.187-301.

Sisler, F.D. (1960). Bacteriology and Biochemistry of the Sediments. Part 4-O., in W. O. Smith, C.P. Vetter, G. B. Cummings, and others, Comprehensive Survey of Sedimentation in Lake Mead, 1948-49. USGS *Professional Paper 295*, pp. 187-193.

Smith, W.O., Vetter, C.P., and Cummings, G.B. (1960). Comprehensive Survey of Sedimentation in Lake Mead, 1948-49. USGS *Professional Paper 295*.

Stevens, J.C. (1946). Future of Lake Mead and Elephant Butte Reservoirs. *Transactions* ASCE, v. 111, pp. 1231-1342.

Swain, R.E. (2008). Evolution of the Hoover Dam Inflow Design Flood – A Study in Changing Methodologies. In *Bureau of Reclamation: Historical Essays from the Centennial Symposium*. Vol.1, pp. 195-207.

Twichell, D.C., Cross, V.A., Rudin, M. J., Parolski, K. F., Rendigs, R. R. (1999). Surficial Geology and Distribution of Post-Impoundment Sediment in Las Vegas Bay, Lake Mead. USGS *Open-File Report 01-070*.

Twichell, D.C., Cross, V.A., Rudin, M. J., Parolski, K.F. (2001). Surficial Geology and Distribution of Post-Impoundment Sediment of the Western Part of Lake Mead Based on a Sidescan Sonar and High-Resolution Seismic-Reflection Survey. USGS *Open-File Report 99-581*.

Twichell, D.C., Cross, V.A., and Rudin, M. (2002). Mapping turbidities in Lake Mead from source to sink. American Association of Petroleum Geologists, *Abstracts with Programs*, Houston, Tex., March 2002, v. 11, p. A179-180.

Twichell, D. C., Cross, V. A., Belew, S. D. (2003). Mapping the floor of Lake Mead (Nevada and Arizona): Preliminary discussion and GIS data release, USGS *Open-File Report 03-320*.

Vandivere, W.B., and Vorster, P. (1984). Hydrology analysis of the Colorado River floods of 1983. *GeoJournal*, Vol. 9:4 (Dec), pp. 343-350.

Vanoni, Vito. (1990). Personal communication.

Westergaard, H.M. and Adkins, A.W. (1934). Deformations of the earth's surface due to weight of Boulder Reservoir. U.S. Bureau of Reclamation *Technical Memorandum 422*.

75 Years of Hydraulic Investigations – Hoover Dam

Philip H. Burgi[1], D.WRE, Hon. M. ASCE

[1]Retired Manager, Bureau of Reclamation's Hydraulic Laboratory and Founding Member Environmental & Water Resources Institute, 3940 Dover St. Wheat Ridge, CO 80033; philipburgi@aol.com

ABSTRACT: Hydraulic investigations for the unprecedented size and scope of the proposed Hoover Dam (also called Boulder Dam) started in 1930 with a small core of Bureau of Reclamation engineers and technicians assigned to the hydraulic laboratory of the Colorado Agricultural Experiment Station in Fort Collins, Colorado. This paper will summarize the hydraulic investigations and their impact on the Hoover Dam design as well as some of the field challenges associated with the construction and operation of Hoover Dam. The salient results of the initial spillway and outlet works model studies conducted at various venues, model scales, and with several design options will be reviewed. More recent hydraulic studies will be presented of the penstock tie rods, intake tower cylinder gate performance, replacement of the internal differential needle valves, and repair and solution for cavitation damage in the elbow of the spillway tunnels. The initial and subsequent studies related to hydraulic design and performance at Hoover Dam have produced standards and criteria that have greatly added to the world-wide knowledge base for today's hydraulic structure design. The design, experimental investigations and construction of the Hoover Dam pushed the science of hydraulics to new frontiers.

INTRODUCTION

At the end of the 19th Century and early in the 20th Century efforts were made to divert flow from the Colorado River into the fertile Imperial Valley near the border with Mexico. In early 1905 with the rise of flow in the Colorado River the entire river was diverted into a newly formed canal which was part of the Alamo River flowing North into the Salton Sea located some 280 ft below sea level. This uncontrolled flow over nearly two years caused a rise of 78 ft in the Salton Sea. The river flow was finally controlled and returned to its natural channel. However, the need to control the Colorado River farther upstream in the canyon country became evident. By 1918 Director Arthur Powell Davis of the U.S. Reclamation Service realized the importance of investigating the possibility of building a large dam in the Boulder or Black Canyons on the Colorado River over 300 miles upstream.

In 1924, the U.S. Bureau of Reclamation (renamed from Reclamation Service in 1923 and relocated from USGS to a separate bureau in the Department of Interior) issued a report recommending construction of a concrete arched-gravity dam in Black Canyon (located 13 miles downstream from Boulder Canyon). In 1928 Congress authorized the Boulder Canyon Project and after the 1929 Wall Street stock market crash appropriations soon followed in 1930 as an unemployment relief measure.

Because of the seriousness of the unemployment problem in 1930, Interior Secretary Wilbur ordered that the U.S. Bureau of Reclamation speed up preparation of plans and specifications so that the contract for construction could be awarded at the earliest possible date. In response to the Secretary's request, engineers at the Denver Office completed and printed specifications for the dam and appurtenant works in December 1930, six months ahead of schedule. Bids were opened at the Denver office on March 4, 1931, with the lowest bid being submitted by Six Companies, Inc., of San Francisco. The contract was awarded to Six Companies on March 11, 1931 and the order to proceed was issued on April 16, 1931 (Boulder Canyon Project, 1941).

HYDRAULIC LABORATORIES

"Models were first used extensively by the Bureau of Reclamation in 1930 in the design of the spillway for the Cle Elum Dam of the Yakima project in Washington. The design of the spillways for Hoover Dam, Madden Dam in the Panama Canal Zone, and Norris and Wheeler Dams for the Tennessee Valley Authority, served as stepping stones in further developing the technique and improving the methods." (Warnock, 1936). These early investigations with hydraulic models started when a dozen engineers, technicians, and mechanics from the Denver Office began working in the hydraulic laboratory of the Agricultural Experiment Station in Fort Collins, Colorado (The 2,600 ft^2 laboratory was built in 1912 under the direction of Ralph Parshall.) The unprecedented height of Hoover Dam necessitated a thorough study of many phases of its design to ensure absolute safety of the structure. The laboratory facilities were adequate for model tests of Hoover Dam spillways and spillway tunnels; tunnel plug needle valves; and the intake towers, with scales from 1:106.2 to 1:64.

With the anticipation of designing Hoover Dam there came the recognition that this structure would impose design and construction challenges well beyond the textbooks and experience of the day. With safety concerns as well as high construction costs associated with this large structure it became evident that careful preliminary studies and model testing were required before one could finalize design and start construction (Burgi, 2002).

Because of the tremendous power of the spillway flows it was felt that larger scale models should be used to develop and confirm the spillway and outlet works design. A model scale of 1:20 was decided upon for the spillway tests and tunnel plug needle valves. This required a model discharge of 112 ft^3/s and a vertical water drop of 30 ft in the model. Since no laboratory was available to provide this size of facility it was decided to build an outdoor laboratory on the Umcompahgre Irrigation Project near Montrose, Colorado. At that time the south canal provided a drop of 50 ft and a discharge up to 200 ft^3/s during the irrigation season. This outdoor laboratory was used for testing the Hoover Dam design features (side-channel spillway and spillway tunnel; tunnel plug needle valves) during the irrigation season in the years 1931, 1932 and 1933.

The laboratory was used to test large 1:20 scale models of the Hoover Dam side channel spillways with drum gates (earlier studies showed the inadequacies of the Stoney Gate and overflow crest for the spillway). A complete sedimentation model of Imperial Dam and its appurtenant works, and a model of Grand Coulee Dam and spillway bucket were also tested at the Montrose laboratory. A total of eight laboratory models were used in the hydraulic design of Hoover Dam with model scales of 1:20(2), 1:60(3), 1:64, 1:100, and 1:106.2

EARLY STUDIES

When the initial appropriation for construction at Hoover Dam was made on July 3, 1930, there were many problems requiring further study in connection with both design and construction. Work to improve the design was under way when, due to the existing general unemployment situation, the President of the United States expressed a desire that actual construction be commenced as soon as possible. Designs were speedily prepared for bidding purposes, based on the results of limited studies then available. At the time of the first Specification (No. 519) in December 1930 the development of the plans showed side spillways with Stoney Gates as the control for spillway flows (Boulder Canyon Project, 1941).

To ensure the 31 million acre-ft storage behind the proposed Hoover Dam, the dam's height would be more than twice as high as Arrowrock Dam, the highest dam in existence at the time (1930). It was obvious to all involved that such a large structure would create new problems and challenges in design and construction never before encountered. The design would have to be thoroughly studied with numerous concrete and hydraulic laboratory investigations before construction commenced.

Several early spillway designs were considered for purposes of cost estimating:

1. Siphon Spillways (1920) 32 -16 ft by 16 ft siphons discharging into diversion tunnels in both abutments.
2. Needle Valves and Slide Gates through Dam (1924) 32 -72-inch Needle Valves and 16- 6 ft by 6 ft Slide Gates through dam itself.
3. Spillway Wash Concept– Several concepts provided reduced spillway velocities (V <100 ft/s) by using multiple tunnels through a saddle in the left abutment. These features would be located entirely on the Arizona side and would not use the diversion tunnels for the spillways.
4. High level abutment spillway tunnels that slowly dropped to the river (V <100 ft/s)

All of these alternatives were abandoned due to the additional costs involved with not using the diversion tunnels ($7-17 million) or because of safety and other technical factors (Steele, 1931).

In the summer of 1928, after the failure of St. Francis Dam in California, a board of engineers and geologists was appointed to review preliminary plans for Hoover Dam (Boulder Canyon Project, 1941). The board recommended:
- Increasing the total spillway capacity from 200,000 ft^3/s to 400,000 ft^3/s
- Add power production to the project
- Place two diversion tunnels in each abutment instead of through the dam
- Use intake towers in the reservoir for power penstocks and outlet works
- Use the two outer diversion tunnels for the spillways (connected by inclined shafts) and the two inner diversion tunnels for power penstocks and outlets

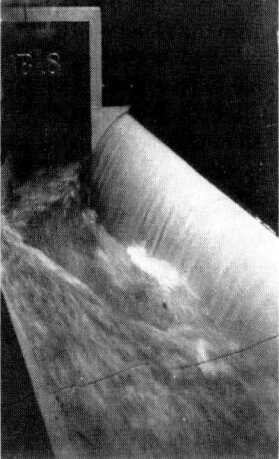

Fig. 1. 1:60 Stoney Gate Spillway Model
(Photograph courtesy of U.S. Bureau of Reclamation - USBR)

SPILLWAYS

The spillways were one feature that demanded especially careful attention. The spillway flows would drop over 500 ft and reach velocities near 175 ft/s. By the start of the hydraulic laboratory studies in 1930 spillway designs were focused on three alternatives:
- Glory Hole spillways on each abutment with vertical drop shafts to diversion tunnels (1928-1930)
- Side channel spillways on each abutment with uncontrolled crests and 50 ft by 50 ft Stoney Gates – inclined shafts to diversion tunnels (1930-32) Figure 1.
- Side channel spillways on each abutment with drum gate controlled crests – Inclined shafts to diversion tunnels (1932-33) Figures 2 and 3

Fig. 2. 1:60 Drum Gate Spillway Model – Fort Collins
(Photograph courtesy of - USBR)

Fig. 3. 1:20 Drum Gate Spillway Model – Montrose
(Photograph courtesy of - USBR)

Once the Hoover Dam construction contract was awarded in early 1931, funding was made available to hire additional design and laboratory personnel. The detailed design studies and laboratory investigations necessary to solve the unprecedented challenges created by mass concrete placement, high energy hydraulic action and the mammoth size of this dam would certainly bring new challenges (Lane, 1933).

By late 1930, the design and laboratory staffs started an intense effort to catch up with the Federal employment relief effort which placed the construction of the Hoover Dam project as high priority. Original specification plans contemplated the construction of a 700 ft long free crest spillway on each abutment, a long concrete-lined channel opening into an inclined tunnel connecting with the outer diversion tunnel. A 50 ft by 50 ft Stoney Gate would be installed at the upper end of each spillway channel as shown in Figure 4 (December 1930). Since it was necessary to make a decision on the general type of spillways without the extensive tests that would follow, the design specifications released in December 1930 reflected this original spillway concept.

Glory-Hole Spillway - The hydraulic laboratory studies at Fort Collins, Colorado which started in October 1930 included studies on Madden Dam (Canal Zone) in addition to Hoover Dam. In order not to delay work on the side channel spillway concept at Hoover Dam, experiments on the glory hole type consisted only of visual observations and photographs using a 1:60 scale model. These cursory studies resulted in two major concerns: (1. The approach flow to the glory hole was not symmetrical due to the fact the spillway was located on the upstream canyon abutment. This caused a concentrated flow down the vertical shaft with very undesirable suction and pulsations; and (2. The very poor flow conditions in the vertical bend at the base of the 600 ft high shaft further lead to the abandonment of the glory hole design. Consequently, work on the glory-hole spillway was discontinued and the side-channel spillway was adopted (Lane, 1933).

Side-channel Spillway - When the total spillway capacity was increased from 200,000 ft^3/s to 400,000 ft^3/s the model studies had already turned to the side channel spillways. This necessitated a redesign and while the office design studies were in progress the hydraulic laboratory studies focused on improvements in side channel spillway efficiency. Two side-channel spillway designs were studied in the laboratory:

- Side-channel 700 ft long ogee spillways with Stoney Gate at the inlet channel (Dec 1930 Specification drawings)

- Side-channel spillways with drum gates on the 400 ft crest (1931)

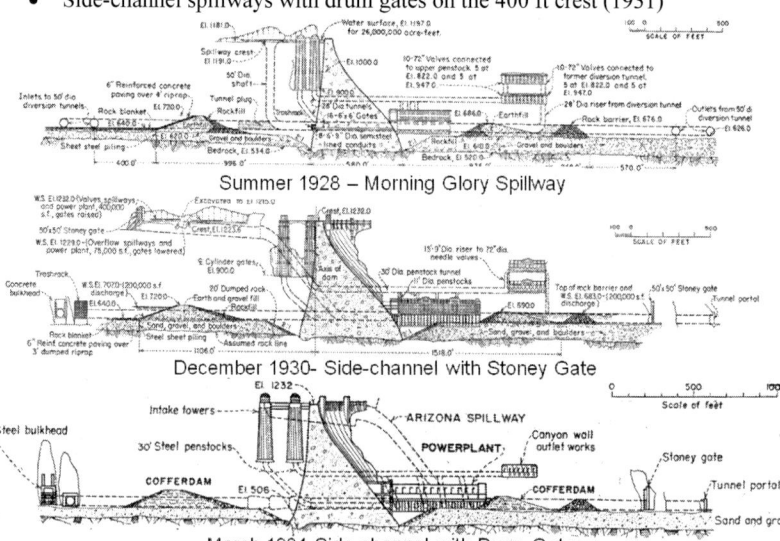

Fig. 4. Spillway Options - 1928, 1930, 1931 (Courtesy of - USBR)

Following this period of research, which included the building of models to scales of 1 to 60 and 1 to 20, the design was changed to a weir with a drum-gate controlled crest equipped with four 16 ft by 100 ft gates, a concrete-lined channel enclosing a stilling basin that was formed by a weir built across the channel at its lower end. The spillway flow exited this "bathtub" through a transition and entered the 50-ft-diameter inclined tunnel to the previously constructed diversion tunnel, Figure 4 (Steele, 1931). This redesign satisfied the requirement of the Colorado River Board that the spillways should have sufficient capacity to carry 400,000 ft^3/s when the reservoir was at the crest elevation of the dam (The Reclamation Era, 1935). The large 1:20 scale model of the drum gate spillway at Montrose, Figure 3, was used to finalize the design of the drum gates on the side-channel spillways at Hoover Dam replacing the proposed Stoney Gates which proved to be unsatisfactory during the model tests.

A significant story not often told about the construction of Hoover Dam was the very successful simultaneous development of the design, laboratory studies and field construction during a very tight schedule that required a large number of U.S. Bureau of Reclamation staff to work long hard hours to keep the design and construction on schedule. The time line shown in Figure 5 illustrates the tight schedule of construction on one side and the design/laboratory studies on the other. Throughout the entire series of tests, close contact was maintained with the design staff in order that the plans developed would be sound not only from the hydraulic performance but also from a structural and construction standpoint. This same concept of laboratory investigations and design collaboration proved to be a very successful strategy throughout the "glory years" of Reclamation's construction program.

Construction of the diversion tunnels started in September 1931. Construction on the spillways and the inclined tunnels that would connect them with the diversion tunnels began in early 1932 with excavation of the spillway channels. The final design of the twin spillways resulted from extensive studies that involved testing of several designs. The final design was subject to numerous tests using models of various scales to determine and verify the hydraulic performance of the structures. The hydraulic model tests were conducted at Reclamation's laboratories in Fort Collins (1:60 and 1:100 scale models) and Montrose (1:20 scale model). Excavations of the spillway channels were completed in March 1933.

The first major improvement resulting from the hydraulic laboratory studies was the increased spillway capacity at Hoover Dam when side-channel spillways replaced the originally planned glory-hole spillway design. Once the side-channel spillway with drum gates was selected as the best design, several additional improvements were made:
- Improving flow conditions in the side channel basin by placing a 35-ft-high weir at the downstream end
- Increasing the length of transition from the 128-ft-wide spillway basin to the 50-ft-diameter spillway tunnel
- Improving the ogee crest (drum gate) shape to maximize flow efficiency

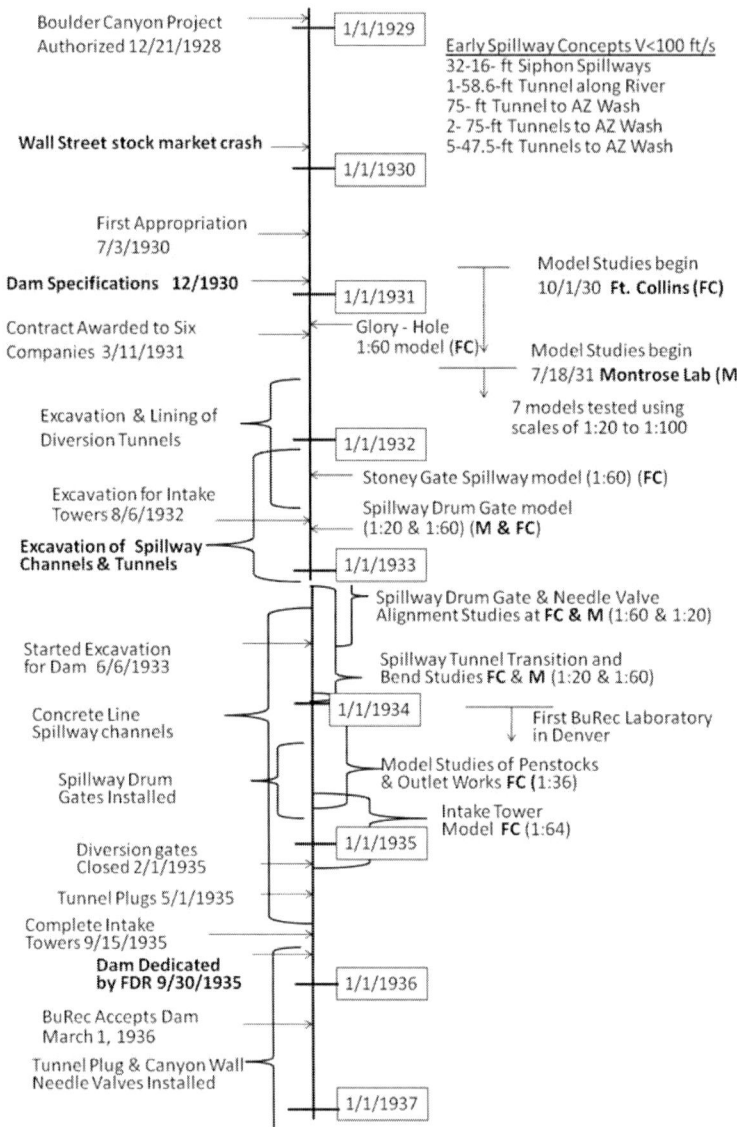

Fig. 5. Hoover Dam Time Line

PENSTOCKS AND OUTLET WORKS

The hydraulic model studies of the side-channel spillways proved to be so valuable that it was decided to test other major features of Hoover Dam by model experiments. The penstock and outlet works models were all conducted at the Fort Collins laboratory with the exception of the 1:20 model of the tunnel plug needle valves studied at the outdoor Montrose laboratory. These studies included:

- Hydraulic losses and pressure variations in the intake towers (1:64 scale model)

- Studies included the junctions from the 30 ft headers to the 13.5 ft penstocks (1:36 scale model)

- Flow characteristics of the needle valves in the tunnel plugs (1:60 & 1:106.2 scale model)

The study results indicated that the proposed set of air vents at the base of each intake tower could be eliminated. In no instance did the study results lead to drastic changes in the design layouts for the penstocks and outlets such as those proposed for the spillways. However, the study results fully verified the adequacy of design (Boulder Canyon Project, 1938).

TUNNEL PLUG NEEDLE VALVES

The Hoover Dam tunnel-plug outlets provided outstanding challenges. Each tunnel had six 72-inch internal type needle valves under pressure heads up to 560 ft and designed to discharge a total of 22,000 ft^3/s to a 50-ft-diameter concrete tunnel. The laboratory model studies included tests at scales of 1:106.2, 1:60, and 1:20 to assure the validity of the design against any scale effects. The final configuration selected represented a distinct improvement over those originally proposed (Bradley, 1936).

A critical design problem realized in the configuration of the tunnel plug outlets was the arrangement of the six needle valves in a very tight pattern which if not properly aligned would damage the tunnel lining and possibly choke off the downstream tunnel. The original design orientation of the valves was level with all located at the same elevation. As the result of the model investigations the six valves were symmetrically placed at different vertical and horizontal angles to achieve the best hydraulic performance in the downstream tunnel. The model studies also investigated a pattern of opening the valves where pairs of valves opposite the centerline are opened sequentially to provide the best performance.

The laboratory tests also showed that a large air vent tunnel originally proposed would not be necessary. This resulted in construction savings of $30,000 (1932 costs). Field tests with the tunnel plug needle valves in December 1938 confirmed there was no need for the air vents (Board of Engineers, 1939).

In the mid-1990s the tunnel plug and canyon wall needle valves were decommissioned and replaced with 8- 68-inch jet flow gates in the tunnel plugs and 4- 90-inch jet flow gates in the canyon walls.

LATER STUDIES

Concerns with Power Penstock System – In the mid-1990s in addition to the needle vale replacement several other features of the power penstock and outlet works system were under a performance and safety review. Concerns were raised over the functionality of the system under emergency conditions. The four intake towers, two on each side of the canyon, are founded on rock benches some 270 ft above the river bed. The towers and their 32-ft-diameter cylinder gates control flow to the power penstocks and outlet works. A header tunnel was excavated from the base of each tower. Tunnels from the upstream towers connect to the inner diversion tunnels and supply water to the power penstocks and the tunnel plug outlet valves. Tunnels from the downstream towers run parallel to and approximately 170 ft above the inner diversion tunnel and supply water to four of the power penstocks as well as the canyon wall outlet valves, Figure 6.

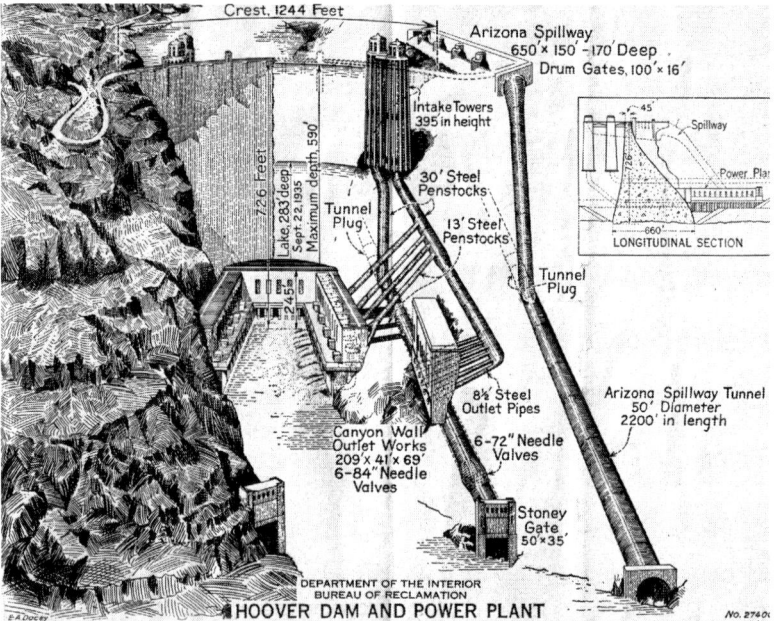

Fig. 6. Isometric Hoover Dam **(Courtesy of - USBR)**

Downpull Forces on Intake Tower Cylinder Gates - The upper and lower 32-ft-diameter and 11-ft-high cylinder gates on each tower were designed in the 1930s and at that time were only expected to operate in the balanced mode (would only be used when there was no flow through the tower). In the early 1950s the decision was made that the cylinder gates should be able to function if an emergency flow closure to the powerhouse or outlets was needed. Field tests conducted in 1950 on the lower cylinder gate of the upstream Arizona intake tower produced severe cavitation and extremely high downpull loading on the gate stem during an unbalanced closure of the gate. By 1953 the gate seal configuration was changed as shown in Figure 7 to remove a significant portion of the gate frame not needed to seal the gate. This decreased the low pressure zone which caused cavitation and greatly increased the downpull forces during gate closure (Todd, 2004). Since 1953 the gate seals on the remaining seven cylinder gates have been modified as shown in the Figure 7a.

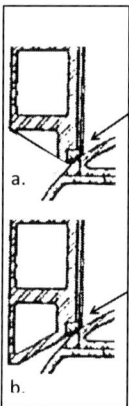

Fig. 7. Cylinder Gate Seals
a. Modified Gate Seal
b. Original Design
(Courtesy of - USBR)

In September 2003 field tests were conducted on the Nevada side downstream intake tower, testing the upper cylinder gate with the new seal arrangement under unbalanced loading. This unbalanced test represented an emergency closure. Two of the three stems were instrumented yielding loads of 19,000 pounds and 35,000 pounds. From these tests it was estimated that the total downpull load would be on the order of 80,000 pounds for the three stems. With the new gate seal shape the downpull load was reduced by 75% in over that experienced with the original seal design from the 1930s. The unequal loading of two gate stems in the 2003 tests leaves some questions as to the definitive demonstration that the remaining gates can operate safely during emergency closure. Although the test results showed that none of the gate system components were over loaded and adequate safety margins prevailed, additional testing of the other cylinder gates was recommended.

Replacement of Needle Valves – There were several catastrophic needle valve failures at dams in the 1980s, causing the deaths of several workers and millions of dollars in damage to hydraulic structures. To provide safe and reliable dam outlet discharge, Reclamation carried out a needle valve replacement program in the late 1990s. This resulted in replacement of the Hoover Dam needle valves with jet flow gates (Figure 8) developed by Reclamation for Shasta Dam in the late 1940s (Lowe, 1946). Jet flow gates can operate at very high pressures, offer excellent control characteristics, operate with no cavitation, and have a high coefficient of discharge.

The twelve 72-inch needle valves in the tunnel plugs were replaced with eight 68-inch jet flow gates, each capable of discharging approximately 3,800 ft^3/s. The twelve 84-inch needle valves in the canyon wall outlets were replaced with four 90-inch jet flow gates, each capable of discharging approximately 5,400 ft^3/s

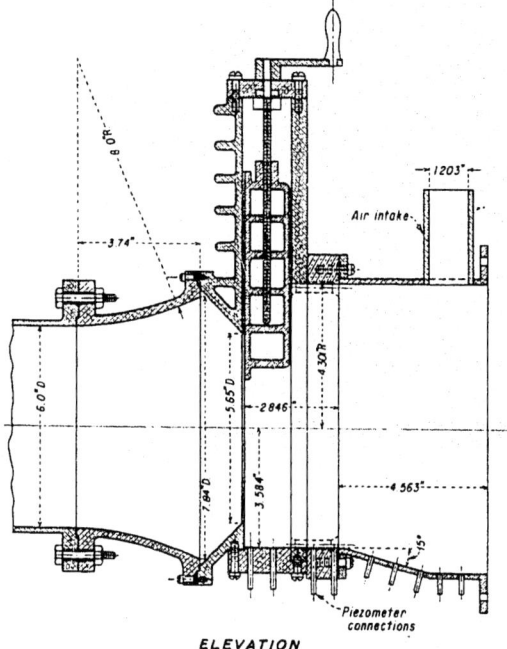

Fig. 8. 6" Model Jet Flow Gate
(Courtesy of - USBR)

The two downstream canyon wall outlets remain operational and four outlets in each tunnel plug remain operational. Figure 9 shows the alignment of the four remaining outlets in the Arizona and Nevada tunnel plugs.

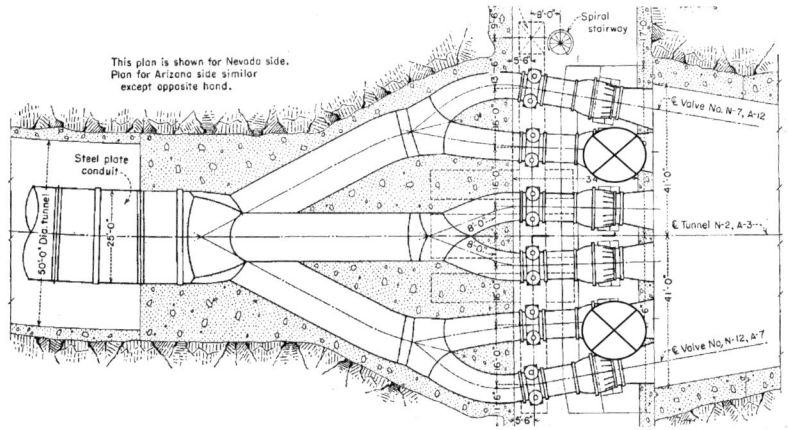

Fig. 9. Configuration of the four jet flow gates in the tunnel plugs
(Courtesy of - USBR)

Penstock Tie Rods – As a result of a recent risk assessment of the Hoover Dam Penstock System a concern was raised over a possible tie rod fatigue failure at any of the 16 penstock junctions. The original design called for two tie rods at each penstock junction as shown in Figure 10. If a tie rod failed resulting in failure of the header-penstock, the cylinder gates would close under unbalanced conditions effectively stopping the flow. Field tests have been performed at three locations to determine the fatigue life of tie rods: Upper Arizona A-1, Lower Nevada N-3, and Upper Nevada N-4. Early study results show remaining service life of from 24 to 400 years (Todd, 2005). For this reason additional tie rod tests are scheduled.

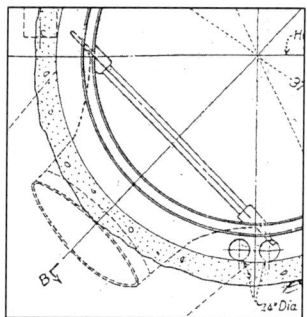

Fig. 10. Tie Rod Upper Arizona Header/Penstock
(Courtesy of - USBR)

Spillway Cavitation Damage (1983-84) – Reclamation's high dam tunnel spillways have proved to be a very economical means to pass large flood discharges in lieu of building large capacity surface spillways and stilling basins on the dam abutments.

However, there have been several incidents of cavitation damage in or near the tunnel elbows. In 1941, the level of Lake Mead reached to within one foot of the top of the spillway gates, and on August 6, the gates on the Arizona side were lowered and water flowed into the spillway for the first time. When the spill was halted in early December, inspection of the spillway tunnel revealed that a 112 ft by 38 ft section of the tunnel lining had been eroded to a depth of 45 ft by the action of the water as it flowed through the tunnel. Repairs to the tunnel lining were completed in early 1943.

Keener describes the damage in a 1943 article in Engineering News Record (Keener, 1943). The damage was thought to have initiated at a "misalignment" of the tunnel invert which led to damage caused by the high velocity flow passing over the roughness and leading to bubble formation in the flow which collapsed sending out high energy shock waves damaging the concrete. This phenomenon is referred to as cavitation formation and damage. Studies were conducted in 1945 in Reclamation's hydraulic laboratory to study means of injecting air into the tunnel spillway (Bradley, 1945). A number of tunnel invert sills and dentates were tested as possible ways to place air along the invert of the tunnel in the area that was damaged. The conclusions were not very promising and the decision was to make the repaired invert as smooth as possible. In the 1940s the damage was repaired by backfilling with river rock and then covering with a thick layer of concrete. A very fine almost terrazzo finish was placed on the concrete surface to prevent reoccurrence of the cavitation and resultant damage.

Tunnel spillways were later constructed at Yellowtail, Flaming Gorge, Blue Mesa, and Glen Canyon Dams. Severe damage occurred at the Yellowtail tunnel spillway in 1967 as a result of a spill and was investigated by the hydraulics laboratory in 1971 (Colgate, 1971). These laboratory studies verified the aeration slot concept similar to that shown in Figure 11. To introduce air into the high velocity flow at Yellowtail Dam tunnel spillway, a 3 ft by 3 ft aeration slot with a 3-inch ramp was used. Subsequent field tests at Yellowtail Dam confirmed the results of the model study relative to the location and configuration of the aeration slot. An examination of the tunnel indicated that sufficient air was being supplied to the flow to protect the flow surfaces downstream from the slot from cavitation damage.

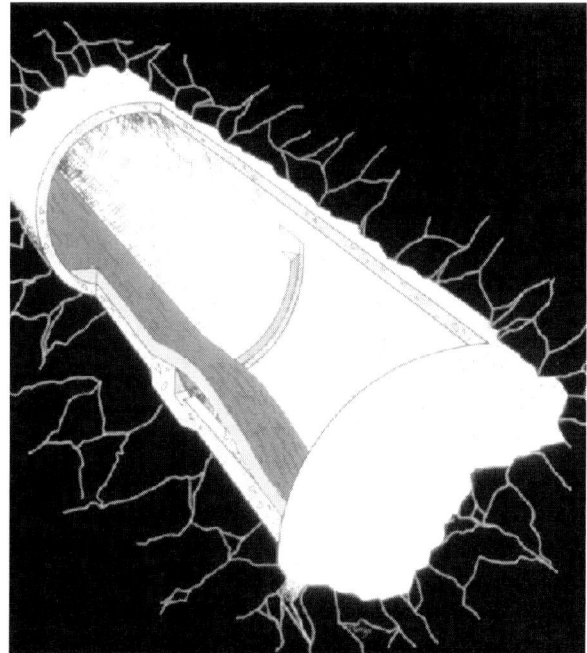

Fig. 11. Tunnel Spillway Aeration Slot Concept
(Photograph courtesy of - USBR)

As a result of intense rain and snow in the Rocky Mountains during the spring of 1983, high runoff in the Colorado River Basin created the need to pass flood flows through tunnel spillways at Blue Mesa, Flaming Gorge, Glen Canyon, and Hoover Dams. The damage was so extensive at the Glen Canyon Dam tunnel spillways that $42 million and a year of reconstruction were required to repair the spillways and install aerators, Figure 12. By early July, Lake Mead was almost at capacity, and on the evening of July 2, water flowed over the Hoover Dam spillway drum gates for the first time in over forty years. The spill continued until September 6 and at its peak, water was more than four- ft deep over the top of the raised gates. The spill caused damage in the tunnel spillways at locations similar to where it occurred during the 1941 spill. The damage was more sever this time on the Nevada side but not to the degree of the damage experienced in the Arizona tunnel in 1941, Figure 13.

Reclamation conducted extensive laboratory model tests to determine hydraulic performance of a unique aerator design for each tunnel spillway at these five Reclamation dams (Pugh, 1988). By 1985 aerators were installed in all five of these high head tunnel spillways in the western United States. In 1990, a comprehensive monograph summarizing Reclamation's experiences and developments in cavitation damage control was published (Falvey, 1990).

Field tests conducted in 1984 on the aerated left tunnel spillway at Glen Canyon Dam proved conclusively the effectiveness of properly placed aerators in preventing cavitation damage in tunnel spillways. An area of eroded concrete near the elbow of the Glen Canyon tunnel spillway was not repaired prior to the testing. The successful outcome of these tests and other experiences with aerators provided a strong basis for relying on aeration as the main defense against cavitation damage at Hoover Dam, instead of the previously used surface tolerance criterion. For Hoover Dam and the other dams with aerators in the spillway tunnels, the decision was made to relax surface tolerances and leave unrepaired many surface roughnesses that exceeded previous surface specifications. The lessons learned through years of experience and research with structures such as Hoover Dam demonstrated the need to update Reclamation's concrete surface specifications and standardize the design process for smooth spillways with high-velocity flow (Frizell, 1991).

Fig. 12. Glen Canyon Dam Left Tunnel Spillway Damage 1983
(Photograph courtesy of - USBR)

Fig. 13. Hoover Dam Tunnel Spillway Damage 1941
(Photograph courtesy of - USBR)

CONCLUSIONS

The design and construction of Hoover Dam in the years from 1929 to 1935 was an extraordinary accomplishment given the state of the economy, the severe conditions where the dam was built and the unprecedented size and scope of the Boulder Canyon Project. It challenged the ingenuity, perseverance and courage of the planners, designers, investigators, and construction personnel of the time.

The hydraulic features of the project as finally evolved represent the results of an extensive process of design, experimentation, and elimination, covering a period of more than four years. The intensive cooperation between design engineers, hydraulic laboratory personnel and construction staff at the dam site was invaluable in leading to a highly successful outcome. The ability of the hydraulic laboratory personnel working closely with design staff even while excavation and initial construction were progressing at the dam site proved to be critical to the early completion of Hoover Dam which was dedicated by President Roosevelt on September 30, 1935. Without the hydraulic model experiments, the same degree of security could only have been obtained by using very conservative designs costing several million dollars more than those adopted.

The results of the hydraulic laboratory studies used for the development of designs for Hoover Dam furnished valuable assistance in designing spillways at other dams. The theory of hydraulic model testing advanced considerably during these years. Tests on models extending over a wide range of sizes showed virtually identical results, a further extension of size to that of the prototype could be expected to produce similar results.

REFERENCES

Board of Engineers (1939), *Tests to Determine Operating Characteristics of Tunnel-Plug Outlet Works at Hoover Dam, Boulder Canyon Project*, u.s. Bureau of Reclamation, Hydraulic Laboratory Report HYD-49, Denver CO, Feb 3, 1939.

Boulder Canyon Project Final Reports (1938), U.S. Department of the Interior, Bureau of Reclamation, Part VI- Hydraulic Investigations, Bulletins 1-3, 1938.

Boulder Canyon Project Final Reports (1941), U.S. Department of the Interior, U.S. Bureau of Reclamation, Part IV Bulletin 1,1941.

Bradley, J.N., and Warnock, J.E. (1936), *Hydraulic Model Experiments for the Design of the Hoover Dam-Model Studies of Appurtenant Structures*, HYD 7.1, Denver CO, June 4, 1936.

Bradley, J.N.(1945), *Study of Air Injection Into the Flow in the Hoover Dam Spillway Tunnels- Boulder Canyon Project,* HYD 186, Denver CO, October 24, 1945.

Burgi. Philip H., *Impacts of Reclamation's Hydraulic Laboratory on Water Development*, U.S. Bureau of Reclamation Centennial Anniversary, Las Vegas, NV, June 17-19, 2002. http://www.usbr.gov/history/

Colgate, D. (1971), *Hydraulic Model Studies of Aeration Devices for Yellowtail Dam Spillway Tunnel*, U.S. Department of the Interior, Bureau of Reclamation, *REC-ERC-71-47*, December 1971.

Falvey, H. T.(1990), *Engineering Monograph No. 42 Cavitation in Chutes and Spillways*, U.S. Department of the Interior, Bureau of Reclamation, 1990, 145.

Frizell, K. H. and Mefford, B. W.(1991), *Designing Spillways to Prevent Cavitation Damage*, Concrete International: Design and Construction of the American Concrete Institute, May, 1991, pp. 58-64.

Keener, A.(1943), Challenge to Hydraulic Designers, *Engineering News Record*, November 18, 1943.

Lane, E.W.(1933), Hydraulic Model Experiments for the Design of the Hoover Dam, HYD 1.3, Denver Co, January 15, 1933.

Lowe, F.C.(1946), *The Hydraulic Design of a Control Gate for the 102 Inch Outlets at Shasta Dam*, HYD 201, Denver CO, March 31, 1946.

Pugh, C.A. and Rhone, T.J.(1988), *Cavitation in Bureau of Reclamation Tunnel Spillways*, International Symposium on Hydraulics for High Dams, Beijing, China, 1988.

Steele, B.W., McConaughy, D.C., and Lane, E.W.(1931), *Structural Design Studies and Model Tests of Hoover Dam Spillways*, HYD – 0.6, Denver CO, October 2, 1931.

The Reclamation Era (1935), U.S. Department of the Interior, U.S. Bureau of Reclamation, pages 30-33, February, 1935.

Todd, Robert V.(2004), *Upper Cylinder Gate Tests for Upper Nevada Penstock,* TSC, Denver, Technical Memorandum No. HV-D8410-04-4, August 2004.

Todd, Robert V.(2005), *Hoover Dam Boulder Canyon Project, Arizona And Nevada – Study of Fatigue Life of Tie rods at A-1 Header/Penstock, Upper Arizona Penstock*, TSC, Denver, Special Study No. HV-D8410-02- 2005, September 2005.

Warnock, J. E.(1936), Experiments Aid in Design at Grand Coulee, *Civil Engineering*, November 1936, pp.737-41.

Performance of Spillway Structures Using Hoover Dam Spillways as a Benchmark

William R. Fiedler[1], M. ASCE, P.E

[1]Technical Specialist, Bureau of Reclamation, Denver Federal Center, Bldg. 67 (86-68130), PO Box 25007, Denver, CO 80225-0007; wfiedler@usbr.gov

ABSTRACT: The designs of the Hoover Dam Spillways were innovative and helped shape the design of future spillways. The hydraulic models studies for the Hoover Dam Spillways and the resulting design details will be discussed and related to present day designs. With the Hoover Dam Spillway designs as a starting point, present day concerns and vulnerabilities of spillway structures will be discussed. Case histories of spillway incidents and failures will be presented. Current approaches for evaluating spillways and identifying potential failure modes related to spillway structures will be discussed, with a focus on: spillway capacity/dam overtopping issues; spillway conveyance capacity; gate performance in non-flood situations; and, performance of spillway linings under high velocity flow.

INTRODUCTION

Hoover Dam is a concrete gravity arch dam, 726-ft-high with a crest length of 1,244 ft. The dam has a side channel spillway regulated by four 100-ft-wide by 16-ft-high drum gates at each abutment. The spillway crest is at elevation 1205.4 and the top of the drum gates in the raised position is elevation 1221.4. Water from the spillway inlet channel flows into a 50-ft-diameter inclined tunnel. A vertical bend connects the end of the inclined tunnel to a near horizontal tunnel. A flip bucket is provided at the downstream end of the tunnel spillway to help dissipate the energy of spillway flows. The current combined capacity of the spillways is about 270,000 ft^3/s. Details of the Hoover Dam Spillways are provided in Figures 1 and 2.

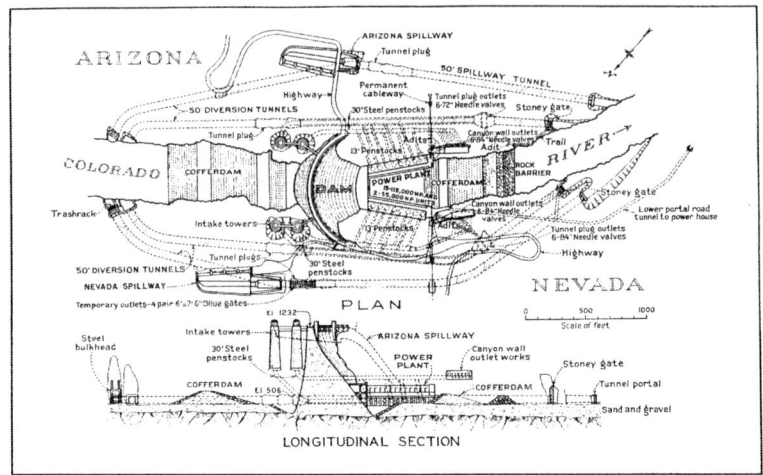

FIG. 1. Hoover Dam Plan and Section (Reclamation)

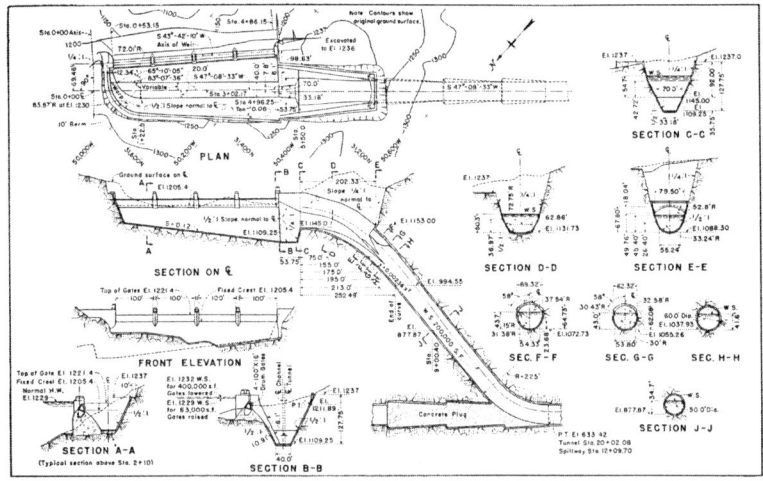

FIG. 2. Hoover Dam – Nevada Spillway (Reclamation)

A major breakthrough in the understanding of high-head, high-velocity spillway designs resulted from the Boulder Canyon Project and construction of Hoover Dam. Between the authorization of the Project in 1928 and the completion of Project documentation in 1948, extensive research was performed which serves as the basis for present-day spillway designs and analyses. The unprecedented size of the spillways (each with design capacity of 200,000 ft^3/s and a maximum average velocity approaching 175 ft/s) for Hoover Dam was the motivation to initiate a comprehensive research program. One of the key focal points was the research and development of methods to design the "ogee" spillway crest, which is still used for spillway designs around the world (Reclamation, 1987).

Prior to this research, methods of estimating the "under-nappe" of a jet of water moving over a sharp crested-weir had been based on approximate observations made by M. Bazin in the late 1800s (Bazin, 1888), which typically used a vertical upstream face on the spillway crest. The shape of the under-nappe defines a minimum shape or profile for the spillway flow surface. Unless the flow surface matches or is flatter than the under-nappe, sub-atmospheric pressure can occur. This can possibly lead to reduced tailwater-induced backpressure, increased cavitation potential, or vibrations. The Boulder Canyon Project hydraulic research expanded on Bazin's methods and developed design tools, which can still be found in Reclamation's Engineering Monograph (EM) No. 9 (Bradley, 1952) and Design of Small Dams (Reclamation, 1987). The design tools provide considerable flexibility and methods to:

1. Determine the spillway ogee shape required to best fit the under-nappe of the overfalling stream for any practical condition of design;

2. Derive the nappe shape due to varying approach velocities;

3. Determine the coefficient of discharge for overfall dams (or spillways) with vertical, sloping, overhanging, and offset upstream faces;

4. Determine effects on coefficient of discharge due to different crest shapes with and without control gates, including the effects of adjacent terrain, piers, and position of gates; and

5. Determine the effects on coefficient of discharge from downstream submergence.

The model tests conducted as part of the Boulder Canyon Project provided guidance on the crest shapes for a variety of conditions but also refined the shape of the ogee crests for the Hoover Dam spillways. Figure 3 shows the ogee crest shape that was constructed for the Hoover Dam spillways.

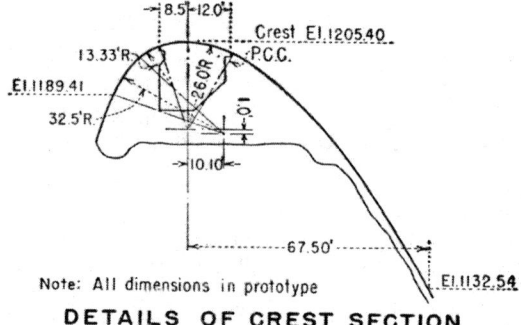

FIG. 3. Spillway Crest for Hoover Dam Spillways (Reclamation)

EVOLUTION OF HOOVER DAM SPILLWAY DESIGNS

All of the spillway designs for the Hoover Dam spillways incorporated the two outer 50-ft-diameter, horizontal diversion tunnels. A spillway inlet structure and inclined tunnel were used to connect to the horizontal tunnels. A number of hydraulic model studies were performed to evaluate and optimize the type and configuration of the inlet structure (Reclamation, 1938).

The initial design for the spillways at Hoover Dam included glory hole spillway inlet structures. The spillways had large diameter, uncontrolled circular crests which transition to a vertical shaft with a smaller diameter. The smaller throat diameter was maintained through the remainder of the vertical shaft, through an elbow section and through the horizontal downstream tunnel. Glory hole spillways have the advantage of providing relatively large spillway discharges at low heads due to the large crest length created by the circular crest. At higher heads, the spillway capacity is restricted, as the hydraulic control shifts from the spillway crest to the throat and releases are limited by the smaller cross-sectional area of the throat.

The preliminary designs of the glory hole spillways at Hoover Dam contemplated a discharge of 100,000 ft^3/s for each spillway with the reservoir water surface at elevation 1234.5. One of the conditions of the spillway designs was that for floods as large as the 1884 flood, the discharges from the dam should be safely carried within the capacity of the downstream river channel. This limited the combined spillway discharge to 62,500 ft^3/s with the reservoir water surface at elevation 1229.

The glory hole spillways designed for Hoover Dam had a crest diameter of 234 feet and a throat diameter of 50 feet. The design of the glory hole spillway was based on the assumption that water would flow over the ogee crest and toward the spillway shaft in radial directions, with equal depths of flow at all radial locations. This condition would produce smooth flow conditions in the crest section and down the vertical shaft. This condition was not achieved in the model, however. Flow concentrated along a radial line from the center to the point at the back side of the spillway formed by the ends of the two approach channels. This concentrated flow jumped across the top of the shaft from the canyon wall to the lake side, and then dropped down the shaft with an irregular distribution. For large flows, this concentrated flow tended to seal the top of the shaft which caused an undesirable suction and pulsation. The vacuum drew water down the shaft. When the vacuum in the shaft was broken by entrance of air from above, the water piled above the throat, the top of the shaft was sealed and then the process repeated itself. Because of concerns over the non-symmetrical flow in the glory hole spillways and the extremely high velocities in the spillways, it was decided to abandon the glory hole spillway alternative and pursue a side channel inlet structure for the spillways.

Several configurations of a side channel spillway were considered for Hoover Dam – a side channel spillway with a free crest, a side channel spillway with a Stoney Gate and free crest and a side channel spillway regulated by drum gates. The basic theory of the side channel spillway was developed by Julian Hinds (Hinds, 1926). The first side channel design for the spillways involved an uncontrolled overflow crest. The designed spillway had a capacity of 31,250 ft^3/s with the reservoir at elevation 1229 and 100,000 ft^3/s with the reservoir at elevation 1234.5. These capacities matched the capacity of the glory hole spillway. With the reservoir at the top of the dam, the unregulated side channel spillway had a capacity of about 140,000 ft^3/s. A 1:60 scale model of the Nevada spillway with an unregulated side channel spillway was built at the Colorado State College laboratory (Reclamation, 1938). Shortly after the model was completed, it was decided to increase the design capacity of each spillway from 100,000 ft^3/s to 200,000 ft^3/s, which required a new design for the spillway inlet structure.

The designs were then modified to include a combination of a side channel spillway with an uncontrolled crest in addition to a Stoney Gate at the upstream channel that allows for releases below the side channel spillway crest (the top of the Stoney Gate, when closed, was at the same elevation as the side channel spillway crest, at elevation 1205.4). The uncontrolled crest portion of the spillway was designed to release 50,000 ft^3/s at a head of 10.7 feet and the flow through the 60-ft-wide by 80-ft-high Stoney Gate was designed to pass 150,000 ft^3/s.

One of the effects that was explored in the model study of the side channel spillway/Stoney Gate combination spillway was the possibility of interference between the two flow regimes. There was concern that drawdown of the flow through the Stoney Gate would reduce the head and the discharge over the side channel ogee crest. There was a similar concern that releases over the ogee crest would cause drawdown that would reduce the head on the Stoney Gate and correspondingly reduce flow through the gate. With a combined discharge of 200,000 ft^3/s the interference of the two spillway releases reduced the total capacity by 5,000 ft^3/s or 2.5%. One of the issues that became apparent with further evaluation of flow through the Stoney Gate was the transition downstream of the Stoney Gate and a horizontal curve in the channel in this same location. The combined affect of these two geometries was to cause a sizable wave on the right side of the right channel wall, downstream of the Stoney Gate. A realignment of the spillway channel would have corrected the problem but the additional estimated construction costs were significant. It was concluded that a side channel spillway regulated by drum gates would eliminate this concern and the designs then focused on this refined concept for the inlet structure (Reclamation, 1938).

The original concept of a drum gate controlled side channel spillway consisted of an ogee crest regulated by four drum gates, each 100 ft long and separated by 10-ft-wide piers. The channel bottom had a slope of 0.232 and a width varying from 26.5 to 52.4 feet. Flow conditions in the model were not favorable due to the steep bottom slope. Because of the significant drop between the spillway crest and the channel invert, a high wave formed on the back wall of the channel which caused considerable impact on the end wall at the portal of the tunnel during higher discharges. Several changes were made to the side channel spillway to improve flow conditions. The changes included flattening the slope of the channel invert, an improved tunnel portal shape to address entrance conditions into the tunnel and the construction of a weir across the downstream end of the channel (Reclamation, 1938).

One of the issues that was explored in the model studies of the final configuration of the side channel spillways was the determination of air entrainment of flow over the side channel. Flow over the spillway crest falls into the water in the channel with considerable turbulence, leading to air entrainment of the flow. This was important in the channel design to ensure that the bulked flow would be contained in the channel. This issue was previously investigated in hydraulic model studies and through observations of the operation of the side channel spillway at Arrowrock Dam (Taylor, 1929), which indicated that the air entrainment could be significant. The air content in the model study was measured by two methods. The first method involved measuring the pressures in the bottom of the spillway channel and estimating the air content by determining the pressure for the same flow if there was no air entrainment and evaluating the difference between the two pressures. The second method involved sampling the flow and then separating the air from the water. The measurements and observations of the spillways indicated that air entrainment could be significant at smaller discharges due to the large drop from the spillway crest to the water in the spillway channel but the entrained air in the flow would be negligible for the maximum discharge. It was neglected in the design of the side channel, because the size of the channel was governed by the maximum discharge case, where the air entrainment was not significant. An allowance of 35 percent air content was made in the tunnel transition to insure free flow conditions at this location.

One of the concerns in the design of the spillways was the high velocities that the tunnels would be exposed to and the potential for erosion and/or cavitation in the spillway concrete. A testing program was initiated in which concrete blocks were subjected to continuous flow from a nozzle that created velocities that ranged from

100 to 175 feet per second. The flow was applied to the concrete surface in a number of ways (through a straight circular hole, following a semi-circular groove, over or entering an open joint, impinging against or over a surface offset, and impinging against a smooth surface) at a variety of angles (from 0 to 90 degrees). There was minimal evidence of wear or erosion of the concrete on any plane or jointed surface subjected to the water jet when the angle between the jet and the concrete surface was small. The critical location in the spillways will be in the lower tunnel sections, where maximum flow velocities were expected to occur. The spillway was designed with the intent of achieving streamlined flow at this critical location. The only tests that were thought to represent flow conditions at the critical location were the cases where water was applied to the concrete block surface at a low angle. For these cases, the worst damage consisted of a slight roughening of the concrete surface after ten days of continuous operation. The conclusion of the potential for erosive damage of the concrete subjected to high velocity flow was:

> "In the regions of the tunnels where the maximum velocities occur, virtually streamline flow is expected. Under this condition, it is concluded from the results of the jet tests that concrete of ordinary proportions and controlled quality will safely withstand any erosive action produced under conditions of maximum discharge in the spillways (Reclamation, 1938)."

PERFORMANCE AND MODIFICATIONS OF HOOVER DAM SPILLWAYS

The hydraulic model studies performed for the Hoover Dam spillways were valuable in confirming the spillway capacity and in evaluating design concerns such as the potential for air entrainment in the flow and the potential for erosion/cavitation damage in the tunnel lining under extremely high flow velocities. While the studies provided a solid basis for the spillway designs, some design vulnerabilities were not apparent until the spillways were constructed and the spillways were put into operation.

The tunnel spillways at Hoover Dam operated for the first time in the winter of 1941. The Arizona Tunnel Spillway operated for about 116 days at an average flow of about 13,000 ft^3/s with a maximum flow of about 38,000 ft^3/s. The Nevada Tunnel Spillway operated for only 20 hours at an average flow of about 8,000 ft^3/s and a maximum flow of about 14,000 ft^3/s. As a result of the 1941 spill, the Arizona Tunnel Spillway suffered major damage but the Nevada Tunnel Spillway only experienced minor damage. Figure 4 shows the damage to the Arizona Tunnel Spillway, which consisted of a hole, 50 feet long, 30 feet wide and 45 feet deep. The damage was apparently initiated by a misalignment in the tunnel invert (see Figure 5). The Arizona Tunnel Spillway was repaired by backfilling and compacting river rock and then covering it with a thick layer of concrete that restored the tunnel lining. Finishing of the concrete surface was accomplished by brushing and wet sandblasting, followed by stoning, and finally by grinding with a terrazzo machine. The result was a very smooth and durable surface.

Other significant modifications (based on a hydraulic model study) were completed on both spillways in 1947 (Walter, 1957). The purpose of these modifications was to address other factors that were suspected to have contributed to the erosion damage that was observed in the Arizona Tunnel Spillway. These modifications included extending the Nevada Tunnel Spillway exit 367 feet further downstream, constructing "simple circular upward deflection elbows" (flip buckets) at the ends of each tunnel spillway (exit portal), and dredging the exit channels immediately downstream of each spillway exit portal. One unintended outcome of these modifications would not be fully appreciated until 1984, when additional hydraulic model studies were undertaken to address cavitation damage that occurred during the 1983 flooding.

FIG. 4. Damage to Arizona Spillway Due to Spillway Releases in 1941 (Reclamation)

FIG. 5. Misalignment in Arizona Tunnel Lining (Reclamation)

In 1983, both spillways operated again for several hundred hours at discharges of about 10,000 ft³/s. The damage in the Arizona Tunnel Spillway was minor but the damage in the Nevada Tunnel Spillway was more substantial (see Figure 6). The damage in the Nevada Tunnel Spillway was believed to have been initiated by a small popout in the tunnel lining.

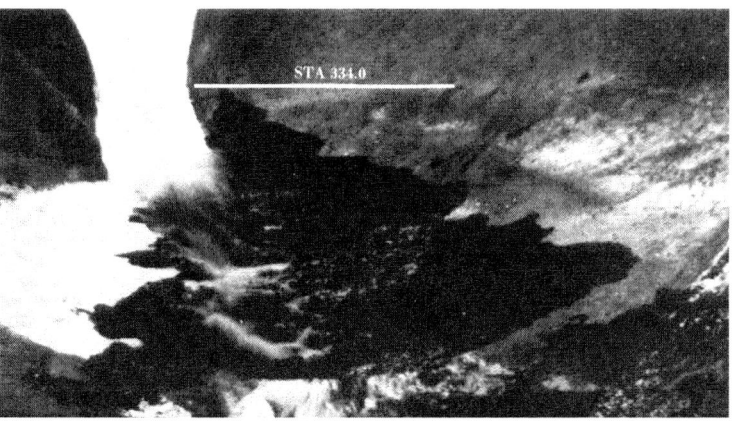

FIG. 6. Damage in Nevada Spillway Due to Releases in 1983 (Reclamation)

As previously noted, a hydraulic model study was used to better understand the hydraulic conditions that lead to the 1983 cavitation damage (Reclamation, 1984a and 1984b). The model studies led to the installation of aeration devices in both tunnel spillways. Since damage occurred in both tunnel spillways, even with exceptionally smooth surfaces, aeration ramps and a downstream offset were installed in both tunnel spillways (see Figure 7). The ramps are 3 feet high on the invert and transition to zero height at 35 degrees on each side of the tunnel crown. The offsets are concentric with the original tunnel diameter. The 5-ft-offset transitions back to the original tunnel diameter in 25 feet. The ramps are located about 256 feet below the inlet to the tunnels. The aeration ramps and offsets were completed in June 1987.

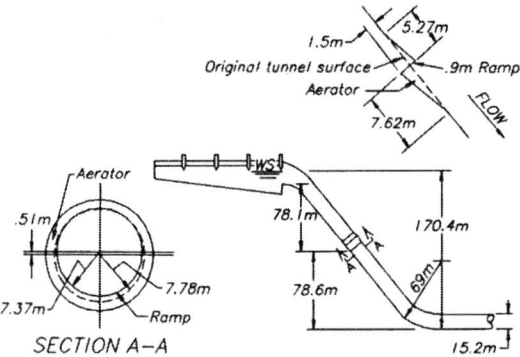

FIG. 7. Aeration Devices for Hoover Dam Spillways (Reclamation: note – metric units used)

The 1984 hydraulic model study also determined that the original design discharge capacity (200,000 ft^3/s) had been affected by the 1947 flip bucket modifications. Results indicated that unstable flow conditions (the hydraulic jump generated very large instantaneous pressure surges in the range of 300 to 400 feet) would occur for discharges between 135,000 ft^3/s and 185,000 ft^3/s, and would shift into pressure flow for discharges greater than 185,000 ft^3/s. A key conclusion of the 1984 hydraulic model study was that the maximum discharge capacity of each spillway had been reduced from 203,000 ft^3/s to 191,000 ft^3/s at reservoir water surface elevation 1232. Recommended options from the study included:

1. Remove the existing flip buckets and replace them further downstream outside of the tunnels to allow free flow for full range of discharges.

2. Remove a portion of each tunnel crown above the flip bucket ramps.

3. Remove a portion of the Nevada tunnel crown (since it has been extended downstream of the tunnel section, such a modification may be less costly),

allowing free flow passage of 200,000 ft^3/s through the Nevada spillway and limiting Arizona spillway to 135,000 ft^3/s.

4. If no modification is undertaken, limit maximum operating discharge in either spillway to 135,000 ft^3/s (until a complete analysis can be done to determine if structural integrity would be compromised, or not).

The modifications to date have been limited to installing the aeration ramps and some repair of the existing concrete flow surfaces. The recommendations previously noted have not been implemented for either spillway flip bucket. Given this, the safe maximum discharge from each spillway is 135,000 ft^3/s or less.

Currently, there are flow surfaces (such as just downstream of the spillway elbows, i.e., bends) that have exposed aggregate (up to 1 inch of concrete paste has been removed). These areas present a "uniformly rough" flow surface but are not considered to create issues associated with cavitation.

CURRENT APPROACHES FOR EVALUATING SPILLWAYS

During the time of the Hoover Dam designs and for decades after that Reclamation was in the mode of designing and constructing numerous dams and spillways. The current emphasis within Reclamation and many dam engineering organizations is on effectively managing their existing inventory of dams. This includes evaluations of dams and spillways to ensure that they will continue to operate safely and not endanger the public. Reclamation's current approach to evaluating dams and related structures for dam safety is to rely on a risk-informed approach. Potential failure modes are identified and the risks are estimated for viable failure modes. The risk estimates are used to help decide if actions should be taken to reduce risk. While the studies performed for the spillway designs at Hoover Dam were invaluable in the design of the Hoover structures as well as at numerous other dams, many of the concerns at the time of the Hoover Dam spillways are still relevant when evaluating spillways today. The remaining discussion will highlight some of the key spillway risk issues that are commonly evaluated and present case histories that illustrate the concerns.

SPILLWAY CAPACITY/DAM OVERTOPPING ISSUES

One of the biggest issues when evaluating existing dams and spillways, especially those that were constructed twenty years ago or more is the spillway design capacity related to current floods that are possible at the site. Spillway capacity was a key design parameter for the Hoover Dam spillways that ultimately impacted the type of spillway inlet structure selected. The required spillway capacity doubled as the spillways were being designed. While some overtopping of Hoover Dam could likely be tolerated, an embankment dam would be much more vulnerable to overtopping (Wahl, 1998). Whether overtopping of a given dam is a risk issue warranting further action will typically be related to the frequency of the flood that initiates overtopping and the downstream consequences but can also be related to operational issues at the dam. Several case histories are presented that relate to dam overtopping.

South Fork Dam, Pennsylvania

South Fork Dam was constructed upstream of Johnstown, Pennsylvania, forming a lake for a fishing club. The dam, as originally constructed between 1840 and 1853, was 72 feet high and over 900 feet long. It was constructed of rolled earth and puddled material, and contained a low-level stone culvert through the dam. In 1862 the stone conduit collapsed and the dam failed through internal erosion. However, there were no significant consequences as the reservoir was low at the time. The low level outlet conduit was plugged and filled in during dam reconstruction. The spillway was 99 feet wide and crossed by a bridge with supports spaced at 6.5 feet. Iron screens were placed across the spillway to prevent fish from escaping. The crest of the dam was lowered 2 feet to widen the roadway on top of the dam. No camber was left in the dam, and the center portion of the dam may have been lower than at the abutments. In May of 1889 a large rainstorm advanced from the west. The large inflows sent debris to the spillway area where it became lodged in the fish screens, plugging the spillway. Overtopping erosion failure of the dam ensued. More than 2,200 people lost their lives (Frank, 1988). This was the second largest catastrophe in the U.S. from a single event (since the 1900 Galveston hurricane flood resulted in more than 6000 fatalities) until the terrorist attacks of September 11, 2001.

Taum Sauk, Missouri

The 2005 failure of Taum Sauk Dam in Missouri brought renewed attention to operational failure modes. Taum Sauk Dam did not include a spillway, and the failure was not related to flood inflows. The dam formed the upper reservoir of a pumped-storage project. The dam was a concrete faced earthfill structure, forming a complete "ring" on the top of a large hill (Figure 8). Water was routinely stored on a 10-foot-high parapet wall above the crest of the dam. After a membrane liner was installed in 2004, the instrumentation for measuring the water level in the reservoir was tied to cables in the reservoir because the warranty for the liner would have been void had holes been drilled through it to secure the instruments. The cables, which were installed near the power intake, loosened due to the hydraulic currents. In addition, settlement of the embankment was not taken into account in re-setting the reservoir level sensors. The reservoir was overfilled due to pumping, overtopped, and failed (see Figure 9). The instrumentation installed to detect and prevent such an occurrence did not provide accurate readings or function as originally intended. A house with five people inside was destroyed in the downstream flooding. Miraculously, everyone in the house was thrown upstream when the water hit, and no one died. Additional information can be found in FERC (2006). While the instrumentation at Taum Sauk Dam was the key contributor to the overtopping at that dam, other operational issues could lead to dam overtopping and failure. These issues include mechanical failure of gates during a flood, loss of power during a flood preventing gate operation and failure to operate the gates according to Standing Operating Procedures during a flood.

FIG. 8. Taum Sauk Dam and Reservoir (USGS)

FIG. 9. Overtopping Failure of Taum Sauk Dam (USGS)

Debris Plugging of Spillways

One of the issues that has gained more attention in recent years is the potential for spillway debris to plug spillway structures, reducing their capacity and potentially resulting in dam overtopping. This was the case at Kerkhoff Dam, a dam owned by the Pacific Gas and Electric Company. The spillway became plugged due to a massive amount of debris, consisting of fallen trees from the drainage basin. The trees became knitted together in a large mass and were pushed against the dam and the spillway and dam by incoming flows. The dam overtopped, but the rock abutments of the concrete dam did not erode and the dam did not fail. Several publications address the potential for debris in reservoirs (Wallerstein, 1997 and Federal Highway Administration, 2005).

SPILLWAY CONVEYANCE CAPACITY

Potential failure modes at a dam may be related to exceeding the design capacity of the spillway conveyance structure. For a tunnel spillway, exceeding the design capacity could result in pressurization of the tunnel, failure of the lining and failure of the abutment which ultimately leads to dam failure. Due to the introduction of flip buckets at the downstream end of the Hoover Dam spillways in 1947, the potential for pressurizing the downstream portion of the tunnels was created and discharges are currently restricted to 135,000 ft^3/s through each spillway. For a concrete chute spillway, exceeding the design capacity could result in overtopping of chute walls which could initiate erosion of the backfill (or adjacent embankment in some cases), undermining of the structure and headcutting leading to a breach of the reservoir. Additional factors that could contribute to chute wall overtopping include air entrainment in the flow (Wilhelms, 2005; Falvey, 2007) and cross waves within the chute (Falvey, 1980). Air entrainment of flow was one of the issues investigated in the model studies for Hoover Dam. The breach of the reservoir at El Guapo Dam in 1999 was caused by inadequate spillway chute capacity.

El Guapo Dam Spillway: December 1999

El Guapo Dam was located on the Rio Guapo, 3 miles south of the city of El Guapo, in the state of Miranda, Venezuela. The reservoir formed by the dam provided potable water for the local area, flood control, and irrigation water. The reservoir volume was 33,000 acre-feet. The dam was constructed from 1975 to 1980. The original spillway at El Guapo Dam consisted of an uncontrolled ogee crest, located on the left abutment of the dam, a concrete chute and a concrete hydraulic jump stilling basin. The spillway had a width of 40 feet, a length of 925 feet and a design discharge capacity of 3,600 ft^3/s. Initial hydrologic studies were based on a similar basin but not the Rio Guapo basin. During construction of the spillway, the chute walls were overtopped, which triggered a new flood study. A tunnel spillway was then constructed through the dam's left abutment, 820 feet from original spillway.

The spillways at El Guapo Dam were severely tested during flooding in December 1999. On December 14, 1999, the reservoir was 3 feet above the normal pool and 17 feet below the dam crest. The radial gate on the tunnel spillway was fully open, both spillways were operating normally. Early on the morning of December 15th the reservoir rose quickly and was 2.5 feet below the dam crest. Early the next morning the reservoir was 8 inches below the dam crest, the spillway chute walls just below the spillway crest began to overtop, and erosion of the adjacent fill initiated. By 4:30 a.m. on December 16th, the cities below the dam were evacuated. At 9:00 a.m. the dam was inspected by helicopter and the reservoir level had subsided (2.5 feet below crest); people believed that flood had crested and the crisis was over. At 4:00 p.m. on December 16th, the reservoir rose again quickly; the bridge over spillway collapsed; erosion of spillway backfill accelerated and the reinforced concrete chute, basin and crest structure failed; but the concrete lined approach channel remained intact and controlled flows through the spillway. At 5:00 p.m. the approach channel failed and the reservoir was breached through the spillway area, resulting in a loss of most of the reservoir. El Guapo Dam never overtopped. Overtopping of the spillway chute walls initiated erosion of backfill behind chute walls and undermining and failure of spillway chute. Headcutting progressed upstream and led to reservoir breach. The spillway foundation consisted of decomposed rock, which was erodible (Villar, 2002).

FAILURE OF SPILLWAY GATES IN NON-FLOOD CONDITIONS

The failure of spillway gates with a full or nearly full reservoir can threaten downstream populations due to the typically large size of spillway gates. Gates can fail under normal conditions if the gate members or components become overstressed or fail due to corrosion or fatigue. During a large earthquake, failure of gates can occur if the additional loads exceed the structural capacity of the gates. Gates that are in the raised position when closed are more vulnerable to failure during non-flood conditions, because there are more mechanisms that can cause the gates to lower and initiate an uncontrolled release. Drum gates have some vulnerabilities as a result of this. The spillways at Hoover Dam are regulated by drum gates. The initial designs of the Hoover Dam spillways involved uncontrolled spillway crests without gates. Cresta Dam had a drum gate that lowered under normal operating conditions.

Cresta Dam, California

Cresta Dam is located on the North Fork of the Feather River, and is a key feature in Pacific Gas and Electric Company's Feather River hydroelectric system. The dam forms the forebay for Cresta Powerhouse and was constructed in 1949. There are two 124-ft-long by 28-ft- high drum gates at the top of Cresta Dam. Drum gates at Cresta Dam are raised to maintain the reservoir level for electric power generation at the Cresta Powerhouse under normal operations and are operated to regulate spills during high river flow conditions. Regulation of the drum gates is accomplished by adjusting the water level in a chamber in which the gates float. On July 5, 1997, the left drum gate at Cresta Dam began to drop uncontrollably. The gate took about 20 to

30 minutes to completely lower. The downstream water level rose from 1.59 feet to 15 feet in approximately 40 minutes. The maximum recorded downstream discharge was 15,140 ft^3/s. No injuries or fatalities occurred as a result of the gate failure.

The left drum gate at Cresta Dam dropped primarily due to a combination of failure of the drum gate drain line and leakage into the drum gate. Failure of the drum gate drain line was likely caused by crimping of the drain line at the downstream stop seal. Water accumulated in the drum gate due to leakage through the check valves as a result of failure of the drum gate drain lines and normal leakage into the drum gate through connections, hatches and the gate skin (Pacific Gas & Electric Company, 1997).

FAILURE OF SPILLWAY LINING DUE TO HIGH VELOCITY FLOW

Many spillway structures are subjected to high velocity flow. High velocity flow combined with irregularities on concrete flow surfaces can contribute to failure of the concrete lining and exposure of the spillway foundation. Under certain conditions, this could lead to erosion of the spillway foundation, headcutting and a breach back through the reservoir. Mechanisms that could cause lining failure include cavitation and stagnation pressures that can fail a floor slab or initiate erosion in the chute foundation and undermine the chute slab. The potential for erosion/cavitation in the concrete tunnel lining at Hoover Dam was recognized during the design of the spillway and tests were conducted to help evaluate the potential for lining damage. While major damage of the concrete lining occurred in the Hoover Dam Tunnel Spillways due to cavitation damage, the spillways at Glen Canyon Dam experienced even worse damage.

Glen Canyon Dam Spillway: June 1983

Glen Canyon Dam is located on the Colorado River in northern Arizona, about 15 river miles upstream of Lees Ferry and 12 river miles downstream from the Arizona-Utah state line. The dam, completed in 1964, is a constant radius, thick-arch concrete structure, with a structural height of 710 feet. Spillways are located at each abutment and each consists of a gated intake structure, regulated by two 40-ft by 52.5-ft radial gates, a 41-ft diameter concrete lined tunnel through the soft sandstone abutments and a deflector bucket at the downstream end. Each tunnel spillway is inclined at 55 degrees, with a vertical bend and a 1,000-ft-long horizontal section. The combined discharge capacity of the spillways is 276,000 ft^3/s, at a reservoir water surface 63 feet above the spillway crest elevation. The spillways experienced significant cavitation damage during operation in June and July, 1983 during flooding on the Colorado River system when the reservoir filled completely for the first time and releases were required. The cavitation damage was initiated by offsets formed by calcite deposits on the tunnel invert at the upstream end of the elbow. Both spillways were operated at discharges up to about 30,000 ft^3/s. Cavitation indices of the flow in the area where damage initiated in the left spillway ranged from about 0.13 to 0.14. The cavitation indices of the deposits along the tunnel (indices at which cavitation was likely to occur) ranged from 0.64 to 0.73. Although flashboards were installed on top of the spillway gates to avoid releases to the extent possible, releases were still made through both spillways. The worst damage occurred in the left tunnel spillway – a hole 35 ft deep, 134 ft long and 50 ft wide was eroded at the elbow into the soft sandstone (see Figure 10, Burgi, 1987). Extensive concrete repair work and installation of air slots were required to bring the spillways back into service and reduce the potential for future damage.

Although Reclamation had investigated cavitation damage and implemented repairs since 1941, the understanding and methodology to adequately mitigate cavitation damage was not developed until after significant cavitation damage occurred at the Glen Canyon Dam and Hoover Dam Tunnel Spillways. Prior to this, standard practice was to specify very stringent concrete "finishes" (at the time, the term "finish" defined both surface texturing and surface offsets) for flow surfaces associated with discharge velocities greater than 75 ft/s. The concrete finishes for these flow surfaces were very difficult to achieve and maintain in the field. A more effective method had actually been employed in 1961 and 1969 with the installation of aerators to address the cavitation damage which occurred at Grand Coulee Dam outlet works tubes and spillway chute, and the Yellowtail Dam spillway. Aerators have since been installed in the tunnel spillways for Glen Canyon, Flaming Gorge, Hoover, and Yellowtail Dams. The aerator ramps were constructed in the Hoover Dam Tunnel Spillways in 1986 and 1987.

Resulting from an eight year effort, EM 42, Cavitation in Chutes and Spillways was published in 1990 (Falvey, 1990), providing common-sense guidance on how to identify and mitigate cavitation potential. Two important developments include: (1) generalized guidelines and tools were developed to assess the potential degree of cavitation, and to develop preliminary aeration designs, and (2) concrete finishes (surface textures) were de-coupled from concrete tolerances (surface offsets and irregularities), recommended surface tolerances were revised to be more achievable in the field, and these tolerances were linked to cavitation indices. These indices are a function of the fluid velocity and pressure, and empirically give an indication of the potential for cavitation.

FIG. 10. Damage Initiated by Cavitation in Glen Canyon Dam Spillway (Reclamation)

CONCLUSIONS

The studies conducted as part of the Boulder Canyon Project were critical to developing the designs of the Hoover Dam spillways. The studies also were expanded to evaluate a broader range of conditions and have provided the basis for the design of many spillways since the completion of Hoover Dam. Many of the issues that were evaluated during the spillway model studies at Hoover Dam (spillway capacity, aeration of spillway flows and the potential for erosion/cavitation on concrete flow surfaces subjected to high velocity flow) are issues that are considered in the evaluation of present day spillways. Some issues at Hoover Dam were not fully appreciated until the spillways operated and further hydraulic model studies were conducted. These issues included the potential for cavitation damage in the tunnel spillways and the tunnel capacity issues related to the introduction of flip buckets at the downstream end of the tunnels.

Current methods for evaluating existing spillways and dams include a risk informed approach. This involves evaluating the designs for potential vulnerabilities, identifying potential failure modes unique to the dam and spillway and estimating risks. Case histories of failures and incidents related to dams and spillways provide insights into the potential failure modes and are useful when estimating risks. Risk estimates are used as key information in dam safety decision making and often result in actions that are taken to minimize the chance of future failures and incidents.

REFERENCES

Bazin, M. (1888). "Recent Experiments on the Flow of Water Over Weirs, Annals des Ponts et Chaussees," (translated by Arthur Marichal and John C. Trautwine, Jr., and published in the *Proceedings of the Engineer's Club of Philadelphia*, Vol. VII, No. 5, 1890, p. 259 and Vol. IX, No.3, 1892, p. 231).

Bradley, J.N. (1952). "Discharge Coefficients for Irregular Overfall Spillways," *Engineering Monograph No. 9*, Bureau of Reclamation, Denver, Colorado.

Bureau of Reclamation (1938). "Model Studies of Spillways," *Final Report, Part VI - Hydraulic Investigations, Bulletin 1*, Boulder Canyon Project, Denver, Colorado.

Bureau of Reclamation (1948). "Studies of Crests for Overfall Dams," *Final Report, Part VI - Hydraulic Investigations, Bulletin 3*, Boulder Canyon Project, Denver, Colorado.

Bureau of Reclamation (1984a). "Hydraulic Model Study Results Hoover Dam Tunnel Spillways," *Report PAP-465*, Denver, Colorado.

Bureau of Reclamation (1984b). "Hoover Dam Spillway Tunnel Aeration Device Design Summary," *Report PAP-473*, Denver, Colorado.

Bureau of Reclamation (1987). *Design of Small Dams*, 3rd Edition, Denver, Colorado.

Burgi, Philip H. and Eckley, Melissa S. (1987). "Repairs at Glen Canyon Dam," *Concrete International Design and Construction*, Vol. 9, No. 3, pp 24-31.

Falvey, Henry T. (1980). "Air-Water Flow in Hydraulic Structures," *Engineering Monograph No. 41*, Bureau of Reclamation.

Falvey, Hank T. (1990). "Cavitation in Chutes and Spillways," *Engineering Monograph No. 42*, Bureau of Reclamation.

Falvey, Henry T. (2007). Discussion on "Gas transfer, cavitation, and bulking in self-aerated spillway flow," *Journal of Hydraulic Research*, Vol. 45, No. 6, pp. 859-860.

Federal Highway Administration (2005). "Debris Control Structures, Evaluation and Countermeasures, Third Edition," *Publication No. FHWA-IF-04-016, Hydraulic Engineering Circular No. 9*.

FERC (Federal Energy Regulatory Commission) (2006). "Report on Findings on the Overtopping and Embankment Breach of the Upper Dam – Taum Sauk Pumped Storage Project," *FERC No. 2277*.

Frank, W.S. (1988). "A New Look at the Historic Johnstown Flood of 1889," *Civil Engineering Magazine,* May 1988, pp. 63-66.

Hinds, Julian (1926). "Side Channel Spillways," *Transactions of the American Society of Civil Engineers*, Vol. 89, p. 881.

Pacific Gas & Electric Company (1997). "Investigation of Uncontrolled Operation, Highway Side Drum Gate, Cresta Dam," Hydro Generation Department, Analysis Group.

Taylor, P. I. (1929). "Arrowrock Dam," *Dams and Control Works*, Bureau of Reclamation, pp 35-39.

Villar, Luis Miguel Suarez (2002). "El Guapo," *Incidentes en las Presas de Venezuela, Problemas, Soluciones y Lecciones*, pp 27.1-27.39.

Wahl, Tony L. (1998). "Prediction of Embankment Breach Parameters, A Literature Review and Needs Assessment," *DSO-98-004, Dam Safety Research Report*, Bureau of Reclamation, Denver, Colorado.

Wallerstein, N., Thorne, C.R. and Abt, S.R. (1997). "Debris Control at Hydraulic Structures in Selected Areas of the United States and Europe," *Contract Report CHL-97-4,* Prepared for U.S. Army Research Development and Standardization Group – UK, London.

Walter, Donald Scott (1957). Thesis, "Rehabilitation and Modification of Spillway and Outlet Tunnels and River Channel Improvements at Hoover Dam," Submitted to Faculty of the Graduate School of the University of Colorado, Department of Civil Engineering.

Wilhelms, Steven C., and Gulliver, John S. (2005). "Gas transfer, cavitation, and bulking in self-aerated spillway flow," *Journal of Hydraulic Research*, Vol. 43, No. 5, pp. 532-539.

Seismic Evaluation of Hoover Dam Powerplant

Adam Toothman[1], P.E., David Gold[2], P.E., Tim Brown[3], P.E., and Mary Beth Schuetz[4]

[1] Structural Engineer, U.S. Bureau of Reclamation, Structural Analysis Group, Denver Federal Center, Building 67, (86-68110), Denver, CO 80225; atoothman@usbr.gov
[2] Structural Engineer, U.S. Bureau of Reclamation, Structural Analysis Group, Denver Federal Center, Building 67, (86-68110), Denver, CO 80225; dgold@usbr.gov
[3] Building Seismic Safety Program Manager, Structural Engineer, U.S. Bureau of Reclamation, Structural Analysis Group, Denver Federal Center, Building 67, (86-68110), Denver, CO 80225; tbrown@usbr.gov
[4] Civil Engineer, U.S. Bureau of Reclamation, Structural Analysis Group, Denver Federal Center, Building 67, (86-68110), Denver, CO 80225; mschuetz@usbr.gov

ABSTRACT: The Hoover Dam Powerplant was designed and built in the 1930s, a time without computers, without modern earthquake engineering, and without established building codes. For the first time since the original design, the Bureau of Reclamation's Building Seismic Safety Program (Program) has conducted a structural seismic evaluation of the both the Nevada and Arizona Powerplant superstructures and the Central Portion superstructure. These structures consist of multi-story reinforced concrete and structural steel systems built integrally with the downstream face of the dam and canyon walls.

This paper illustrates the Program's usage of ASCE 31 and the finite element modeling techniques employed during the Powerplant evaluation.

The Bureau of Reclamation owns 67 powerplants and over 200 pumping plants that are located throughout the western United States. The Program is tasked with assessing the structural and nonstructural seismic risks in these plants and providing recommendations and/or considerations for mitigating identified risks to the decision-makers for the facilities.

INTRODUCTION

This paper presents the Bureau of Reclamation's Building Seismic Safety Program (Program) seismic evaluation of the Hoover Dam Powerplant (FIG. 1). It describes the Program's overall approach to the evaluation, details regarding the finite element modeling techniques used to facilitate the analysis, and overall building performance. The analysis was completed using a Tier 3 evaluation, as defined in ASCE 31-03, *Seismic Evaluation of Existing Buildings*. The earthquake magnitude used was based on a recurrence period of 1,000 yrs, and the expected performance level was Immediate Occupancy.

FIG. 1. Hoover Dam Powerplant (Original Photo).

Hoover Dam Powerplant, located directly downstream of the dam, is located on the Colorado River approximately 36 miles southeast of Las Vegas, Nevada. With a structural height of 726.4 feet and hydraulic height of 592 feet, Hoover is still the highest concrete dam in the United States. There are two powerplants at the Hoover facility, one on each abutment. Because the dam is located on the Arizona-Nevada border, the plants are named after the state in which they reside. Together, there are 17 turbine generators and two Pelton Waterwheel station service units with a total plant capacity of 2,079 MW.

The scope of the seismic analysis of the powerplants was limited to the primary buildings that are essential for the continued operation of the facility. The primary buildings included the Arizona and Nevada Powerplants, as well as the Central Portion. The Central Portion is a multi-story reinforced concrete and structural steel system built integrally with the downstream face of the dam.

The powerplants consist of a three-story, mass-concrete substructure supporting an 85-foot tall superstructure. The superstructures are identical for both plants and consist of reinforced concrete cantilever column bays spaced at 20 feet on center. Reinforced concrete walls span between the columns in the longitudinal direction. The columns support a massive structural steel truss roof system. The roof trusses span approximately 65 feet between the columns in the transverse direction (FIG. 2).

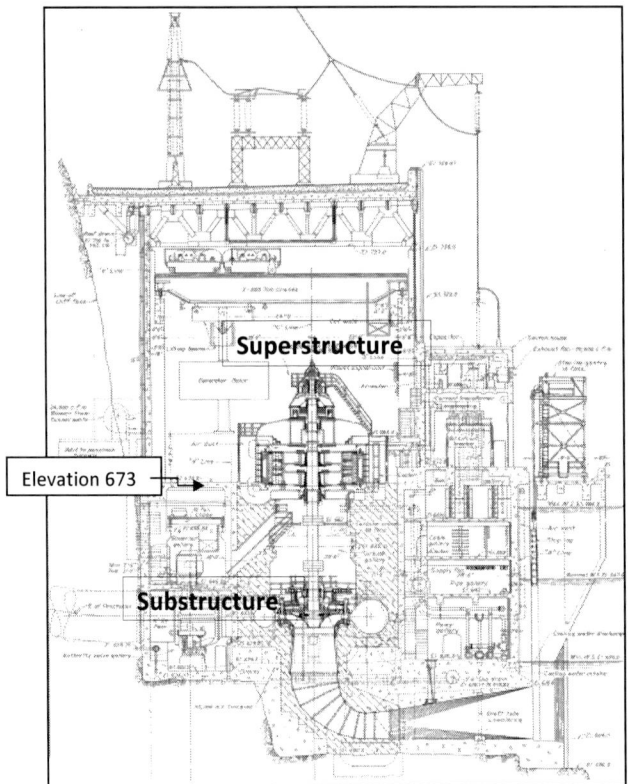

FIG. 2. Typical cross section of Hoover Dam Powerplant (Figure courtesy of the Bureau of Reclamation).

DECOUPLED ANALYSIS APPROACH

For the evaluation of the powerplants, a single three-dimensional finite element model of the entire powerplant substructure and superstructure system was not practical. The structures are complex and would require extensive modeling time to process data. The evaluation of the Central Portion has the same issue; a single model for the entire building would also be computationally expensive.

Balancing the need for accurate analysis and efficient minimization of modeling time, a decoupled seismic analysis approach was conducted. Separation of the buildings at logical locations resulted in smaller and less complex models. First, the powerplants were separated at the expansion joints spaced at every third bay, or approximately 60 feet on center. Next, the models' bases were chosen at a location where the stiffness and mass changed significantly. Each powerplant was constructed in a similar manner with very stiff and massive substructure supporting a relatively flexible superstructure. This point of separation between the substructure and superstructure was the same for all models; elevation 673.0. FIG. 2 illustrates the massive substructure below the flexible superstructure. Lastly, four superstructure models were identified to adequately model the powerplant's multiple structural conditions: two three-dimensional superstructure models captured the Central Portion's significant structural features; one two-dimensional model analyzed the typical bay, and one three-dimensional model analyzed the end bay of the powerplant.

Division of the Central Portion into two separate three-dimensional models required extensive review of the building's design and connections. Some of the expansion joints in the roof of the Central Portion did not extend the full depth of the building. The Central Portion construction also included different model building types separated by full height expansion joints with shared, common foundation elements. The expansion joints were initially identified as potentially good separation locations. However, the foundation elements are not entirely rigid and separating those models might not fully capture the load transfer forces between the two systems. Final model geometry of the Central Portion included a model of the Center Section between building 6 ½-Line and 12 ½-Line and a model on the Nevada side between 1-Line and 6 ½-Line (FIG. 3).

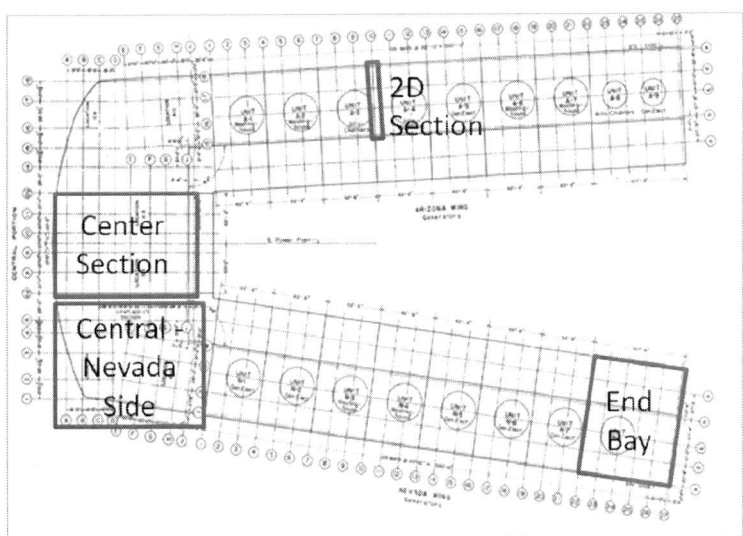

FIG. 3. Plan view of powerplant (Figure courtesy of the Bureau of Reclamation).

SEISMIC DEMAND AND PERFORMANCE LEVELS

The seismic demand considered at the Hoover Dam Powerplant is a function of various parameters and has evolved over the years as engineers, scientists and the like have come to a deeper understanding of the behavior of structures subjected to seismic ground motions. Beginning in the early 1970s, the Bureau of Reclamation (Reclamation) began using the Maximum Credible Earthquake (MCE) as the demand for its critical structures as a replacement for a standard 0.1g acceleration, which had been considered as a design basis loading. This MCE earthquake would produce the most severe ground motion capable of being produced at the site under the presently known tectonic framework. The maximum credible earthquake is considered a rational and believable event that is in accord with all known geological and seismological facts, but provides little regard to its probability of occurrence. In 1976, Reclamation adopted the use of a Design Basis Earthquake (DBE) and an Operating Basis Earthquake (OBE) as design loadings for certain types of its structures, acknowledging the relationship between service life of the plant and the probabilistic likelihood of the seismic event. This relationship is carried through in this evaluation and helps define the level of earthquake hazard to be considered. Recognizing the importance of the Hoover Powerplant to generate power and its impact on the public safety as well as the economic implications, it was deemed that a DBE may be too small. In general a DBE, as specified within the seismic design provisions of the International Building Code (IBC), is based on an earthquake hazard level equivalent to 2/3 times the MCE. In the Western United States, the DBE is typically, but not always, equivalent to an earthquake with a 10% probability of being exceeded within 50 years (10%/50) or a recurrence period of approximately 500 years. Many of Reclamation's powerplants including Hoover Powerplant have exceeded or will soon be approaching the 50-year benchmark, and are expected to have a much longer service life. A service life on the order of 100 years is common for such critical public works projects. Given the longer service life and the essential function of the Hoover Powerplant, a seismic hazard with 10% exceedance probability in 100 years (10%/100) or a return period of approximately 1,000 years was used during the evaluation.

Response Spectrum

Once the seismic event has been defined, a general mapped response spectrum can be determined based on the provisions of ASCE 31 and the MCE as defined by the United States Geological Survey (USGS). The response spectrum procedure considers a modal analysis used to calculate the linear response of complex, multi-degree-of-freedom structures. Through superposition of the responses of the individual natural modes of vibration of the Hoover Powerplant, each mode responding with its own particular pattern of deformation (mode shape), with its own period, and with its own damping, seismic demand is determined. Although the seismic demand as determined by the response spectrum procedure is not as simple as saying the horizontal acceleration is 0.1g, as was done in the past, it is clear from the graph below (FIG. 4) that the overall demand can be 2-4 times larger than the original design of the Hoover Powerplant assuming the identical effective seismic weight.

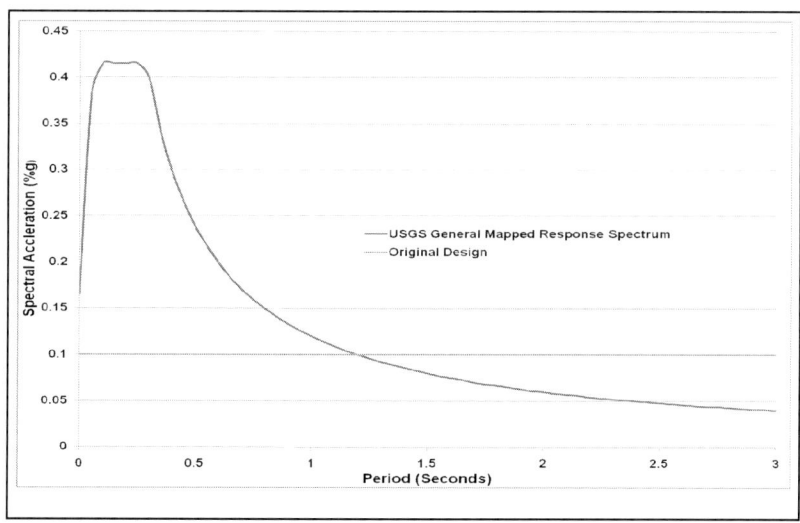

FIG. 4. Hoover Dam Powerplant general mapped response spectrum (10%/100years) (Original Figure).

Powerplant Loads

The effective seismic weight used to evaluate the Hoover Powerplant superstructure was predominately a function of the self-weight of the structure. This includes a large superimposed roof dead load that is made up of several layers of sand and gravel fill with a 2-inch concrete protective slab as shown in FIG. 5. Most of the Powerplant's operational equipment, such as pumps, turbines, etc., were not considered due to being located in the substructure. However, the Central Portion of the Powerplant does house numerous controls and offices that add to the seismic effective mass. These additional loads were considered in the models as the live load specified by the original structural design data drawings and were verified by inspection. Additionally, the self-weight of the two large 300-Ton overhead cranes that run the length of the Powerplant wings were considered.

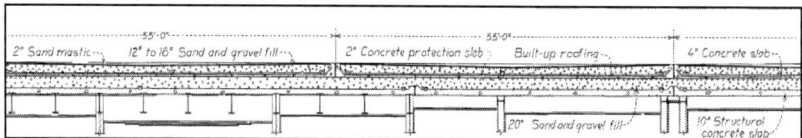

FIG. 5. Hoover Dam Powerplant Roof Detail (Figure courtesy of the Bureau of Reclamation).

Performance Level

The performance of structures can be broken down into two distinct categories as per ASCE 31; Immediate Occupancy and Life Safety. The purpose of the performance level is to define the level of damage that one is willing to accept given the previously defined seismic demand. The Hoover Powerplant is considered to maintain an Immediate Occupancy performance level due to the critical nature of the functionality of the Powerplant. This translates to a post-earthquake damage state where the Powerplant remains safe to occupy, essentially retaining the pre-earthquake design strength and stiffness. Minor damage to structural and nonstructural components during the seismic event is acceptable, such that partial or total collapse of the Powerplant does not occur and nonstructural collapse is neither life-threatening nor causes loss of safe egress for occupants.

Unlike new design, the evaluation procedure as defined by ASCE 31 considers the Equivelent Displacement Approach. This approach considers a demand applied to the structure that results in 'actual' displacements. In order to obtain 'actual' force demands, a modification factor, m-factor, is used for each independent ductile component instead of applying the global ductility related R-factor to the applied loads, as is the case with new design. This component demand modifier accounts for the expected ductility of specific elements of the lateral force resisting system by taking the component demand and dividing it by the m-factor. It is through these m-factors that one ensures that the level of damage, as defined through the ascribed performance level, is satisfied. It should be noted that the performance level is directly related to ductile or deformation controlled behavior. Components, such as connections or braced frames, with little ductile behavior, are required to consider the full demand.

MODELING TECHNIQUES

Frame and Shell Elements

All beams and columns were modeled as frame elements, and all walls and floor and roof slabs were modeled as shell elements. Beams framing into columns and beams framing into beams were modeled as pinned connections because the original drawings typically showed shear tab connections only.

Concrete columns were modeled as fixed to the substructure because the reinforcing steel is continuous from the columns into the superstructure. Concrete shear walls were modeled as pinned to the substructure based on its small relative stiffness in the weak axis.

Material Properties

Material properties used in the evaluation were based on the original design data sheets and default lower-bound material strengths from ASCE 31 given the era of construction. The compressive strength used to model concrete members was 2,500 lb/in^2 and the unit weight was 155 lb/ft^3. The yield strength of the reinforcing steel is 40 $kips/in^2$. The lower-bound yield strength for structural steel members was taken as 33 $kips/in^2$ and the tensile strength was 60 $kips/in^2$. Steel bolts were also assumed to be 33 $kips/in^2$.

Boundary Conditions

Modeling the boundary conditions for the connection between the powerplant and the dam face was a crucial component for the Central Portion. The type of connection was determined from the original design drawings (FIG. 6). Historical photographs during construction were found to verify that the conditions shown on the drawings matched the actual construction (FIG. 7 and 8).

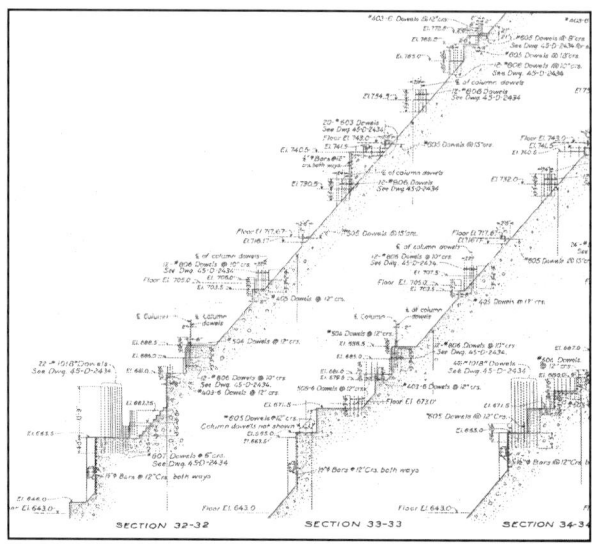

FIG. 6. Design drawings of protruding rebar for floor slab connection (Figure courtesy of the Bureau of Reclamation).

FIG. 7. Construction photo – Floor slab dowels in downstream face of Hoover Dam (Photograph courtesy of the Bureau of Reclamation).

FIG. 8. Construction photo illustrating shear wall dowels in downstream face of Hoover Dam (Photograph courtesy of the Bureau of Reclamation).

From the original dam drawings, it was shown that floor slabs and beams are bearing on a ledge formed into the dam. Vertical and/or horizontal reinforcing steel protrudes from the dam face, and as a result, the beams and slabs in the Central Portion were modeled as pinned connections when attached to the dam face. Due to having one end of the structure fixed (dam face) and the other free to move (downstream end), the structure racked when subjected to an earthquake load resulting in large tensile and shear forces at the dam face connections.

Steel column base plates were modeled as moment restraints in the strong axis (upstream/downstream direction) and pinned in the weak axis direction. The base plates, which are on average 3 inches thick, were originally designed to resist small moments. During the analysis, if the moment demands were too large for the base plate alone to resist, two 1½-inch-diameter anchor bolts in the connection were relied upon to resist the additional moment. The shear force is resisted by bearing friction between the base plate and the concrete (FIG. 8).

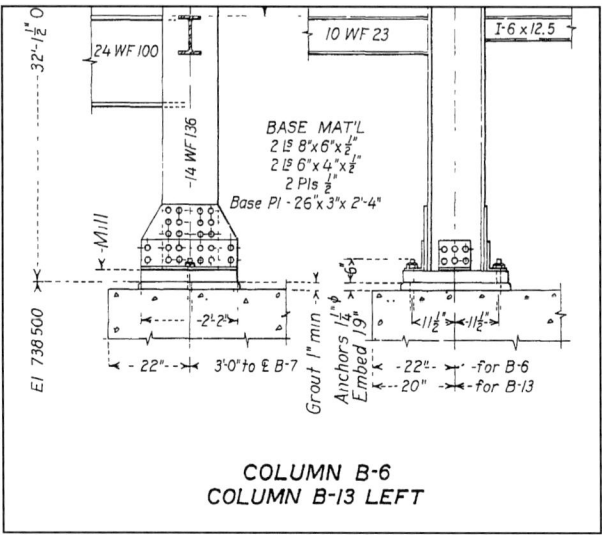

FIG. 9. Typical steel column base plate detail (Figure courtesy of the Bureau of Reclamation).

Expansion Joints (Interaction of Structures)

As previously discussed, the Hoover Powerplant is separated into numerous structures by ¾-inch expansion joints. Although the Central Portion is "wedged" into the canyon, the deflection of each structure was less than ¾-inch, so load transfer between the structures was not considered. It is uncertain if the original design allowed for the structures to transfer load, but for this evaluation, the structures were analyzed independently with no load transfer allowed.

While most expansion joints were continuous through the entire superstructure whereby completely isolating it from the adjacent structure, the Central Portion also had a roof expansion joint (FIG. 10). This created a discontinuous diaphragm at the roof level only. The roof expansion joint was modeled so that the shell elements on each side of the expansion joint were not connected allowing the roof to displace independently at this location. For the Center Section, this resulted in the lateral loads from the roof diaphragm getting transferred to different lateral force resisting systems in the structure.

FIG. 10. Roof expansion joint separating truss from moment frame (Original Photo).

Counterforts

Modeling the counterfort walls also presented some unique challenges during the evaluation. The counterforts are located at each column line along both the Arizona and Nevada Powerplant Wings and essentially act as a strut between the powerplant and the canyon rock wall (FIG. 11). The connection of the counterfort to the canyon rock could not be determined from the design drawings so it was assumed that there is no reinforcing steel embedded into the canyon rock. This presented a modeling concern because the lateral loads will get transferred differently if the load is into the canyon wall or if the load is in the opposite direction, pulling away from the canyon wall. This boundary condition becomes like a compression spring in that it can take large loads when in compression, but cannot resist tensile load.

A 2-D model of a typical powerplant column line was used to investigate the behavior of the counterforts. The structural analysis software has certain restrictions when it comes to using compression springs while running a response spectrum analysis and therefore, two separate models were created to account for the different boundary conditions of the counterforts depending on the direction of the earthquake. When the earthquake load is in the direction of the canyon, it was modeled as a pin and transfers lateral loads directly into the canyon wall. However, because there is no reinforcing steel connecting the counterfort to the canyon rock, when the earthquake load is in the opposite direction, the counterfort was modeled as free so that it could move laterally and rotate with the rest of the structure.

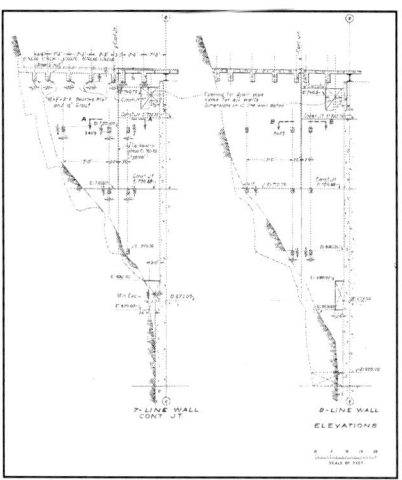

FIG. 11. Elevation view of counterforts and canyon wall (Figure courtesy of the Bureau of Reclamation).

BUILDING PERFORMANCE

Central Portion – Nevada Side

The analysis of the Central Portion – Nevada Side proved to be a complex model to analyze with half of the structure orientated orthogonal to the river direction and the other half in-line with the wings of the Powerplant rotated 8° from the upstream/downstream axis (FIG. 12). The model itself, isolated by a ¾-inch expansion joint, has a lateral force resisting system composed of various shear walls and concrete or steel columns tied together by steel or concrete floor diaphragms. While many of the individual lateral force resisting elements, such as columns and shear walls, do not appear to be unique in type and use, it is their location, orientation and associated boundary conditions that separate this plant from others. One such distinction is that the floor diaphragms and adjacent shear walls were tied directly to the dam as previously mentioned. The boundary condition of the roof and other floor diaphragms being restrained against translation at the dam interface, results in an extremely stiff structural response. This boundary condition alone limits the amount of demand within the individual lateral force resisting elements as the movement in the river direction is restrained against translation at the dam interface and the movement in the cross-canyon direction is resisted primarily through the counterforts, downstream wall and the connection to the dam through shear-friction of the reinforcing steel. As a result, a maximum high roof displacement in the river direction and cross-canyon direction was less than 1/8-inch. Further highlighting the influence of the boundary conditions at the dam interface, the seismic accelerations due to the modal analysis reflected an extremely stiff dynamic response and required multiple modes to ensure an effective mass participation. While this may not be considered a typical response as it relates to buildings at large, the boundary conditions along the height of the model require higher modes to engage the mass at the various floor elevations as previously described.

FIG. 12. 3-D finite element model of Powerplant Central Portion - Nevada Side (Original Photo).

Central Portion-Center Section

The Center Section is "wedged" between the Nevada and Arizona Central Portions, although isolated by a ¾-inch expansion joint. As previously stated, the finite element model focused exclusively on the superstructure beginning at elevation 673.0 and extending up to the roof at elevation 775.0. The structure consists of concrete columns and shear walls on the downstream portion, and steel beams and columns attached to the sloping face of the dam (FIG. 13). The lateral force resisting system for this structure was unique. In the cross canyon direction, the lateral loads are transferred through steel columns and concrete columns. The concrete columns then transfer loads to the shear walls and into the substructure. A portion of the lateral loads get transferred directly into the dam face. In the river direction, lateral loads are transferred either into the steel columns or through the connection to the dam face between building "a" Line and "e" Line. Located in the middle of the structure, four small shear walls terminate at elevation 733.0, not extending to the structure's roof. The lateral loads from the building "e" Line to "j" Line are resisted by concrete columns spaced nearly 66 feet apart.

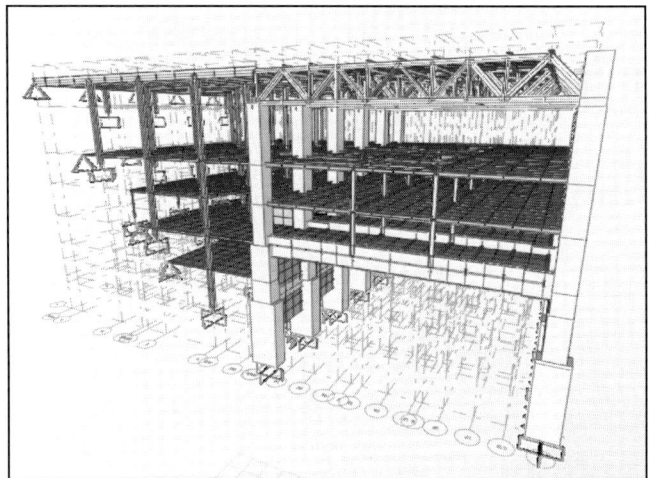

FIG. 13. 3-D finite element model of Powerplant Center Section (Original Photo).

Due to the type of lateral force resisting system and the connection to the dam face at each floor elevation, the Center Section is a very stiff structure. Typical displacements were 0.3 inch or less. However, a maximum displacement of 0.7 inch occurred at the downstream portion of the roof where the discontinuous roof diaphragm exists. Similar to the Nevada Side of the Central Portion, the influence of the boundary conditions at the dam interface reflected an extremely stiff dynamic response in the modal analysis and required multiple modes to ensure an effective mass participation.

The connection of the floor and roof slabs to the dam face was a critical component. The connection to the dam restrains that end of the structure while the other end (large exterior columns) is free to move causing the structure to rack. This results in large loads pulling away or pushing into the dam.

As mentioned previously, the steel truss at roof level acts as a horizontal brace between the concrete columns spanning nearly 66 feet. During a seismic event, the roof dead load of more than 400 lb/ft^2 causes large tension and compression in the steel truss which translates to large shear forces on the columns and truss-pin assemblage.

Powerplant (Nevada Side) – End Bay

Unlike the models of the Central Portion, the Powerplant End Bay is more 'representative' of typical powerplants found throughout the country (FIG. 14). Once again the superstructure above elevation 673.0 was considered during analysis, focusing on the critical elements of the lateral force resisting system. The components of the lateral force resisting system were well defined with shear walls in the river direction and concrete columns tied together by a large roof truss in the cross-canyon direction. In addition to the concrete columns, the end shear wall located at building 27-Line provides additional lateral stability in the cross-canyon direction. The decision to consider a three-dimensional finite element model for the End-Bay was ultimately decided to ensure that the end wall does not produce excessive demands on the adjacent concrete columns due to potential racking at the roof. Since the stiffness of the concrete columns is much less than that of the building 27-Line shear wall in the cross-canyon direction, horizontal torsion at the roof resulted in a differential displacement of approximately ½-inch due to an earthquake in the cross-canyon direction. Furthermore, the dynamic behavior of the structure did not produce any results that were not expected. The fundamental period in the cross-canyon direction was 0.42 second, while the fundamental period in the stream direction was 0.19 second. Sixty-five modes were required in order to establish a mass participation greater than 90% per code requirements. Although the racking had a significant impact on the demand of the individual lateral force resisting elements, all components considering ductility maintained a demand-to-capacity ratio of less than one.

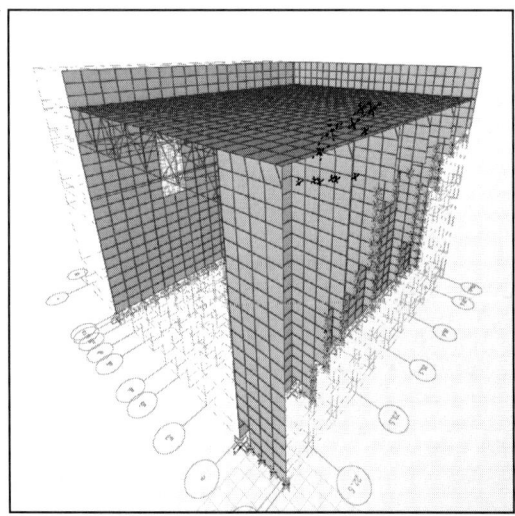

FIG. 14. 3-D finite element model of Powerplant Nevada Wing, End Bay (Original Photo).

Typical Powerplant Section

A 2-D finite element analysis was performed for a typical column line for the Powerplant wing portion. The model consisted of "b" and "e" Line columns connected by a steel truss. The counterfort at "e" Line was also modeled to analyze how the counterforts affect the structure with different boundary conditions. This was discussed previously in the modeling section.

As opposed to the Central Portion structures, the 2-D model of a typical powerplant section was much more flexible when the counterfort boundary condition was free to translate (FIG. 15). The fundamental period of the structure was 0.93 second, which resulted in a seismic acceleration of 0.11 g. As can be seen in the response spectrum curve in FIG. 4, a period of 0.93 second is a long period that correlates to acceleration on the downslope of the response spectrum curve. The maximum deflection was 1.2 inches at the roof level.

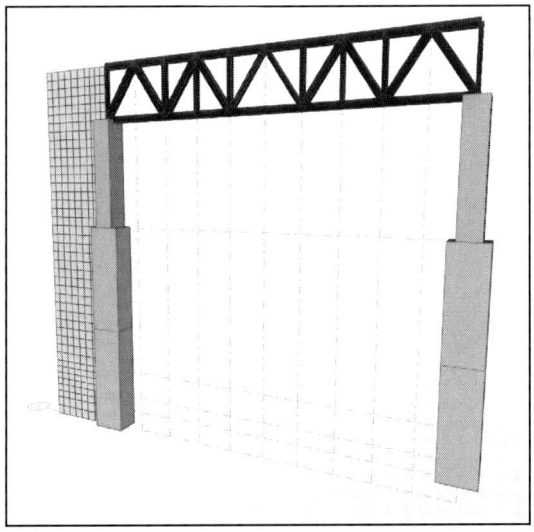

FIG. 15. 2-D finite element model of a typical Powerplant wing section (Original Photo).

CONCLUSIONS

Structural seismic evaluation of the Hoover Dam Powerplant was a unique and challenging process. Accurate analysis required researching and understanding historical construction practices, materials, photographs and drawings to fully comprehend the structure's load and force distributions. Knowledge and understanding of the Powerplant's interaction of different structural sections and connections to the dam and canyon walls were required to dynamically decouple it into efficient sections for analysis. Full understanding of the historical design was then incorporated into today's technology of finite element modeling using the current ASCE 31 seismic code to analyze the Hoover Dam Powerplant. Analysis regarding the Hoover Dam Powerplant remains ongoing. Due to the uniqueness of the Powerplant, primarily associated with the boundary conditions and loading parameters, further investigation is required to ensure that the behavior exhibited accurately reflects the anticipated response of the structure.

REFERENCES

SAP2000 – Static and Dynamic Structural Finite Element Analysis Software. Computers and Structures, Inc., 2008.
ASCE standard ASCE/SEI 31-03. "Seismic Evaluation of Existing Buildings." American Society of Civil Engineers, Reston/VA.

Frank Crowe: General Superintendent of the Six Companies, Inc. Hoover Dam Project

Philip Dunn[1], Jr. P.E., M. ASCE

[1]Associate Professor, Construction Management Technology, School of Engineering Technology, College of Engineering, University of Maine, 5711 Boardman Hall, Room 119, Orono, ME 04469-5711; philip.dunn@umit.maine.edu

ABSTRACT: Part of the success of the Hoover Dam project is directly attributable to the work of Francis Albert Trenholm (Frank) Crowe, the general superintendent of Six Companies, Incorporated. Frank Crowe led the effort to pioneer construction techniques to make the project succeed in the isolated area of the confluence of the mighty Colorado River. To make these construction decisions with the associated risk, Crowe maintained an even and skilled hand based on the 20 years of experience he had had with the Bureau of Reclamation. This paper will provide a biographical sketch of Frank Crowe through articles and information gathered about one of the famous alumni of the University of Maine. A 1905 graduate of the University of Maine, Francis Crowe began his career just before his formal graduation; Crowe's career led him to the development and construction of 19 dams. He maintained a tireless work ethic to build the best structure possible within the established budget and schedule. His style gave him the respect of all who worked with him and the reputation of a Master Builder. The paper will conclude with a discussion about the UMaine Francis Crowe Society that is established to recognize the unique heritage of engineers who graduate from the University of Maine and who are engaged in engineering in Maine. The legacy of Frank Crowe lives beyond his short time here on Earth.

INTRODUCTION

The Hoover Dam was one of the largest and most successful public works projects of the 20th Century. At the time of its construction, this massive concrete dam was the largest manmade structures ever attempted. Many modern construction practices were pioneered and were perfected here. These practices and the overall drive to build this project resulted in the Hoover Dam being delivered well ahead of schedule and at the bid estimate. Because of the success of the project, the Hoover Dam project was considered to be one of the seven civil engineering wonders of the world. (Wofat, 1955)

The successful development of the Hoover Dam can be attributed to many factors not the least of which included the people who worked to bring the project to reality. During the height of construction, as many as 5,000 people worked on the project. One hundred plus workers lost their lives during the construction of the project and many sustained injuries. The physical environment was harsh, but people persevered to complete the task.

Through the efforts of several principal individuals involved with the Hoover Dam project, the dam was completed on February 26, 1936, two years ahead of schedule. The construction company that built the dam was formed through a partnership of several companies to cover the required bond. The combined resources and leadership of the partnership created a company well suited to handle such a large project. The Six Companies, Inc. partnership was formed from the Utah Construction Company, the Morrison Knudsen Company, the Pacific Bridge Company, the J. F. Shea Company, the MacDonald and Kahn Construction Company, the Bechtel Company, the Henry J. Kaiser Company, and the Warren Brothers Paving Company. Frank Crowe worked for the Morrison Knudsen Company. He had had substantial experience in building large dams throughout the West and had worked for Morrison Knudsen since 1925. Because of his experience and reputation, the principals of the Six Companies, Inc. unanimously chose Frank Crowe to be the General Superintendent of the project. (Wolf, 1996)

This paper will give a biographical sketch of Frank Crowe and will focus on his work at the Hoover Dam project. The paper will present Frank Crowe through both factual and anecdotal information. It will conclude by describing the legacy of Francis Crowe at his alma mater the University of Maine, Class of 1905.

EARLY YEARS, 1882-1905

Francis Albert Trenholm Crowe was born on October 12, 1882, in Trenholmville, Quebec, Canada the second child of John and Emma Crowe. John Crowe was a woolen mill operator and moved the family to various locations as opportunities arose. During Frank Crowe's teens, the family moved to Byfield, Massachusetts. He entered Dummer Academy in 1899, a boarding school. In this high school setting, Crowe discovered that he had talents in both math and science. He contemplated future education in ministry or medicine. Because of the costs associated with schools to get this type of education, Crowe could not afford to attend. His older brother Joe had entered the University of Maine in the Electrical Engineering program and spoke about the engineering programs to Frank. In 1901, Francis Crowe entered the engineering program at the University of Maine. (Rocca, 2007)

While at the University of Maine, Francis Crowe developed his mathematical skills and interest in engineering. After his first year, Crowe worked for the summer doing various jobs in the greater New England area. He liked the outdoors and soon became interested in the highways and bridges that he saw. Upon entering his sophomore year, Francis Crowe transferred into the Civil Engineering Department. He developed his talents in taking mathematics and conceptual ideas and transforming them into mechanical drawings. Crowe did well in his Civil Engineering courses and was able to relate the application of these subjects into practice. (Rocca, 2007)

In 1904, Frank E. Weymouth, an 1896 University of Maine graduate, came to the University of Maine to speak about the new U. S. Reclamation Service. Weymouth described the application of irrigation in the Western United States and the importance of the work for the development of the area. Frank Crowe attended the lectures and was interested in the opportunities that the Reclamation Service could offer. He approached Weymouth about a summer position with the Reclamation Service. Weymouth offered Crowe a position for the summer and Crowe worked on the Lower Yellowstone River Project. This position in the summer of 1904 energized Crowe for a career in the Reclamation Service. Crowe returned to the University of Maine in the fall of 1904. In the spring of 1905, Weymouth offered Crowe a full time position with the Reclamation Service that would begin on May 15. Crowe received permission to take his final exams early to be able to accept the position. (Rocca, 2007)

PROFESSIONAL EXPERIENCE, 1905-1930

Upon completing his education at the University of Maine, Francis Crowe embarked on a career that would develop his skills to become the master builder of the Hoover Dam. He learned organizational skills and pioneered many technical practices that made the Hoover Dam one of the most successful engineering projects of the 20th Century. Throughout the period leading up to the Hoover Dam project, Frank Crowe networked and worked with many of the personalities that would later be part of the organization and team used in building the dam.

In his first job, Frank Crowe worked as part of a survey party in the Lower Yellowstone River Project laying out irrigation canals. By 1906, rumors started about the bad financial situation that threatened the Reclamation Service and impending shake-ups. Crowe had met up with James Munn, a contractor that had gained a contract with the Lower Yellowstone River Project. Crowe decided to join Munn and to work in construction. He was put in charge of constructing the lateral canal in Montana. He did other projects as were available through Munn. In 1908, the Reclamation Service received solid financial support and Frank Weymouth transferred to the Minidoka project in Idaho. Crowe left Munn and rejoined the Reclamation Service in late 1908 and followed Weymouth to Idaho. By 1910, Crowe was assigned his first project as supervisor to build the dam at Jackson Lake. He set up a working camp and organized the construction to sequence as many operations as possible with the given resources to complete maximum amounts of work. He maintained work through the winter and completed the project in 1911. In 1911, he moved onto the Arrowrock Dam where some of the techniques used at Hoover Dam were perfected. A 487-foot diversion tunnel was excavated by dynamiting in established patterns and excavating the blasting. Air compression and pneumatic tools were used for drilling. Cableways were employed to transport materials and supplies from the top of the canyon to the bottom. These cableways were located to allow coverage over the entire project. Telephone communication systems were installed to assist in directing the operation of the cable system. The riverbed was mucked using a 2-1/2 cubic yard orange-peel bucket developed by Crowe.

With the construction well underway at Arrowrock, Crowe was transferred back to survey the Pioneer Drainage Project and eventually back to the Jackson Lake Dam. In

1916, he reported to the Flathead Project. Due to the sudden death of project superintendent Ernest Tabor, shortly after Crowe's arrival at Flathead, Frank Crowe was appointed to be the construction engineer. He rapidly learned the details of bureaucracy and the protocols of administration. Crowe soon became disenchanted with his role as a non-technical administrator. By January of 1920, Crowe resigned from the Reclamation Service to become a partner in the small start-up construction company of Rich, Markhus, and Crowe. The small company had secured three road building contracts in Montana. Though he had not worked in such projects, Crowe left the Reclamation Service with the option of remaining as a consultant to the Reclamation Service. After six months, Crowe resigned the new business venture to rejoin the Reclamation Service to be the superintendent of the Tieton Dam in the Yakima Valley.

At the Tieton Dam, Crowe soon encountered the local interest of the townspeople in the project. He met with the local media to explain the approach for the project and how local people would be employed in the construction. He became involved as "city manager" for the construction town that developed and eventually had as many as 1,000 occupants. Crowe became part of the operations beyond the actual building of the dam. Work progressed despite Crowe becoming more involved with visitors and other public parts of the project. During the time of the Tieton Dam, Crowe assisted in gathering information and estimating a possible future project for a dam in the Colorado River in Boulder Canyon. At the end of the project, Crowe accepted a promotion to the position as General Superintendent for the 17 western states headquartered in Denver. In late 1924, the U.S. Bureau of Reclamation changed its operation to no longer build dams and irrigation projects, but rather contract these projects out to companies. Crowe spent his time reviewing requisitions and other administrative duties. He did not like this type of work and soon concluded "never my belly to a desk." By May 1925, Frank Crowe resigned from the Reclamation Service for the last time to work for the Utah Construction Company on the Guernsey Dam project in Wyoming.

Within his first couple of weeks at the Guernsey Dam, Crowe had the project town under construction, a railroad spur under development, and a couple hundred local men hired. Through his organization, Crowe developed the support systems to build the dam. By mid-1927, the Guernsey Dam was completed. Crowe then built the Van Giesen and Deadwood Dams through the Utah and Morrison Knudsen Companies. (Rocca, 2007)

Crowe had acquired extensive experience with the Bureau of Reclamation from basic field operation skills to logistics of staffing and preparing large jobsites. Under the tutelage of Frank Weymouth, Crowe was given an opportunity to experience the untamed West and to develop solutions to irrigating the vast lands. He had opportunities to mobilize the needed resources to create the necessary stage for completing construction project work specifically needed in building dams. He was able to develop new methods of construction such as cableway systems and pneumatic systems. Frank Crowe learned the importance of working with the stakeholders on the project: the locals, the politicians, and varied personalities that enter the project site. These skills were all brought to the Hoover Dam project.

HOOVER DAM

By the time the United States committed to constructing a dam along the Colorado River in Boulder Canyon, Frank Crowe had been successfully working for the Utah Construction Company and with Morrison-Knudsen. He had been aware of a possible project in Boulder Canyon since the early 1920s and was interested in being part of the project. Crowe knew that this project would be the largest to build and he was confident that his experience would be suited for construction of a dam of this scale.

Requirements for the construction included a $5,000,000 bond. The magnitude of this bond was beyond the capability of western contractors. Crowe convinced his bosses with Morrison-Knudsen and Utah Construction Companies to develop a consortium to raise the bond. The partnership of the Six Companies, Inc. was created to bid this dam. The bids for the project were opened on March 4, 1931. The winning bid was prepared by Frank Crowe on behalf of Six Companies, Inc. at $48,890,955.00, only $24,000.00 above the engineer's estimate.

SITE INNOVATION

As the General Superintendent for Six Companies, Inc. Frank Crowe was responsible for all of the site logistics of the construction operations of the Hoover Dam project. Though he had done many of the tasks needed on 15 previous dam projects, the scale of the Hoover Dam was much greater. The physical environment associated with the project was harsh and isolated.

The site for the project was over 24 miles from the next nearest town of Las Vegas, Nevada. The road to the site was not developed and utilities did not exist to the area. One of the first challenges that Frank Crowe encountered upon arriving at the project site was the large itinerant population who had come there to seek work on the project. A make shift community developed in the valley near the project site. Because of the nationwide Great Depression, potential workers camped out in the heat with their families seeking any work. Many of these people were not prepared to deal with the extreme heat conditions of the desert.

Crowe was tasked with providing adequate housing by building workers' dormitories near the construction area. He had to set up a company mess hall, central office, and transportation system for supplies. All of these items were established from scratch. With six months of starting, Crowe had set up a town with family cottages and single men's dormitories in Boulder City. Crowe had established communities before, but this one was established for 5,000 workers and associated families.

Materials for the project were another challenge and Crowe needed to build a railroad spur from Las Vegas to the project site. Warehouses had to be built to store supplies needed for both the project and the community of Boulder City. Crowe also established aggregate processing facilities to make concrete at the site and the Babcock-Wilcox Company established a fabrication facility to make the large machined parts needed for the tunnel stops and mechanical operations.

Crowe had the work begin with putting in 4 bypass tunnels to divert the Colorado River and dewater the riverbed for construction. As he had used in other projects, the tunnels were drilled and dynamited with the broken rock being removed after each blast operation. To expedite this operation, large staging was mounted on the frames of International Trucks to allow driller crews a circumferential access to the tunnel simultaneously. This apparatus was called a Williams' Jumbo. A 30-person crew could drill on the inside of the tunnel and place explosives at the same time. The jumbo could then be driven out and the rock blasted. After the loose rock was removed, the jumbo could be driven back in to the tunnel. The implementation of the jumbo saved significant time in excavating the diversion tunnels. The jumbo allowed drilling and blasting operations without the erection and dismantling of scaffolding for each cycle of tunnel excavation.

Frank Crowe had used many cableway systems on previous projects. The use of cableway systems on the Hoover Dam allowed the efficient transport of workers and supplies into and out of the canyon and from one side of the canyon to the other. He also devised the systems such that concrete buckets could be precisely placed to allow a five foot block pattern in interlocked columns for the concrete placement. The stopping systems and communications systems had been perfected on earlier projects. In March 1934, the system was operating concrete buckets to deliver concrete each 78 seconds to at total of 264,000 cubic yards that month. The buckets and other transporting devices could travel along the cableway lines at 1,200 feet per minute and then drop the loads precisely as needed. (Tobin, 2001)

Another innovative practice used by Crowe on the project included pumping concrete to areas as needed. He also employed cooling tubes to cool the concrete as it was setting up after placement. Estimates had been that a mass pour would take over 125 years to properly cure the concrete used in the dam. By developing a series of 5-foot blocks that were set in alternating columns, air assisted in cooling the concrete. Tubes were run through the bottoms of subsequent forms allowing cooling water to be pumped through. When the concrete was set, the tubes were cut off and the pipes were grouted and sealed. (Tobin, 2001)

Crowe had the largest equipment available at the time. He ordered larger trucks to carry out the large excavation and materials within the construction site. Beds were deepened and more heavy duty springs were installed to allow 50 ton loads. Crowe met with officials of the Ford Motor Company to devise a way for the trucks not to overheat as much. A 4 blade fan with a heavy duty radiator cooling system was developed to prevent the continued overheating in the extreme heat of the area. (Rocca, 2007)

PROBLEMS

The Hoover Dam project presented some unique problems to the General Superintendent Frank Crowe that he had to resolve to make the project continue to operate. These situations illustrate how Crowe moved the project to be on time.

The Six Companies, Inc. organization was formed through a partnership of several construction companies. Each company was lead by strong personalities that had built their respective companies and continued active operation in the respective company. Each of these personalities felt that they could easily run the project. After much debate, a Board of Directors was formed and each member was given a title and position. Unfortunately, these board members started offering advice and other pressures to Crowe in operating the project. After three months, Charlie Shea got the board to relinquish control and form an executive committee of four who alone would interact with Frank Crowe. Charlie Shea informed Crowe of the arrangement and eventually became the major contact. (Rocca, 2007)

By August 1931, working conditions at the site had become intolerable for the workforce. The extreme temperatures were causing many workers to collapse due to heat exhaustion. Workers complained about poor food and lack of suitable drinking water. They complained of poor safety practices and the Six Companies announcement to cut back the wages of the "muckers." The workers called a strike. Crowe believed that the strike was the result of agitators and did not give into the strike. Security fencing was installed at the site and workers were locked out and fired. They were transported to the limits of the project and let go. New personnel were hired for the project. Because of the Depression, hundreds of people were available to replace the workers. This strike lasted for several days before workers were replaced. Some accounts indicate that many of the original workers were rehired, but those identified as agitators were arrested and jailed in Las Vegas.

A second strike occurred in July 1935 when shift times were changed. Three hundred carpenters went on strike. Crowe negotiated with the workers giving in on the demands to return to the original shift times, 6-day weeks, and not to maintain a "black list" toward those on strike. They also wanted to have a wage increase from $0.75 to $1.00 per hour. He held on the wage increase and eventually workers returned one by one to working.

Under Nevada law, mining operations did not allow motorized equipment in confined spaces due to potential carbon monoxide poisoning. The tunnel work required for the bypass tunnels used motorized equipment. Six Companies argued that the excavation operations within the tunnels was not mining and continued the excavation operations. Several workers became ill. The illnesses were diagnosed as "pneumonia." No deaths were attributed to gas. The State of Nevada sued Six Companies to stop the practices. The litigation went on for several years and was found in favor of Six Companies. The tunnels were completed by the time of the ruling. Several of the impacted workers sued Six Companies for injury. Evidence shows that Six Companies bribed many sources and settled out of court with all of the Plaintiffs in 1936. (Hopkins and Evans, 1999)

PERSONAL

Francis Crowe was physically described as standing over 6 feet tall and approximately 205 pounds. He had a weathered complexion, stood erect, and was "distinguished looking." In a Hoover Dam Oral History Interview on June 2, 1986, Robert Parker described Francis Crowe.

> "Everyone who worked down here knew who Frank Crowe was. He was all over the job. His workmen, he knew them by their first name, nearly every one of them. Even we who did not work for him, he knew who we were and he always called us...once he learned your name, he never forgot it. Frank Crowe, his trademark was a white shirt. He came to work every morning wearing a brand spanking newly-ironed white shirt. His daughters told me when they were here in 1950 that that's one part that their mother played in the building of Hoover Dam. She never let him go out of the house on any morning without a

clean white shirt on, and that was his trademark...that and a large Stetson hat. That was his trademark. Any time you saw a white shirt and a large Stetson hat coming you knew that it was Frank Crowe, even though you couldn't see who it was. That was Frank Crowe. Very tall and erect, kind of a stately-looking person, a very likeable fellow. But he never forgot you if you ever crossed him. Men that had worked for him fifteen, twenty years, before this dam started, if they ever did anything working, he knew it. He remembered it, he never forgot. He had an elephant like memory, Frank Crowe."

In a March 4, 1975 interview, Curly Francis said

"Frank Crowe was a very truthful man. He treated everybody on his level, or on their level. All the time that I known him, I can't ever remember him eating anybody out or making a person feel uncomfortable...He was a man that didn't talk very much, but nothing, I don't think, ever missed his observation of looking at something and understanding. I think he understood every aspect of construction: concrete, and rigging, and hydraulics, and the works. I think he was a very outstanding man. Everybody liked him. Everybody had respect for his ability to make decisions. I think that this is the reason that the dam was built so successfully without too much of the engineering problems because Frank Crowe's word was usually final on any decision."

In a June 23, 1975 interview, Walker Young (Construction Engineer) said of Frank Crowe,

"He was actually my boss. He was the man that I obeyed. He was a wonderful man with long, long experience in the Bureau of Reclamation. I want to comment on him, his characteristics. He was very strict in running his office. Nothing would go wrong, but soon you would hear from him. He'd call you in and ask you about it. Maybe that wouldn't be the end of it. He came to see us during the investigations when we were drilling. He was the chief engineer for the personnel who were drilling. At no time did he ever criticize anything that he saw wrong on the work of the entire project. He knew everything. No matter what was going on. You might be down there and see a derrick tipped over, or a truck in the river, or a train off the tracks. He never even noticed it."

In an April 14, 1986 interview, Marion Allen spoke about Frank Crowe,

"I only knew him for twenty or twenty-five years, and I didn't know him. He was that type of a person. He was a very deep person. Even his girls, Pat and Betty, said that their father was so bashful that he hardly ever kissed their mother in front of them. Now this was the type of man he was...He was always right there. Boy on the other hand, he'd get you a job, like he got my job here, but if I'd have got fired or laid off the next couple of days, it would have been too bad. And that's the type he was. He'd give you a boost, but you'd better look out for yourself."

Frank Crowe was a quiet man who had few close friends that he developed through his various years of work with the Bureau of Reclamation. These friends remained loyal to him throughout his career as they travelled to various sites. One friend from the beginning was Bob Sass who later became Crowe's brother in-law. Due to the isolation of the construction camps, close knit friends often shared enjoyable evenings with Frank and his family. Crowe maintained these friendships outside of the work place, but, during working hours, he made it a point to maintain professional relationships so that he would not be accused of favoritism. On his 50th Birthday, October 12, 1932, Crowe's wife gave him a surprise party that included long time construction friends. Though he was embarrassed by the special attention, he enjoyed himself.

Crowe worked hard and kept his family life private. He married Marie Sass on September 9, 1909. Marie supported her husband in his construction career living with him in the remote camp at Jackson Lake. Tragically, Marie Crowe died on October 17, 1911 from complications of pregnancy. Soon after his wife's death, Frank buried himself into his work. Crowe remarried on December 9, 1913, to Linnie Korts. Linnie also supported her husband's work and remained his confidant for the remainder of his life. Sadly, tragedy continued with this marriage. The first child of this marriage, Frank, Jr. died at 5 days old in October of 1914. The second child of the marriage, John, was born on May 15, 1918. Frank and Linnie were very happy and a third child Patricia was born on April 16, 1922 followed by Elizabeth in January of 1925. However, once again tragedy would visit the Crowes when 5 year old John died suddenly from stomach problems on August 8, 1923. Frank Crowe maintained his construction responsibilities during his lowest points in life demonstrating his keen sense of duty and fortitude.

During his early career, Frank Crowe maintained a close relationship with his family often writing letters. In 1908, Crowe sent $600 to his mother to help the family and later he would send his sister money during various times of need. He remained loyal to his family and later, upon his death, he left monies to his brothers and sister as part of his final estate.

Frank Crowe was a quiet man who did not like to get up in front of a group. He avoided public speeches and didn't care for his oratory class in college. He would not accept public speaking engagements and would provide brief statements to news media of the time. He referred questions to his assistants whenever possible. He did not enjoy long written communications and rarely wrote anything beyond a page in length. He was succinct and brief. When Dr. Charles Crossland of the University of Maine's Alumni Association asked Crowe for a story about himself and the Hoover Dam project, he wrote on March 16th, 1936, "It is impossible for me to write anything worthwhile for the 'Alumnus'. However, I am sending you an editorial from the Engineering News-Record: of March 5th, 1936, which describes the situation at Boulder Dam. Also, I am forwarding you a few photographs which might be of interest to you..." Dr. Crossland eventually gathered information but did it through Crowe's personal assistant.

Frank Crowe was known to appear throughout a project, at any time, with no set routine. Crowe enjoyed interacting with the men and considered himself a "construction stiff." He viewed the construction stiff as the hard working individuals that made a project work. Saul "Red" Wixson followed Crow from job to job for years. He noted that "(Crowe) had an awesome job and if he wasn't in his office he was down at the dam. It'd never surprise me to see him down there at 2 o'clock in the morning looking around...if something went wrong he was there...to explain what was wrong, fix it. He was there to help you, not to fire you." (Hopkins and Evans, 1999)

FRANCIS CROWE SOCIETY

In 2000, the College of Engineering Dean's office instituted the Francis Crowe Society at the University of Maine. This organization recognizes graduates of the UMaine engineering programs and their unique link to famous engineering graduate Francis Crowe. Graduating seniors are presented with a commemorative medallion with a portrait of Francis Crowe and a certificate marking the induction into the society. These symbols provide a reminder of Crowe's legacy of achieving the greatest possible, no matter what the adversity.

The Francis Crowe Society also recognizes "distinguished" members who have contributed to engineering through their achievements in professional practice. Many practicing engineers with heritage to UMaine are recognized for their contributions to society. Like Crowe, these engineers did not seek recognition, but rather look for the best possible solutions in solving problems.

Additionally, the Francis Crowe Society recognizes "honorary" members for their contributions toward engineering and betterment of society. Such members include John Glenn, Astronaut, and the Herbert E. Sargent family who built many miles of WPA roads, interstate highway, and earth projects in the greater New England Area. These individuals impact the lives of many and don't seek any recognition for their achievements.

The legacy of Francis Crowe extends beyond the great monuments that he created. It is the pursuit of perfection and the betterment of mankind through knowledge that speaks of his legacy. (for information on the Francis Crowe Society, see http://www.engineering.maine.edu/home/francis-crowe-society)

CONCLUSION

Francis T. Crowe was a visionary engineer who had a passion for his work and love for building dam projects. He was a quiet driven man who expected the best from his people and delivered the best project. Crowe believed in what he did and had confidence to tackle seemingly impossible tasks. He was nicknamed "Hurry Up" Crowe because of his expectations to complete projects on time with accuracy. He did not like to be idle and wanted to be part of the working operations of all of his projects.

Though he was a shy man, he liked to interact with his workers. He made himself visible and was an active participant of the projects that he built. He was engaged in the detail and took the time to think through the processes required. People either liked him or hated him, but all thought him fair. He was able to deliver what he promised accepting the responsibilities of these promises.

The Crowe legacy is seen in the monuments that he created that continue to serve mankind. His legacy also is continued in the young charges that graduate and are challenged to follow Francis Crowe and achieve great things.

REFERENCES

Dunar, Andrew L. and McBride, Dennis (1993). *Building Hoover Dam: An Oral History of the Great Depression*. University of Nevada Press, Reno, Nevada.

Dutemple, Lesley A. (2003). *The Hoover Dam*. Lerner Publications Company, Minneapolis, Minnesota.

Fredrich, Augustine J. (1989). *Sons of Martha: Civil Engineering Readings in Modern Literature*. American Society of Civil Engineers, New York, New York.

Great Basin Chautauqua News (1993). "Frank Crowe: Applying the Jeffersonian Vision – with Concrete," Fred Krebs, Volume 2, Number 2, July 1993.

Hopkins, A.D., Evans, K.J., Editors, *The First 100: Portraits of the Men and Women Who Shaped Las Vegas* (1999) "Frank Crowe Crowning Achievement (1882-1946)" Dennis McBride, Huntington Press, Las Vegas, Nevada.

Reader's Digest (1955). Ira Wofert, "The Seven Wonders of American Engineering," November 1955, pgs. 123-129.

Rocca, Al M. (2007). *America's Master Dam Builder: the Engineering Genius of Frank T. Crowe*. Renown Publishing, Redding, California.

Stevens, Joseph E. (1988). *Hoover Dam: An American Adventure*. University of Oklahoma Press, Norman, Oklahoma.

Tobin, James (2001). *Great Projects: The Epic Story of the Building of America, from the Taming of the Mississippi to the Invention of the Internet*. The Free Press, New York, New York.

Wolf, Donald E. (1996). *Big Dams and Other Dreams: The Six Companies Story*. University of Oklahoma Press, Norman, Oklahoma.

Engineering and the Sculptural Program of Hoover Dam

Alfred Willis [1], Ph.D.

[1] Alfred Willis, Assistant Director for Collection Development, Harvey Library, Hampton University, Hampton, VA 23669 alfredwillis@yahoo.com

ABSTRACT: Hoover Dam's sculptural program, worked out by Oskar J. W. F. Hansen, relates in three significant ways to the dam's engineering. First, Hansen engaged not only in producing fine and decorative art but in the small-scale engineering of innovative pieces of furniture, a fact that made him especially suitable as an artistic collaborator with a team of engineers. Second, the scale of Hansen's interventions at Hoover Dam required the engineering of his works of art themselves. The art-workers on the project realized that they were dealing with engineering problems when executing Hansen's designs, and Hansen himself reflected on the engineering problems he had to tackle as an artist. Third, Hansen's work on Hoover Dam is not merely decorative or commemorative but moreover points up the representational qualities of the structure as a work of engineering. Those qualities simultaneously formed an integral part of the dam's engineering and transcended its practical goals. By externalizing and thus revealing the subjective essence of what was, after all, merely a dense and opaque mass of purely objective shape, Hansen's sculptures made Hoover Dam more than just a feat of human ingenuity. They made it magical.

INTRODUCTION: ART AND ENGINEERING

Engineering can be defined as an art, "the art of directing the great sources of power in nature for the use and the convenience of humans" (Parker, 1993). Everyday perceptions, however, see art and engineering as contrasting and even mutually opposed endeavors. Art (more especially, plastic art) is associated with handicraft, with small-scale production, with the exploitation of subjective experience and choices that may well be irrational. Engineering, on the other hand, involves mental prowess, the creation of works on a large scale in accordance with objective criteria by rational methods. While both art and engineering must take into account properties of materials and physical constraints, and both require the application of creativity to designing and making things, art seems to fulfill some higher, metaphysical purpose while engineering solves practical problems. Art might occasionally embellish a work of engineering, but art is unlikely ever to underpin it. Hence, in cultural terms, engineering appears necessary in a sense that art does not.

As the scale of a work of art increases, however, so does its dependence on its apparent opposite: engineering. This dependence is most obvious in the case of architecture (the art of creating aesthetically refined buildings) which always requires engineering calculations and in many cases actual engineering services. But it happens as well in large-scale sculpture. As is well known, for instance, Frederic Bartholdi's colossal statue of *Liberty Enlightening the World* (1870-86) in New York Harbor depends for its stability on a framework engineered by Gustave Eiffel. But that example is far from unique, and less familiar examples can easily be multiplied (Schodek, 1993).

Architecture by Gordon B. Kaufmann (1888-1949) brought artistic distinction to the external appearance of the Hoover Dam by adjusting the proportions and detailing of certain of its elements and surfaces (Wilson, 1985). Recognizing the subordination of his work to that of the dam's engineers, Kaufmann made artistic adjustments notably to the parapets and towers atop the structure and to the facades of the powerplants at its base ("Boulder Dam Architecture," 1938). His architecture also incorporated decorations in both two and three dimensions. Allan Tupper True (1881-1955), a mural painter, designed colorful terrazzo floors throughout the dam complex ("Boulder Dam Architecture," 1938; Nelson, 1938). Meanwhile, the sculptural ornamentation of Hoover Dam proper formed part of a larger sculptural program. That program extended beyond the dam itself to the dam's approaches on both the Nevada and Arizona sides. Oskar J. W. F. Hansen.(1892-1971) both devised that program and created its constituent relief and free-standing sculptures, as well as the special setting in which the latter were installed. The general effect of Kaufmann's architecture and Hansen's sculptures was to enhance the dam's "beauty and grace and fitness"; to achieve in "an engineering feat, ... architecture on a grand scale" (Kaufmann, 1936). Thus it became possible to assimilate Hoover Dam to such monuments as the "Leaning Tower" at Pisa (Simpich, 1940) or the Great Pyramid at Gizeh ("Boulder Dam and the Great Pyramid," 1938) claimed as monuments by the history of architecture as much as by that of engineering.

OSKAR, III, JOHAN WALDEMAR F. HANSEN

Oskar Hansen, an ethnic Norwegian, was born into Swedish nationality 9 March 1892, immigrated to the United States in 1910, and (according to his World War I draft registration card) had become a naturalized citizen by 1917. As a youth he had been a sailor, and become fascinated by Greek sculpture on Mediterranean voyages. Shortly before settling in America he had the opportunity to spend eight to nine months in the Meudon, France, studio of Auguste Rodin (Søyland, 2005). He studied at Northwestern University (1914-15) and after service as a noncommissioned officer in the U.S. Army (infantry) opened a decorating business in Evansville, Illinois. Over the 1920s he focused his interest increasingly on the fine arts, especially carving, and signaled his professional identity as a sculptor by exhibiting work at the Art Institute of Chicago in 1927, 1928, and 1931. His fascination with oversized sculpture manifested itself no later than 1929, when he created "Wings" (Opitz, 1984) for the lobby of the Rand Tower, Minneapolis. The durability of that fascination found literary expression later in essays on "The Colossi of Memnon" and "The Rock Temple of Abu Simbel" (Hansen. 1964) and, of course, plastic expression at the Hoover Dam. Following his success there,

Hansen went on to create a number of other large public sculptures, notably the large granite figure of Liberty on the Victory Monument, Yorktown, Virginia, and busts of the Wright Brothers for the monument to their achievement at Kitty Hawk, North Carolina. In 1942 he moved to an estate near Charlottesville, Virginia, where he died in 1971 ("Oskar Hansen, Virginia Sculptor, Dies at 79," 1971). Over all of his adult life Hansen had exemplified a polymath's range of interests including, beyond decorative and fine arts, creative writing, archaeology, religion, philosophy, politics, and astronomy.

Between 1932 and 1938 Hansen contributed a number of furniture designs to *Popular Mechanics* magazine. The designs reveal a strong interest in combining materials, structural innovation, and the integration of structural, functional, and decorative qualities in single works. One of Hansen's two 1932 designs for built-in tables used the cantilever principle (Hansen, 1932). A coffee table design of 1935 demonstrated his familiarity with diagonal bracing (Hansen, 1935a). His design for a combination dressing seat and clothes hamper of 1936 depended on understanding tension connections and sheer in wood assemblies (Hansen, 1936). His serving tray of 1935 (Hansen, 1935b) and indoor aviary of 1937 (Hansen, 1937) showed the artist's ability to marry decorative elegance with straightforwardness in construction. Finally, his 1938 design for a rolling buffet (Hansen, 1938a) demonstrated his virtuosity in not only cantilevering but also kinetics and dynamic loading, albeit at domestic scale. With such experiments in furniture fabrication behind him or underway in the mid-1930s Hansen at Hoover Dam would not be exploring the challenge of resolving objective and subjective considerations in design for the first time; rather, he would only be scaling them up.

With only slight license, Hansen in the mid-1930s could be described as "a man of many interests – sculptor, engineer, astronomer and mathematician" (Hilton, 1940). His cultivation of both art and science, combined with an interest in technology, had made him especially well suited to join, alongside Kaufmann, a team of engineers when he began his work in the Black Canyon of the Colorado River after reportedly winning a competition ("Boulder Dam Architecture," 1938). Undoubtedly he was not only an accomplished artist but also a practical man who could understand engineers' viewpoints and appreciate their expertise in matters of structure, materials, function, and economy.

HANSEN'S SCULPTURES AND THEIR PLACEMENT

As Kaufmann stated, "throughout the entire project the only pure ornament comprises two sculptured panels ... emphasizing the entrances to the passenger elevators." Each of Hansen's models for these exterior, cast concrete reliefs were originally intended to "represent Irrigation, Navigation, Power Development, Flood Control and Water Supply" on the one hand (the elevator tower on the Nevada side) and "the seals of the surrounding seven states" on the other (Kaufmann, 1936). Ultimately an alternative iconography was adopted for the tower on the Arizona side, perhaps in part because of the difficulty in either arranging seven seals in five panels (to achieve symmetry) or else balancing one five-panel relief against a corresponding seven-panel relief. The alternative iconography evokes not the historical constructs of certain modern political entities within the United States but instead the ahistorical essence of the geography underlying them. That essence Hansen symbolizes through references to the region's Indian tribes. The particular motifs of Hansen's Indian-themed panels, which include depictions of manpower and horsepower used in war and spiritual power used to achieve peace correspond to his industry-themed panels depicting mankind's mastery of water and the harnessing of water-power for purposes of commerce, agriculture, and generating electricity. Thus instead of a didactic program Hansen substituted a symbolic one that emphasized the idea of correspondence.

Hansen further developed the idea of correspondence in his arrangement of the flagpole plaza near the Hoover Dam's Nevada abutment. The terrazzo pavement of this polygonal plaza he treated as a star map, thus making it an earthly correspondent of the skies above in the same sense that a floor might correspond to a ceiling. Indeed, because the surface of the flagpole plaza counts as an element co-extensive with the earth's own surface, Hansen's composition suggests the more general correspondent of the earth and heaven. To one side of this pavement, on whose perimeter appear the state seals previously contemplated for one of the elevator panels, he placed an octagonal platform whose upper surface is treated as a compass. Around the 'compass' he set twelve bronze reliefs depicting the signs of the zodiac (*American Architect and Architecture,* 1938). The zodiacal motif here suggests that the correspondence between heaven and earth is to be seen in something other than prosaic terms. The reliefs inset into the platform's upper surface Hansen carved in a way that evokes ancient gems, hence yet another correspondence between macrocosmic and microcosmic phenomena. Flanking the flagpole, which Hansen made the pivot around which he balanced his entire plaza composition, he placed two colossal bronze figures on diorite bases.

These, "The Winged Figures of the Republic," are clearly symbolic – visual correspondents of something that otherwise would remain invisible -- but exactly what they allegorize is left ambiguous. According to one interpretation they "symbolize the height and depth and strength of a Republic conceived in liberty, dedicated to the principle that all men are equal and that nothing is impossible to men with the tools and with the will and the skills and the ability to use them" (Dwyer, 1941). Thus the paired statues would evoke the ideals of the democratic United States in which free enterprise, epitomized by engineering, can flourish. This interpretation, while feeding the patriotism of the World War II era, fails to account for Hansen's provision of two (rather than one or three figures). Its rhetoric, furthermore, insists not upon dualities but upon triads (height, depth, strength; liberty, equality, potential; will, skills, ability) that relate not so much to any political conception as to the reality of Hoover Dam. According to a second, less straightforward or vulgar, interpretation Hansen's two winged figures would represent "the spiritual and physical unity of man" (Bragdon, 1942). Extrapolating from this line of thought, the Republic would be the projection of a perfect idea – the essence of a physically and metaphysically unified, and therefore perfected, humanity – into that material reality where engineering can reproduce a similar correspondence of project to product. The pairing of winged figures would itself point to a correspondence within the Republic: perhaps a correspondence of what is perfectible to that which has been perfected; perhaps that of a constitution to the ongoing effort required properly to implement it. In any event the pairing as such would reproduce the structure of a state of affairs without which it would have been impossible to bring either the United States or its Hoover Dam into practical existence. Like a team of engineers (or, better, of engineers and artists), the United States is a unity and at the same time something more than a unity. Precisely because the country is and represents just that sort of unity that is also something more, it can accomplish great works.

Completing Hansen's sculptural program at Hoover Dam was a bronze plaque set into the rock wall of Black Canyon on the Arizona side. This plaque memorializes those "who died to make the desert bloom" (Hansen, 1968). Its iconography adapts that of the ancient Egyptian myth of Osiris, whose murdered corpse had been cast into the Nile before being resurrected out of it in an instance of cosmic rebirth. The connection of this myth to agriculture in arid regions, whether of Egypt or the American southwest, is obvious. On the Hoover Dam plaque, a perfected body (the spirit of the construction workers who lost their lives in building the structure) rises out the waters of Lake Mead and spreads out from its open palms plentiful sheaves of grain.

It cannot be without significance that in Hansen's own explication of his sculptures at Hoover Dam he emphasized the free-standing winged figures and their setting in the flagpole plaza to the almost total exclusion of a discussion of his reliefs over the elevator entrances. He clearly meant to privilege a symbolic understanding of what he accomplished at the dam in executing its sculptural program as a whole. At the same time, however, Hansen called attention to the technical difficulties that had to be overcome in fabricating and installing that program's manifold parts (Hansen, 1968).

FABRICATION AND INSTALLATION

In 1938 a commentator for *Architectural Forum* remarked on "the staggering problem of putting sculpture on Boulder Dam. Imagine the conditions: the Dam is an arc on top of which is a forty-foot roadway; there is a half-mile drop into the canyon; sheer cliffs rise at either end" (Saylor, 1938). Although this commentator went on to emphasize the daunting aesthetic challenges posed by the situation thus described, the same situation clearly impinged on the technical means of meeting them. They were engineering as much as artistic challenges. The immense scale of the dam itself meant that any works of art placed on it or alongside of it would themselves have to be of very large scale, in order to remain in proportion. They would have to be not just crafted but engineered. Thus the 30-foot, bronze winged figures on the dam's flagpole plaza would almost have to be "the largest single cast figures of the kind in history" (Ainsworth, 1940). The Roman Bronze Works of Long Island City, New York, cast these figures in four pieces each. Their fabrication on site involved fitting the male into the female joint of each piece, and riveting them together (Dwyer, 1941). These were heavy-construction techniques, not studio techniques. Installing the figures on their bases was a construction operation in and of itself, and required a crane (Ferrence, 2008). Similarly, the placement of the heavy bases was a feat of construction engineering that, according to the official government report, was accomplished by them being "first centered and rested on blocks of ice, the gradual melting of which permitted their being lowered into precise position" (*Boulder Canyon Project*, 1948).

Hansen recorded in some detail what he called "the mechanics of monument making" at Hoover Dam (Hansen, 1968). He described the fabrication of sand molds from which to cast his winged figures; the casting process; the contributions of the many artisans and technicians involved other than himself; the installation of their diorite bases; the insertion of the flagpole into a specially prepared hole in its plaza; and the polishing of surfaces. Hansen clearly recognized the engineering aspects of his art.

Oral history shows that construction workers on the dam project also evinced an awareness of those aspects, though they probably also identified Hansen as quite definitely an artist rather than an engineer (Dunar and McBride, 2001). Bob Parker recalled Hansen's interaction with construction workers setting the terrazzo of the flagpole plaza, an activity at once artistic and technical but more importantly one that integrated the construction of artwork into the execution of an engineering work. Harry Hall remembered how Hansen collaborated with a surveyor, Leland Robinson, to lay out the constellations of his star map. Other workers remembered Hansen's artistic temperament, but of course they could not have remembered him at all had he not been working alongside and with them, an artist among laborers, a sculptor among engineers.

MATERIALISM AND IDEALISM AT HOOVER DAM

Being a feat of engineering, Hoover Dam has practical purposes and a massive presence as a (literally) concrete artifact. It belongs fundamentally to the realm of matter. Completing it entailed not only mental powers applied to its design but moreover great physical efforts applied to its construction. It asserts itself as a tangible thing with serious economic, as well as environmental, consequences. All that is evident from any number of appreciations of the dam published during and since its erection (e.g., Young, 1932). At the same time the Hoover Dam also has a representational value, due in part to its status as an investment with expectation of return (thus the representational value of a promise), but more especially because of its appearance as a wondrous creation of man in a natural setting. It thus represents what mankind can do, mankind's mastery over nature.

Hansen's sculptural program at Hoover Dam plays up the structure's inherent representational value. It idealizes that value. It makes the dam something other than a mere fruit of labor, something greater than an economic instrument, something higher even than a feat of humanity. As its engineers had positioned Hoover Dam in a specific mundane context, Hansen positioned it in a specific cosmic context. To do so he deployed both artistic iconography and artistic style.

Hansen was essentially an academic artist whose style owed little to 20th-Century Modernism ascendant in the 1920s and 1930s. His work of those decades remained conservative enough that it could easily have been made years before. Hardly any of it would have been out of place in Lorado Taft's survey of American sculpture of the first two decades of the 1900s (Taft 1921, 118-146). Taft mainly noticed portraits and allegories, the same categories that account for nearly all of Hansen's oeuvre. He pointed out the Idealism of many of the allegories he illustrated, an Idealism evident (for example) as the "'portraiture of human fears suddenly hushed in the presence of eternity'" in Karl Bitter's "Kasson Memorial" in Utica, New York (cf. Taft, 1921). Such sculpture, with its conceptual roots in late 19th-Century Symbolism, strives to make visible, the invisible ideas hidden within the deepest recesses of the human psyche. It is easily distinguished from didactic allegory, such as that of A. M. Calder's "Triumph of Energy" at the Panama-Pacific Exposition of 1915 (cf. Taft 1921). That sort of piece imparts its symbolic message by accumulating denotative parts rather than by capturing the significance of the parts' interrelationship in one singularly striking emblem. The creation of an emblematic image is, by contrast, the aim of Idealism.

Hansen's work at Hoover Dam includes both didactic and Idealist imagery. But it is the Idealist allegories of the flagpole plaza, with its "Winged Figures of the Republic" enthroned upon a map of the universe, that are most prominent. Indeed Hansen's sculptural program as a whole there reflects an Idealist conception whose poignancy is pointed up by the materialism of the dam it decorates. Hansen at Hoover Dam was, above all, an Idealist.

ENGINEERING MAGIC

The Idealism of Hansen's creations predating his involvement with the Hoover Dam made his work especially appealing to Claude Bragdon (1866-1946), America's leading Theosophical artist and art critic of the first half of the 20th Century. Bragdon found in Hansen's artwork and discourses "the sense of an unsuspected esotericism in the sculptor's art" (Bragdon, 1929). That Idealism probably reflected both the artist's taste and his religious views. Although nominally a Lutheran (Lueker, 1975) Hansen described himself as a "mystic" on the dustjacket of his self-published anthology of writings by and about himself (Hansen 1964). A critic in 1931 had described the figures Hansen exhibited that year at the Art Institute of Chicago as "mystical and semi-abstract" ("Thirty-fifth Chicago Exhibition," 1931).

The sculptor repeatedly expressed mysticism in his work by using images of the human body to convey aspects of the human soul. He intimated through such figures as his "Victory" for the Hinsdale, Illinois, Memorial Building of 1928 the transcendence of corporeality and spirituality brought together in a mystic marriage. "*E Pluribus Unum*," inscribed on the hilt of the figure's sword, underscores the mystery of such a marriage even as it makes reference to the United States. Although ostensibly secular, its hieratic pose commemorating Hinsdale's casualties in World War I has strong undertones of religiosity. Hansen's winged figures on the flagpole plaza of Hoover Dam, as stiff and frontal as any statue surviving from ancient Egypt, also pose as hieratic figures and exude religiosity. Religiosity, or at least a confidence in the possibility of externalizing spiritual identity in bodily expression, recurs in Hansen's portraits. The latter are formally quite close to his images of classical gods and Christian characters alike. A mahogany carving of "Madonna" shows the Virgin and her offspring unified within an envelope of drapery into an indivisible, divine whole. A similarly composed, and similarly allegorical image of mother and child appears in one of Hansen's concrete panels for the Hoover Dam as the personification of peace between former enemies, thus quintessentially the unification of opposites.

In a number of his works Hansen used triangular geometry to reinforce the religiosity of a figural depiction. The triangle possesses symbolic or religious value in many cultures, from ancient Egypt to the Christian West, where trinity is the key motif of the prevailing theology. In a low relief entitled, "And God said, 'Let there be light: and there was light'," Hansen enclosed a face in whose image mankind's own face had been made within an enfoldment of triangles. The triangles in this composition resemble nothing so much as works of *origami* being unfolded to reveal the esoteric nucleus of exoteric appearance. At Hoover Dam, the Nevada-side concrete relief of "Power" shows a figure whose triangular torso reveals itself in a middle ground as, in the foreground, triangular screens are seemingly pulled away by unseen forces to either side. The unveiling of personified Power in this relief transcendentally reproduces the function of Hoover Dam as a hydroelectric installation that "reveals" the electrical power latent beneath the aqueous surface of Lake Mead. The corresponding Arizona-side relief of a "Buffalo Hunt" makes a similar though more subtle use of triangles in its composition to reveal how an exercise of human power can bring forth the nutrient value of animal bodies that, until they are butchered to expose the flesh beneath their hides, appear not as food but as fauna.

The most dramatic and significant use of geometry in the sculptural program of Hoover Dam appeared in the star map of the flagpole plaza. Hansen applied his knowledge of astronomy and physics to devising this heavenly map and the precessional diagram it incorporated as a means of locating Hoover Dam not only in space but time. The chronological location provides a key to understanding the spatial location. Hansen's precessional diagram measures time in Platonic Years, each equivalent to some "25,694.8 or our ordinary years" (Hansen 1942). This "universal clock" thus measures not historical time but cosmic time. Similarly, the star map does not locate the Hoover Dam in geographic space but rather in cosmic space. At the same time it locates the dam's visitors in that same cosmic space. It returns man, and man's great works, to the center of the cosmos. The goal thus achieved recalls that of the sacred landscapes of prehistoric Great Britain, where the placement of stones and monuments upon the ground reproduced stellar patterns in the skies above (Pennick, 1982). Hansen in his "Wings" of 1929, installed in Minneapolis before a background mural of the solar system, aimed at a similar goal. But at Hoover Dam, on the flagpole plaza, the effect is stronger because not only an artwork ("The Winged Figures of the Republic") and hence what that work symbolizes (the very idea of one Republic arising from the marriage of many) are centralized. So, too, is the spectator of that work together with the larger work (Hoover Dam) in which both the artwork and the spectator participate.

Hansen's zodiacal figures around the plaza's 'Compass' seem to link Hansen's composition with modern occultism. Astrology, having been superseded in the early modern period by astronomy, had by the 19th Century lost all claim to inclusion among the proper sciences. Only occultists still accorded it any serious respect. It offered nothing to empirically based enterprises like engineering. The appearance of astrology in a place of prominence at one of modern engineering's most prominent triumphs, therefore appears at first blush anomalous. Perhaps its apparent anomaly is what motivated Hansen to explain astrology's inclusion in the Hoover Dam's sculptural program in an article of 1938. There he noted that "it was the realization that man had to perform a mental journey among the stars before he could go back and forth on the earth which caused me to embody the ancient symbols of the zodiac as part of the sculptural decoration for Boulder Dam" (Hansen, 1938b). "The signs of the zodiac … have no particular scientific value," he went on. "However, they represent the early dream of man concerning the world of which he was a part" (Hansen, 1938b). His inclusion of these zodiacal motifs he thus explains away in historical terms; they represent a pre-scientific stage in mankind's development toward the scientism of the modern age.

However rooted in science and rationalism may be Hansen's scheme with its star map and zodiacal reliefs, those features of his program still point up the complementarity of his art and the engineering of Hoover Dam. There Hansen's Idealist art provides an experience that the dam, for all its massiveness and objectivity, cannot. Hansen's art locates man between a heaven of aspirations and earth of achievements. It helps man to recognize his own physicality in the dam's materiality but also his spirituality in the dam's potential. It simultaneously allowed him to see his human perfectibility in the perfection of "The Winged Figures of the Republic" on its plaza.

There is also the possibility that the layout and orientation of Hansen's star map could have invested it with a propitious or geomantic power.

Hansen's sculptural program at Hoover Dam thus adds magic to what otherwise would be a wonder. This magic is, of course, not the miraculous magic of pre-scientific times and their associated belief systems. Rather, it is the magic understood by modern anthropology to be aimed at arousing "emotion in the group and to make such roused emotions effective agents in the practical life of the community" (Read, 1966). Hansen did not need to be a Modernist to understand such magic, which was invoked by the much publicized work of the Surrealists of his time and therefore permeated the art-thought of the 1930s.

For all its rationality, or indeed because of it, the Hoover Dam transcended rationality by doing something that the rational mind is inclined to see as impossible: stopping up the Colorado River with a barrage of colossal scale. Hansen's sculptural program for the dam matched that scale while complementing the dam's rationality.

CONCLUSIONS

Engineering, by privileging objectivity as the key to solving problems and creating designs, was among the chief means by which the modern world was disenchanted. Restoring magic to a disenchanted world became the aim of a certain humanistic tendency in art of the late 19th and early 20th centuries. To the mathematical perfection of the Hoover Dam, which as a work of engineering was noting more than a correctly dimensioned mass, the architecture of Gordon Kaufmann added a higher, aesthetic, thus ultimately subjective and hence humane perfection. By adjusting the dam's lines and proportions he transformed the dam into a metaphysically perfected body. The Idealism of Oskar Hansen's work at the dam underscored the significance of Kaufmann's interventions. By bringing out the magical qualities latent in an unmagical thing, the sculptures Hansen planned and executed at Hoover Dam helped to re-enchant a world that, during the Great Depression of the 1930s, was anything but enchanting for most Americans.

REFERENCES

Ainsworth, E. (1940). "Along the Camino Real," *Los Angeles Time*s (3 May), p. A10.
American Architect and Architecture (1938), Vol. 52 (February): 4
Andersen, A. W. (1975). *The Norwegian-Americans*. Twayne, Boston, Mass, 274 p.
Boulder Canyon Project Final Reports: Part I, Bulletin I, General History and Description of Project (1948). Bureau of Reclamation, Boulder City, Nev., 158 p.
"Boulder Dam and the Great Pyramid" (1938). *Architect and Engineer*, No. 132: 63-64.

"Boulder Dam Architecture" (1938). *Engineering News-Record*, Vol. 121 (9): 277.
Bragdon, C. (1929). "A Modern Medusa." *Outlook and Independent*, Vol. 151 (8): 289, 308.
Bragdon, C. (1942). *The Arch Lectures*. Creative Age Press, New York, 239 p.
Dunar, A. J. and McBride, D. (2001). *Building Hoover Dam: An Oral History of the Great Depression*. University of Nevada Press, Reno and Las Vegas, 350 p.
Dwyer, P. (1941). "Huge Bronze Figures Ornament Boulder Dam." *Foundry*, Vol. 69 (10): 62-63, 140-143.
Ferrence, C. (2008). *Around Boulder City*. Arcadia, Charleston, S.C., 127 p.
Hansen, O. (1932). "Inexpensive Built-in Tables for Your Home." *Popular Mechanics*, Vol. 57 (2): 321-322.
Hansen, O. (1935a). "Modern Coffee Table." *Popular Mechanics*, Vol. 64 (3): 455-457.
Hansen, O. (1935b). "Two Simple Serving Trays." *Popular Mechanics*, Vol. 63 (3): 427-429.
Hansen, O. (1936). "Dressing Seats and Clothes Hampers." *Popular Mechanics*, Vol. 65 (6): 907-908.
Hansen, O. (1937). "Indoor Aviary." *Popular Mechanics* 68 (1): 127-128.
Hansen, O. (1938a). "Modern Rolling Buffet Has Swinging Trays." *Popular Mechanics*, Vol. 69 (3): 423-425.
Hansen, O. J. W. (1938b). "Signs of the Zodiac." *Coronet*, Vol. 4 (4): 171-175.
Hansen, O J. W. (1942) "A Split Second Petrified on the Face of the Universal Clock." *Reclamation Era*, Vol. 32 (3); 57-59.
Hansen, O. J. W. F. (1964). *Beyond the Cherubim*. Vantage, Washington and Hollywood, 224 p.
Hansen, O. J. W. (1968). *The Sculptures at Hoover Dam*. Bureau of Reclamation, Washington, D.C. 20 p.
Hilton, J. W. (1940). "He Built a Monument in the Nevada Desert." *Desert Magazine*, Vol. 3 (12): 29-30.
Kaufmann, G. B. (1936). "The Architecture of Boulder Dam." *Architectural Concrete*, Vol. 2 (3): 1-5.
Lueker, E. L., ed. (1975) *Lutheran Cyclopedia*. Concordia, St. Louis, Mo., 845 p
Nelson, W. R. (1938). "Ornamental Features of Boulder Dam." *Compressed Air Magazine*, Vol. 43 (6): 5615-5618.
Opitz, G. B. (1984). *Dictionary of American Sculptors*. .Apollo, Poughkeepsie. N.Y., 656 p.
"Oskar Hansen, Virginia Sculptor, Dies at 79" (1971). *Washington Post* (2 September): B4.
Parker, S. P., ed. (1993). *McGraw-Hill Encyclopedia of Engineering*, 2nd ed. McGraw-Hill, New York, 1414 p.
Pennick (1982). *Sacred Geometry: Symbolism and Purpose in Religious Structures*. Harper & Row, San Francisco, 159 p.
Read, H. (1966). *A Concise History of Modern Sculpture*. Praeger, New York, 310 p.
Schodek, D. L. (1993). *Structure in Sculpture*. MIT, Cambridge, Mass., 312 p.
Simpich, F. (1940). "Seeing Our Spanish Southwest." *National Geographic*, Vol. 77 (6): 711-756.

Søyland, C. (2005). *Written in the Sand: Fragments of the Emigrant Saga.* Tr. R. K. Swensen. Western Home Books, Minneapolis, 253 p.

Taft, L. (1921). *Modern Tendencies in Sculpture.* University of Chicago, 152 p.

"Thirty-fifth Chicago Exhibition" (1931). *Bulletin of the Art Institute of Chicago*, Vol. 25 (2): 38-39.

Wilson, R. G. (1985) "Machine-Age Iconography in the West: The Design of Hoover Dam." *Pacific Historical Review*, Vol. 54 (4): 463-493.

Young, W. R. (1932). "Hoover Dam." *Scientific American*, Vol. 147 (3 and 4), 134-138, 222-223.

Megaproject Success: Hoover Dam Construction and Pre-Construction Management Ingenuity

John Walewski[1], Ph.D., M. ASCE and Hessam Sadatsafavi[2], M. ASCE

[1]Assistant Professor, Department of Civil Engineering, Texas A&M University, College Station Texas, 77843-3136; jwalewski@civil.tamu.edu
[2]Graduate Student, Department of Civil Engineering, Texas A&M University, College Station Texas, 77843-3136; hessam@neo.tamu.edu

ABSTRACT: Construction of the Hoover Dam was the largest public project in the United States to date and ushered in an age of infrastructure megaprojects by federal agencies. Confronting technical, organizational, and physical risks, the Hoover Dam was delivered two years ahead of schedule and within the budgeted cost for construction. The technical challenges and successful results for this megaproject are well documented but the pre-construction planning and construction management innovations have received little attention. In retrospect, the Hoover Dam remains unique as it created the mold of structuring the relationship between government requirements and private-sector expertise to successfully deliver on megaproject objectives, cost, and schedule. Modern megaproject construction, pre-project planning, and construction management often do not follow the Hoover Dam best practices that incorporated ingenuity and creativity. The best practices and lessons learned during the planning and construction of the Hoover Dam can and should be applied to current and future federal megaprojects.

INTRODUCTION

Built between 1931 and 1936, Hoover Dam was the first of the great multipurpose dams that set records for height, volume, and power production of such structures. Although surpassed by later dams, these attributes make Hoover Dam one of the monumental engineering achievements of the twentieth century. President Roosevelt promoted the dam as a federal government mega response to combat the Great Depression and tame the arid West. Outstanding aesthetic features and its economical achievements have characterized the Hoover Dam as an expression of America's ingenuity and determination (Dunar and McBride, 1993).

The construction of the Hoover Dam is well documented; however, little has been published on the front end planning and construction management to successfully deliver this megaproject. In spite of the technical, organizational, and geographic difficulties, Hoover Dam was delivered two years ahead of schedule and within the budgeted cost for construction. As such the Hoover Dam project is an excellent example for studying the management innovations specifically with respect to structuring the relationship between government requirements and private-sector expertise in a federal megaproject.

This paper, after providing a brief description of the major steps in planning, design and construction of the Hoover Dam, identifies issues and challenges of the project along with its successful outcomes, followed by explaining practices exercised by the project team for overcoming identified issues and challenges.

HOOVER DAM PROJECT PROCESS

Construction Project Lifecycle

According to the Construction Industry Institute (CII), the project lifecycle can be broken into four distinct phases: business planning, pre-project planning, project execution, and facility operation. Although certain project types may delineate additional or fewer phases or give them different names, they are broadly similar in that they divide the project into distinct segments where each has a predetermined purpose and an identifiable series of tasks. In many cases decision points are identified at the conclusion of each phase where progress and risks are assessed and mitigated.

The first function in the project lifecycle, perform business planning, is the strategic planning which focuses on the goals and objectives of the organization as they relate to the project. The owner/investor is typically responsible for performing this phase.

The second function in the project lifecycle is pre-project planning, defined as the process of developing sufficient strategic information for owners to address risk and decide to commit resources to maximize the chance for a successful project (CII, 1999). This process begins with an identification and validation of the project concept during the business planning phase and ends with the decision to proceed with detailed engineering design and facility construction (Cho, 2000). A number of research studies conducted and published by CII have investigated and documented the importance of the pre-project planning phase of the construction lifecycle (Gibson and Hamilton, 1994). The facility owner is typically in charge of this project phase with help from consultants.

The third phase in the project lifecycle is the execution of the project that encompasses the detailed design, procurement, construction, and startup for a facility. During project execution, the decision to proceed is transformed into a completed project that typically encompasses a facility that has no missing or deficient systems or components. The detailed design, construction, and startup processes are typically outsourced contractually by the owner.

The fourth and final function in the project lifecycle is operation of the facility. In this function, the completed project is operated by the owner during the economic life of the facility.

Understanding the relationship between risk management and project phases for large capital projects can be a difficult task. Megaprojects are often first- or one-time efforts where project progress and phasing decisions can be isolated from risk management. For most megaprojects, different participants are responsible for and control the various phases of a project's lifecycle. In many cases, the project owner is largely responsible for business planning and pre-project planning, a third-party is often hired to manage and control design and engineering to meet the constraints set by the owner, and a contractor is hired to construct the project, who turns the results over to the owner for operation or production.

Structuring projects with distinct phases and responsibilities can increase risk by isolating the project participants in such a manner that minimal attention is given to overarching project concerns. Individual project participants become concerned with only their own project risks and either willingly or unwillingly attempt to transfer these risks to other project participants (Walewski, 2005).

Project Planning and Success

Dumont, Gibson, and Fish (1996) found that success during the detailed design, construction, and startup phases of a project highly depends on the level of effort expended during the scope definition phase as well as the integrity of the project definition package. Gibson et al. (1997) identified several issues including standardizing the pre-project planning approach, having the proper expertise, appropriate individuals, and end users involved as critical success factors for better pre-project planning. Griffith and Gibson (2001) also found that the level of team alignment during pre-project planning positively contributes to the ultimate success of the project.

In an extensive review of the literature on defining project success and measuring factors that enhance the probability of meeting project objectives, Cho (2000) found almost unanimous agreement that the success of a project highly depends on the level of effort expended during the early stage of the project. To study the steps undertaken for the project planning, design and construction of the Hoover Dam, we use the framework developed by the Department of Energy (DOE), which acknowledges the following four major phases for the planning and construction of its projects (NRC, 1998):

- **Preconception phase** where activities take place before a project is formally defined and include identifying ideas, making preliminary evaluations of their feasibility, and documenting the need for the project.
- **Conceptual phase** where technical and project requirements are defined and necessary resources are identified. In this phase of the project, conceptual design report and project execution plans are prepared, both of which are critical in setting the scope, cost, and schedule baselines. Costs for the conceptual phase of the project are included in the total project cost (TPC).
- **Execution phase** that consists of design and construction of the project and the transition to start-up and acceptance.
- **Closeout phase** that generally takes place at the conclusion of construction. Generally by the time of the closeout phase, the project has been completed and turned over for operation. Closeout can also be the termination of an incomplete project or the retirement of a facility at the end of its lifecycle.

DOE's portfolio of projects has historically been large, complex, and sophisticated. Many projects are one of a kind, involving unique systems, processes, and technical challenges and meet the megaproject threshold. Because of post Cold War underperformance issues, Congress mandated a series of reviews and assessments of DOE project management procedures. This review was conducted by the National Research Council's (NRC) Board of Infrastructure and the Constructed Environment (BICE) beginning in 1999. The investigations by BICE found that delivering projects of this magnitude that meet baseline costs and schedules is a constant challenge that requires excellent management (NRC, 1999). An appendix to the original report to Congress was entitled "Characteristics of Successful Megaprojects or Systems Acquisitions", and has become a standalone publication (NRC, 2000) and an often cited source on structuring government megaprojects. Using the NRC and DOE megaproject framework, the project processes to construct the Hoover Dam are categorized and analyzed below.[1]

Preconception

First used as a commerce route for transporting supplies to the Black Canyon area, the Colorado River was used for irrigation purposes which, despite the legislative difficulties in 1890s, was pursued by various land promotion companies and materialized by building a canal to irrigate part of the Imperial Valley in 1901. Operational problems of this canal, such as lack of an appropriate system for controlling the high flow of water in the river caused by torrential rains as well as the rapid rise of heavy silting disturbing the normal stream of the water in the canal, convinced the local and federal officials including the United States Reclamation Service that there was an essential need for a stronger flood control program. In the Fall-Davis report of 1922, the Reclamation Service, which then was a part of the Interior Department, brought the necessity of constructing a dam on Colorado River to the attention of the congressmen and other interested parties. The report was accompanied by an abundant amount of technical information supporting the recommendation.

Conceptual Phase

The recommendation made by the Reclamation Service was then followed by a course of action including the following:
- Agreeing on the amount of water to be apportioned to the seven Basin states affected by the project included Arizona, California, Colorado, Nevada, New Mexico, Utah, and Wyoming. The agreement signed by six of these seven states (Arizona signed in 1944) in November 1922 is known as the Colorado River Compact.
- Studying the eight candidates initially proposed for the location of the dam with respect to the geological and topographical features of each alternative,

---

[1]. Project Processes mentioned at this section are extracted from the following reference:
Joseph E. Stevens, Hoover Dam: An American Adventure, Norman, Okla. University of Oklahoma Press, 1990.

water and silt storage capacity of the reservoir, location of the site in relation to the railroad, and market for the hydroelectric power. After eliminating six of the alternate locations and by further analysis of the remaining two candidates being Black Canyon and Boulder Canyon, the final location was determined to be in Black Canyon, the current location of the Hoover Dam. The report at the end of this stage was received favorably by the interior secretary and congressmen.
- Specifying a method the federal government gets reimbursed for funding the project. This was a key feature of the legislation and was determined to be secured by executing contracts for the sale of the hydroelectric power generated at the dam over a fifty-year period at the rate determined by the interior secretary.
- Preliminary engineering design of the dam, including study of the various dam types and load analysis of the selected type by the Bureau of Reclamation (formerly Reclamation Service) with the help of University of Colorado in Boulder and under supervision of a board of consulting engineers that had been appointed by Congress in 1928 to monitor the design effort and approve the final design.
- Agreeing on dividing the power equitably among competing bidders. After 7 months of study and analysis, the interior secretary decided to divide the power between Metropolitan Water District (36 percent), City of Los Angeles (13 percent), Southern California Edison Company (9 percent), and States of Nevada and Arizona (18 percent each). The total value of contracts was higher than $327 million.

Finally, in December 1928 and after 4 years of study and review, the fourth version of the Boulder Canyon Project Act, which consisted of rough plans, cost estimates and two hundred pages of supportive information about the Hoover Dam, was introduced to the floor of both houses of Congress and then passed by Senate. In June 1929, the Boulder Canyon Project Act was declared effective in the proclamation signed by President Herbert Hoover. Afterward, in July 1930 and by approval of the requested funding, the interior secretary ordered Bureau of Reclamation to commence the construction phase of the project.

Execution Phase

In order for the Bureau of Reclamation to get the approval of the Boulder Canyon Project Act, part of the design related activities of the project had to be conducted concurrent to the activities described in the previous section. The design work continued until January 1931 when the bid documents were made available and continued until November 1932 when the final design was approved by the consulting board after updating the design and incorporating the changes required for addressing unforeseen conditions of the project. At the same time, the Bureau of Reclamation initiated pre-construction activities, including work on building Boulder City to accommodate construction workers, on the railroad spur linking Las Vegas and Black Canyon, on access to the site, and on the site communication line. In March 1931, the Six Companies, Inc., which was a consortium formed by six smaller general contractors, was announced as the winner of the dam construction contract, the largest single contract ever let by the United States government.

The Six Companies, Inc. soon started the mobilization and logistics activities and started operations related to constructing the four diversion tunnels, the mixing plant and high scaling the canyon walls. Construction of the railroad spur connecting Las Vegas to the jobsite in Black Canyon by Union Pacific, and a 10.3 mile spur from Boulder City to the canyon by Lewis Construction Company, were also undertaken. This first portion of the construction activities was completed about one year ahead of the schedule and made it possible for the contractor to reroute the Colorado River in November 1932. In the mean time and eight weeks before completion of the diversion tunnels, construction of the upstream and downstream cofferdams was started to help begin of the work on the foundation of the dam as soon as preparation work was completed. Work on the foundation was also accomplished one year ahead of the baseline schedule and was followed by the concrete work on the body of the dam.

Technical and productivity innovations contributed to the accelerated schedule. The use of electric lighting throughout the site enabled a 24 hour a day work schedule using three 8-hour shifts per day. A series of aerial tramways transporting huge steel buckets for pouring massive amounts of concrete were also deployed (Dunar and Mc Bride, 1993).

Beside construction of the main structure of the dam, operations on other elements of the dam complex included penstocks, spillway tunnels, powerhouse foundation, and intake tunnels which were also underway at this time. As a result of the increase in the number of construction activities, the number of workers on the job reached its peak in July 1934 when 5,251 workers were employed (Stevens, 1990).

Closeout Phase

In early 1936, after installing hydroelectric power equipment and transmission lines, Six Companies and the federal government reached final agreement on the fulfillment of the construction contract and resolution of the disputes that had arisen. The dam and power house were turned over to the Interior Department in February 1936, 26 months ahead of schedule (Stevens, 1990).

Project Challenges and Performance

Hoover Dam is known as one of the seven modern wonders of the United States and as one of the monumental civil engineering achievements of the twentieth century. The outstanding aesthetic features of the dam as well as its economic functionality have also been acknowledged. However, review of the project history reveals the other dimensions of its prosperity which are highlighted when project challenges and issues are brought into consideration. Apart from the technical and technological difficulties involved with the design and construction of a structure with the characteristics of the Hoover Dam, there were significant managerial challenges that included:

- Ensuring the profitability of the project due to uncertainty in availability of buyers for the generated hydroelectric power;
- Determining the hydroelectric power rate in order for the project to be able to compete with other sources of electricity and be attractive for potential buyers while ensuring profitability for the government;
- Dividing the water and power equitably between the seven Basin states and other potential buyers of the project products;
- Ensuring accuracy and correctness of the design and engineering aspects of the project. To address this, a board of consulting engineers was assigned by Congress in 1928 to advise the Bureau of Reclamation during the design process. This board was in charge of approving the final design of the project and any changes in the design during construction;
- Supporting the construction activities in Black Canyon, which was located in a remote area with harsh climatic conditions, making housing, feeding and general care for the workers as well as transportation and supply of equipments, water, and electricity difficult;
- Safety and health issues of the construction workers in Black Canyon, which were intensified by the extensive number of operations needed to be undertaken at the same time;
- Unusual size of the project, which made participation in the project impossible for many construction companies and contractors. This fact is reflected in the very high amount of bid and performance bonds requested by Government to be furnished by the bidders. For example, many companies were not able to afford the five million dollars performance bond and were ruled out of the bid process. That was the main reason six different companies joined and established Six Companies, Incorporated.

Because of all these challenges and other unique characteristics associated with the Hoover Dam project, the feasibility and success of the project were questioned by different individuals and companies in the first place. However, construction of Hoover Dam turned out to be one of the most successful experiences for the Owner and Contractor of the project, setting a great example for future megaprojects.

The construction phase was completed on budget and two years ahead of schedule, leaving a huge amount of profit for the contractor followed by collaborating on a series of similar projects with the Bureau of Reclamation. All of these accomplishments happened during the tough economic situation of the Great Depression. On the other hand, success of the project brought significant advantages to the Bureau of Reclamation, including employment opportunities, reclamation and irrigation, power supply, an effective flood control mechanism, and most importantly, high publicity of the project. Additionally, the idea of building Boulder City to accommodate the project's workers proved very successful and helped the Government and other interested parties recognize all dimensions of the benefits of megaprojects (Stevens, 1990).

INNOVATIVE MANAGERIAL STRATEGIES AND PRACTICES

In this section, to uncover the innovative and creative practices applied in the Hoover Dam project, lessons learned from analyzing the project processes described above are compared to the general characteristics of the DOE projects. To this end, general conditions for success of projects identified and listed in the report prepared by the committee in charge of assessing the policies and practices of the DOE's construction projects in 1999 are utilized (NRC, 1999).

- **Condition Cited as Essential to Success by DOE**: Project sponsors know what they need and can afford, where they want to locate the project, and when it must be ready for use or otherwise completed. The project has a purpose, and the benefits are clearly defined and understood by all participants.
  - **Comparable Lesson Learned from Hoover Dam Project**: The essential need for flood control as a part of the idea supporting construction of Hoover Dam was well understood by both public and private agencies leaving no doubt about approving the project idea. Later activities related to project development were also supported by extensive studies and effort during the 4-year period the legislation was under review in Congress. All the revenue and benefits generated by the project including water and power products were divided long before the operation of the dam in the way agreed by all the interested parties.
- **Condition Cited as Essential to Success by DOE**: The project has a champion in the owner's organization whose position and influence enable him or her to affect behavior and performance in the owner's organization that would benefit the project.
  - **Comparable Lesson Learned from Hoover Dam Project**: The Bureau of Reclamation, as the government agency in charge of the project, maintained a close relationship with the parties involved with different phases of the project and helped the entire project team overcome serious challenges, such as debates and critics during review of the legislation in Congress, labor strikes during the construction phase, lobbying to secure the adequate annual funds, and resolving the situations in which the project team had to address the conflict between public against private interests and state against state benefits.
- **Condition Cited as Essential to Success by DOE**: Open communications, mutual trust, and close coordination are maintained between owner/users and project management during planning, design, construction, start up, and turnover of the completed project to the owner.
  - **Comparable Lesson Learned from Hoover Dam Project:** One of the factors key to the success of the Hoover Dam project was the constructive relationship between the contractor of the project and governmental agencies involved with the project lead by the Bureau of Reclamation. Cooperation between Six Companies' project personnel and the resident construction engineer assigned by the Bureau of

Reclamation helped the project team overcome various technical and operational difficulties of the project. As mentioned above, a good relationship also existed between the contractor and federal officials in the executive level from early after the contract was awarded until the end of the project.

- **Condition Cited as Essential to Success by DOE**: Project managers (in owner's as well as contractor's organizations) are experienced professionals dedicated to the success of the project. Each demonstrates leadership, is a project team builder as well as a project builder, possesses the requisite technical, managerial, and communications skills, and is brought into the project early.
  - **Comparable Lesson Learned from Hoover Dam Project**: Apart from the interior secretary who paid principal attention to the overall progress of the project and helped in addressing financial and political considerations of the project, and apart from the commissioner of the Bureau of Reclamation who was playing the role of project champion for the owner from the early days of the project through construction, a resident construction engineer was assigned to the project by the Bureau of Reclamation as soon as the construction phase started. Additionally, after commencement of construction operations, to strengthen the chain of command between site superintendent and company's senior management, the Six Companies Inc. assigned an executive committee consisting of four members.
- **Condition Cited as Essential to Success by DOE**: Contract incentives are clear and unambiguous, appropriate to the performance objectives, and adequately compensate the contractor for the use of resources, risks, and performance contribution to the owner's objectives.
  - **Comparable Lesson Learned from Hoover Dam Project**: One of the driving forces of the Hoover Dam project, specifically during the construction period, was completing the project on time. To guarantee the timely execution of the project, the construction phase had been divided into major sub-phases and specific deadlines were set for each one of them. To make sure the contractor was meeting the deadlines, a particular clause was incorporated in the contract specifying a penalty for each day the contractor violated the deadline. For instance, the deadline for diversion of the Colorado River was set as October 1st 1933. The fact that the contractor had to pay $3,000 for each day passing this deadline persuaded the contractor to work on the four diversion tunnels at the same time during winter 1931-1932 and finish the job before spring when the water level rises. This strategy also contributed to the project team's early stage acceleration of the project and finishing the project two years ahead of the schedule.
- **Condition Cited as Essential to Success by DOE**: The half life of the political sponsors that decided to proceed with the project exceeds the half life of the project. Thus, there will be no change in the political will during the execution of the project.

- **Comparable Lesson Learned from Hoover Dam Project**: The Hoover Dam project team enjoyed a supportive relationship with the federal government during the course of the project from the initiation phase to the final stages of the construction. However, in 1933 when the Hoover administration gave way to the Roosevelt administration, the Hoover Dam project experienced significant challenges including the nullification of immunity from state taxes which was established by the Hoover administration. This gave the state of Nevada the right to collect taxes on the contractor's property within the Boulder Canyon Project site as a result of which Six Companies Inc. was required to pay $182,000 (Stevens, 1990).

The most important characteristics of the project, which are believed to have the highest contribution to the success of the Boulder Canyon Project, can be summarized as follows:

- Accuracy and adequacy of project development activities including feasibility study, site selection, and conceptual design essential for satisfying legislative requirements as a result of which project mission, scope, and challenges were clear for all the parties involved with the project and helped them overcome project issues;
- Close relationship between project participants specifically Bureau of Reclamation and Six Companies, Inc. both at the field level and the executive level;
- Ensuring the accuracy and correctness of design and engineering activities by assigning a design review board and implementing effective change management processes which minimized rework and delay during construction;
- Establishing a clear chain of command in the owner and contractor organizations to adjust relationships both internally and externally; and
- Supporting the project by securing adequate annual funding and relevant legislative and regulatory facilities.

CONCLUSIONS

With respect to unique characteristics of the Hoover Dam project, including the location of the site, its geologic and topographic features, and the unusual size of the dam structure which necessitated use of new construction technology and equipment, the project team encountered considerable technical and managerial difficulties in the planning, design and construction phases. Despite all these challenges, construction of the Hoover Dam turned out to be a grand achievement and brought significant benefits to the owner, contractor, and other interested parties involved with the project. Review of the project history reveals the fact that to overcome project challenges, the project team came up with a number of strategies and practices which are comparable to practices recommended as critical success factors for today's megaprojects. Constructive relationships between project participants, effective project development and change management practices, and commitment of the project owner to support the project are among the key factors and innovative practices used by project participants.

ACKNOWLEDGMENTS

The authors appreciate the support and comments of Richard Wiltshire, Bureau of Reclamation (retired), and the assistance from the Bureau's Lower Colorado Regional Photo Lab staff regarding construction era photographs of the Hoover Dam.

REFERENCES

Cho, C.S. (2000). "Development of the Project Definition Rating Index (PDRI) for Building Projects". Ph.D. Thesis, University of Texas at Austin, Austin, Texas.

CII (Construction Industry Institute). (1999). Development of the Project Definition Rating Index (PDRI) for Building Projects. Research Report 155-11. Authored by Chung-Suk, Cho, Jeffrey Furman and Edward Gibson, Jr. Austin, Texas.

Dunar A.J. and McBride, D. (1993). Building the Hoover Dam: an Oral History of the Great Depression. University of Nevada Press, Reno, Nevada.

Dumont, P.R., Gibson, G.E., and Fish, J.R. (1996). "Scope Management Using Project Definition Rating Index." *ASCE Journal of Management in Engineering*, 13(5), 54-60.

Gibson, G.E. and Hamilton, M.R. (1994). Analysis of Pre-Project Planning Effort and Success Variables for Capital Facility Projects – Source Document 105. Austin, Texas: Construction Industry Institute, The University of Texas at Austin.

Gibson, G.E. Lao, S., Broaddus, J.R., and Bruns, T. (1997). The University of Texas System Capital Project Performance 1990-1995. OFPC Paper 97-1, Office of Facility Planning and Construction, The University of Texas System, Austin.

Griffith, A.F. and Gibson, G.E. (2001). "Alignment During Preproject Planning." *ASCE Journal of Management in Engineering*, 13(2), 69-76.

NRC (National Research Council). (1998). Assessing the Need for Independent Project Reviews in the Department of Energy. National Research Council, Board on Infrastructure and the Constructed Environment. Washington, D.C.: National Academy Press.

NRC. (1999). Improving Project Management in the Department of Energy. National Research Council, Board on Infrastructure and the Constructed Environment. Washington, D.C.: National Academy Press.

NRC. (2000). Characteristics of Successful Megaprojects or Major Systems Acquisitions. National Research Council, Board on Infrastructure and the Constructed Environment. Washington, D.C.: National Academy Press.

Stevens. E.J. (1990). Hoover Dam: An American Adventure. Norman, Oklahoma. University of Oklahoma Press.

Walewski, J.A. (2005). "International Project Risk Assessment". Ph.D. Thesis, University of Texas at Austin, Austin, Texas.

Construction Management of a Mega Project

Charles R. Parrish[1], L.M., ASCE, Civil Engineer, P.E. Inactive

[1]Member, ASCE-Southern Nevada Branch, Life Member Forum, P.O. Box 36696, Las Vegas, NV 89133; charandmar@aol.com

ABSTRACT: Mega projects and small projects may be managed well by using the processes derived in 1910 by world class dam builder, Frank T. Crowe, while working on a small project in Wyoming and subsequently used by him on several successful projects during the following years including Hoover Dam–a mega project requiring over 1,500,000 cubic yards of rock excavation and over 3,400,000 cubic yards of concrete constructed on the Colorado River in the early 1930s.

INTRODUCTION

Good management practices for a small construction project are the same as for a mega project according to Crowe. The steps he set out are as follows: (Rocca, 2001)

1. Study, understand and plan each task as completely as possible before starting work
2. Schedule as many processes to be done concurrently as possible without causing physical interference
3. Keep in daily contact with project associates. Ask for and listen to input regarding improvements to current processes and those to come. Keep all parties informed
4. Make adjustments quickly for unforeseen obstacles and for new and better ideas

This paper will validate the efficacy of these steps by looking at some of the actual management practices of Frank T. Crowe while he managed the building of Hoover Dam in the early 1930s, certainly a mega project of its time.

CROWE'S HISTORICAL PRACTICES

Crowe, after graduating from the University of Maine and serving 5 years with the Reclamation Service by 1910, had attained the rank of full Engineer while working on the Jackson Lake Dam Project located in Wyoming just south of Yellowstone National Park. He eventually supervised 400 people on the project and arrived at a management strategy that would underlie his work on all subsequent projects and would become the foundation of his success. He simultaneously sequenced as many of the job operations as possible, thereby employing a maximum number of workers to complete the maximum amount of work during any given period of time. Regular staff meetings ensured that all obstacles to progress were discussed and that foremen knew exactly what the short and long term work objectives were. Although Crowe frequently talked with field workers only the foremen made assignments in the field thus maintaining a firm chain of command. Crowe used this process on numerous successful projects while working both for the Reclamation Service, later known as the Bureau of Reclamation, and for private contractors.

Crowe seriously began to coax Harry Morrison and Morris Knudsen, owners of the Morrison-Knudsen Company (MK), in 1928 to consider preparing a bid for the dam construction project in southern Nevada. Crowe inspected the project site prior to bid on four occasions and floated down the Colorado River through the canyon twice while preparing the detailed bid estimate for the project. (See FIG. 1.)

FIG. 1. Black Canyon, ca. 1929. (Photograph courtesy of Special Collections, UNLV Libraries)

In order to afford the large start up cost of a mega project like Hoover Dam, MK, Utah Construction, Charles A. Shea, Pacific Bridge Company, and San Francisco hotel builders Felix Kahn and Alan MacDonald, came together in 1929 and pooled their resources. They were still short of the anticipated $5-million needed to fund the field operations until progress payments began to flow. Then in 1930, San Francisco contractor William A. Bechtel joined the group bringing along Henry J. Kaiser from southern California and Warren Brothers Construction Company of Massachusetts. The group calling themselves Six Companies, Inc. bid and received the award, March 4, 1931, to construct Hoover Dam for the Bureau of Reclamation.

When he was put in charge as project superintendent, Crowe had intimate knowledge of the task ahead as well as a wealth of similar dam building experience although on a smaller scale. It is unlikely that anyone at the time had experience to exactly meet the demands of a project involving 119 separate work phases and eventually requiring over 5,000 workers. The two most critical phases focused on excavating 1,563,000 cubic yards of rock and placing over 3,400,000 cubic yards of concrete. To insure understanding of the work by the joint venture partners during the bid phase, Crowe went to the extent of constructing a wooden scale model of the dam complete with movable parts for demonstrating crucial operations.

The first order of business for Crowe, after getting the camp facilities for workers underway, was the excavation of over 15,000 linear feet of 56-foot-diameter solid-rock diversion tunnel–two tunnels located on each side of the river. (see FIG. 2)

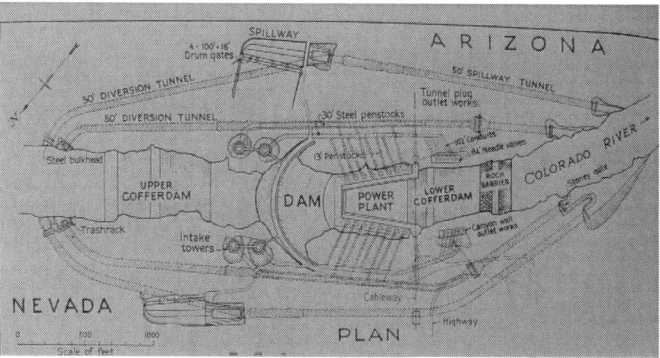

FIG. 2. Plan view showing Diversion Tunnels. (Figure courtesy Boulder City/Hoover Dam Museum)

A contract clause would penalize Six Companies $3,000 per day after October 1, 1933, if the tunnel phase remained unfinished. Tunnel work began on May 12, 1931, on the Arizona side of the river. Working with the Bureau of Reclamation engineers and his staff, Crowe developed a plan to drive two 10 X 8-foot adits from the canyon walls perpendicular to the main tunnel lines at a point midway between the inlet and outlet portals. Once these small adits were completed they added eight more faces from where tunnel excavation could proceed, bringing the total to 16 faces that could be worked simultaneously. This effort commenced with men working from barges floated down the river and moored to the rock face of Black Canyon.

At the time few, if any, people had experience in excavating a 56-foot-diameter tunnel through solid rock. Initially the work proceeded by drilling and blasting a 12 X 12-foot heading near the top of the tunnel, then the flanking wings were chipped out, followed by excavation of the main 30-foot-wide bench and finally removing the remaining sides and the invert. Clearly a very detailed and time consuming process.

Crowe put out the word that, somehow, they had to speed up the tunnel excavation work. He expected everyone to give this some thought and to make suggestions.

Crowe announced in February 1932 that diversion tunnel No. 2 was holed through and he disclosed that a new "drilling and trimming jumbo" now in use had enabled the work crews to set a world record pace. Crowe gave most of the credit for the jumbo to tunnel superintendent Williams who put together a three-tiered wooden drilling platform complete with compressed air and water connections to simultaneously power 26 pneumatic drills. (see FIG. 3) Williams then mounted the whole platform on a large truck frame. When maneuvered into place, it allowed drilling and loading of blasting materials over the entire 56-foot-diameter tunnel face at one time instead of blasting and mucking the rock out in layers. This greatly increased the rate of tunnel progress. The initial drilling platform worked so well, Crowe ordered four more to be built, this time using steel materials.

FIG. 3. Drilling and Trimming Jumbo, ca. 1931. (Photograph courtesy Morrison-Knudsen Co., Inc., Boise Idaho.)

In mid-October 1932, almost a year ahead of the contract required date, the first phase of the Colorado River diversion started with the construction of a dike up the center of the stream splitting the river flow. On November 13, 1932, at 11:30 a.m. the cofferdam was blasted away in front of Tunnel No. 4. Next, using a fleet of dump trucks, rock was placed across the river from a temporary wooden bridge in order to turn the flow into the tunnel. (see FIG. 4) Work continued until about 7:00 a.m. the next morning when the newly formed rock barrier began to appear above the river surface and water began to flow out of the diversion tunnel outlet. Although much work remained, a portion of the river had been diverted.

FIG. 4. Colorado River Diversion, ca. 1932. (Photograph courtesy Boulder City/Hoover Dam Museum.)

Newspapers across the nation reported the event. Engineers from all over the world besieged the State Department with pleas for extended visits to the site. They all wanted to see and talk to Crowe. In 1933, Crowe met with engineers and statesmen from: Scotland, Japan, Canada, England, Switzerland, China, Mexico, India, Austria, Germany, Belgian Congo (now Zaire), South Africa and Italy.

CONCLUSIONS

Crowe developed his management theory on a small project and continued to use it successfully through the years and eventually on the mega project of its time–constructing Hoover (Boulder) Dam. As noted above, in accordance with his first management step, Crowe studied the project extensively before starting the work by personally examining the work site and by preparing a detailed review of the plans and specifications during the bidding process. Since there were four diversion tunnels to construct offering eight locations where work could progress, in keeping with step two of his management plan, he found a way to double the work areas available allowing work simultaneously at sixteen headings in order to speed up this time sensitive work. Crowe kept in daily contact with his staff in accordance with step three of his management plan by walking around the site accompanied by Walker R. Young, the Bureau of Reclamation's Construction Engineer. They observed work progress and frequently talked with workers during their rounds. (see FIG. 5)

FIG. 5. Frank T. Crowe at left and Walker Young, c. June 2, 1934. (Photograph courtesy of Florence Lee Jones, Water, A History of Las Vegas.)

Crowe asked people for input regarding improvements to the work plan. And, as expected by step four of his management plan, his staff came up with a new innovation, a drilling and trimming jumbo that greatly increased the speed of the tunnel excavation phase. Although he had already revised the tunnel excavation plan once by doubling the available work sites, Crowe quickly changed to this new and better idea for hard rock tunnel excavation thereby gaining almost a year in the project schedule. Ultimately, Crowe's management practices succeeded in bringing the Hoover Dam project in ahead of schedule and within the budget, the real goal of every small and mega construction project management team.

REFERENCES

ASCE Southern Nevada Branch, Life Member Forum (2009). *"From the Spanish Trail to the Monorail - A History of Civil Engineering Infrastructure in Southern Nevada."* InstantPublisher.com, Collierville, Tennessee.

Rocca, Al M., (2001). "America's Master Dam Builder – the Engineering Genius of Frank T. Crowe," University Press of America, Lanham, Maryland.

The Construction of Hoover Dam: A Case Study from a Builder's Perspective

Tamiko Powell-Melhado[1], M. Arch., Michael Hein[2], P.E. M.ASCE, and Linda Cain Ruth[3], A.I.A.

[1] Graduate Student, Masters of Building Construction, Auburn Univ., Auburn, AL 36849; tfp0002@auburn.edu
[2] Professor, McWhorter School of Building Science, Auburn Univ., Auburn, AL 36849; heinmic@auburn.edu
[3] Associate Professor, McWhorter School of Building Science, Auburn Univ., Auburn, AL 36849; ruthlin@auburn.edu

ABSTRACT: One of the mechanisms of progress in modern societies is their investment in physical assets, which can provide the goods, services and symbols a society needs. Hoover Dam was the largest construction of its kind and came about during the Great Depression in the U.S. It served not only as a means of flood control, irrigation and hydroelectric energy, but also as a Mecca for the jobless and hope for progress. The task of organizing all resources available to realize such aspirations might have seemed insurmountable, but the builders were able to achieve them. While the historical record is filled with information about the social, cultural and design achievements of masterpieces of the built environment, much less light has fallen on the innovation required of and summoned by constructors in creating masterpieces such as Hoover Dam. A closer look at the innovative contribution of constructors to major constructed works is appropriate and long overdue. This paper attempts to reveal a few of the milestones in construction innovation that were born in the process of building Hoover Dam. Examples such as scheduling and management strategies, materials and methods in concrete, and equipment such as the 'drilling jumbo' are explored.

INTRODUCTION

The character *Bob the Builder* from the British children's television program is a building contractor who, along with his team, helps with renovations, construction, and repairs, and with other projects as needed in each episode. The children's show stresses that coordination of labor and other resources, and co-operation among the members within the construction team are important in getting the job done. As it is in art, so it is in life; the construction industry is a project-oriented trade and effective project management is imperative to the successful completion of complicated projects (Chan, Scott, & Chan, 2004; Hubbard, 1990).

Construction projects commonly experience uncertainty because of shortages in resources and the nature of the project (Jaselskis & Ashley, 1991). Construction management decisions are generally made based on schedules that were developed during the early planning stages of projects. There are however several possible scenarios that could transpire during construction and builders are responsible for taking action (Castro-Lacouture, Süer, Gonzalez-Joaqui, &Yates, 2009).

Hoover Dam, which is toured by close to one million people each year, is one of the icons in civil and water resources engineering history. Like the Eiffel Tower or the ancient pyramids, this hydroelectric dam, by its sheer magnitude, leaves visitors in awe of the obstacles that were overcome for its construction, and the immense efforts being made today to preserve it. The dam stands as testament to the triumphant efforts of effective construction management procedures and building techniques that were employed to optimize productivity.

BACKGROUND

Constructed by the Bureau of Reclamation in the depths of the Great Depression in the 1930s, Hoover Dam was the largest federal project of its time. Years after proposals were made to build a dam, then Secretary of Commerce, Herbert Hoover, served on a commission charged with finding an impartial way to divide the waters of the Colorado River among the seven basin states: Wyoming, Colorado, Utah, New Mexico, Arizona, California, and Nevada. The result of this commission's discussions, known as the Colorado River Compact of 1922, finally cleared the way for the dam to be built. Congress and President Calvin Coolidge authorized the Boulder Canyon Project in 1928.

The dam, which was referred to as Boulder (Canyon) Dam in the 1920s, was publicly called Hoover Dam for the first time on September 17, 1930, by Interior Secretary Ray Lyman Wilbur. On May 8, 1933, President Roosevelt's new Interior Secretary, Harold Ickes (who disliked Mr. Hoover) decided that the name should be reverted back to its original moniker, Boulder Dam. The name Hoover Dam was finally restored by a resolution signed by President Truman on April 30, 1947 (Mann, 2001).

The construction, which began in 1931 and was completed in 1936, was nothing short of a monumental civil engineering achievement, sitting high above the Colorado River between Arizona and Nevada. The Hoover Dam project was fashioned from 4.4 million cubic yards of concrete, 3.25 million cubic yards for the dam alone. It was a welcome income source for thousands of unemployed men and destitute families affected by the great economic depression that occurred in the United States following the 1929 stock market crash. For the West, this structure was a beacon of hope - a major structure that would symbolize the nation's technological competence (Haussler & Rekentbaler, 1999).

Hoover Dam would serve several purposes:
1. The flooding along the Colorado River as it made its way to the Gulf of California had to be controlled.
2. The water flow had to be harnessed to provide much needed water to irrigate the fertile, yet arid agricultural areas of California and Arizona; and

3. Hydroelectric energy was to satisfy the requirements of millions of people in adjacent regions.

Several construction companies submitted bids to undertake the task of constructing the dam. It was however the conglomeration of six construction firms called the Six Companies, Inc. that submitted a competitive proposal to build Hoover Dam, and as the lowest qualified bidder at $48,890,995, was awarded the contract. The Six Companies, Inc. consisted of: (1) Morrison Knudsen Company in Boise, Idaho, (2) Utah Construction Company, (3) J. F. Shea Company of Portland, (4) Pacific Bridge Company, also of Portland, (5) MacDonald and Kahn, a San Francisco company, and (6) San Francisco-based, W. A. Bechtel Company. Francis Albert Trenholm (Frank) Crowe, having had a twenty-year career history with the Reclamation Service and considered the most renowned construction engineer in the country, was especially skilled at devising timesaving, efficient construction methods. Crowe had pioneered the use of numerous pieces of new equipment, and developed a reputation as the government's best construction man. With an impressive track record, he was appointed as the General Superintendent. "In directing the construction of the dam, Crowe set basic patterns and helped define procedures utilized in most future Six Companies, Inc. projects. It was soon realized that an understanding of how government works was as important to large-scale construction as knowing the details of dam or bridge building", (Dunar & McBride, 1998).

The building of Hoover Dam, which was deemed by many as being an impossible feat, was made possible through the resilience of over 5,000 workers who endured harsh conditions and extreme dangers to complete the dam almost two years ahead of schedule. Today, Hoover Dam is a project of the Department of the Interior's Bureau of Reclamation, whose mission is "…to manage, develop and protect water and related resources in an environmentally and economically sound manner for the benefit of the American public" (Bureau of Reclamation). Hoover Dam is also an ASCE National Historic Civil Engineering Landmark (Stevens, 1988) and more recently (2000) was selected as the Monument of the Millennium dam.

PROJECT MANAGEMENT

On March 1, 1931, the composite bid of $49 million from Six Companies, Inc. was accepted and construction began on April 30, 1931. The construction of Hoover Dam was not only one of the biggest engineering projects at that time, but also the greatest testimony to functional organizations and old fashioned management control techniques. Active management was left to four people: Henry J. Kaiser, Charles A. Shea, Felix Kahn and Warren A. Bechtel. Frank Crowe, who worked under Shea, acted as the point man between the Board of Directors and the operations personnel. It was Shea's responsibility to carry out the construction on time and on budget.

TIME MANAGEMENT

The massive construction of Hoover Dam was split into five categories, with performance timelines:

1) River diversion (cofferdams and four large diameter tunnels)
2) Concrete gravity arch dam
3) Side channel spillways and tunnel chutes
4) Outlet works and valve houses
5) 'U' shaped concrete power house

COST MANAGEMENT

The country was experiencing a period of great economic depression and therefore financial resources had to be stringently managed. The Boulder Canyon Project cost a total of $165 million of which - $49 million went to build Hoover Dam. At the dam site unskilled labor was paid $4/day. Expenses incurred for the construction project were to be paid back over a period of fifty years to the Federal Treasury by selling electricity.

Under-balanced Bid

Early in the project, Six Companies, Inc. intentionally submitted a high bid for rock excavation work - $8.50 per cubic yard, asking $13,285,000 for 1,563,000 cubic yards of tunnel excavation. Later, to balance this act, they bid the concrete placement well below market price; at only $2.70/yd^3: requesting $9,180,000 for the dam's mass concrete and $3,432,000 for lining the diversion tunnels. (See Figure 1) An under-balanced bid guaranteed cash up front to finance the project.

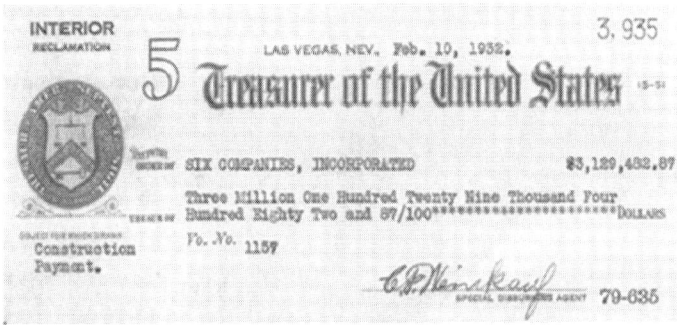

FIG. 1. Check of $13,129,482.87 made out to Six Companies, Inc. on Feb. 10, 1932 (USBR).

The excavation of the four diversion tunnels cost $13,285,000, which was 27% of the project cost. The concrete lining cost another $3,432,000. This provided a great deal of cash income up front, to balance out the $5 million performance surety secured by the partners at the beginning of the job.

RESOURCES

Hoover Dam was the first round-the-clock federal public works project, using three shifts per day, and seven days a week. Five million barrels of cement were used in the concrete; 9,000 tons of structural steel components; and 44,000 tons of large-diameter steel pipe and fittings and giant cooling towers.

The Government would provide for the construction project:
a) All materials, except concrete aggregate.
b) Railroad spur and highway to crest of gorge.
c) Construction of Boulder City, which provided housing for 80% of workers.
d) Assumption of flood damage liability after cofferdams accepted.
e) Turbines and machinery for hydroelectricity.

PIONEERING TECHNOLOGY

The Jumbo Drill

Frank Crowe had perfected the art of scheduling using the critical path method. The biggest challenge of this construction project was completing the four diversion tunnels by May 1, 1934, ahead of the spring run-off.

At 56 feet in diameter and averaging 4,000 feet long, these were the largest diversion tunnels ever constructed at the time. In order to complete excavation of the diversion tunnels, the world's first Jumbo Drill was created. This came in the form of large 10-ton trucks, modified to support platforms holding thirty drills with water and compressed air lines attached to operate its battery of thirty drills. (See Figure 2) The excavations were done in stages: a pilot bore at the crown of the tunnel and the tunnel invert excavated last. (See Figure 3)

FIG. 2. The Jumbo Drill (USBR)

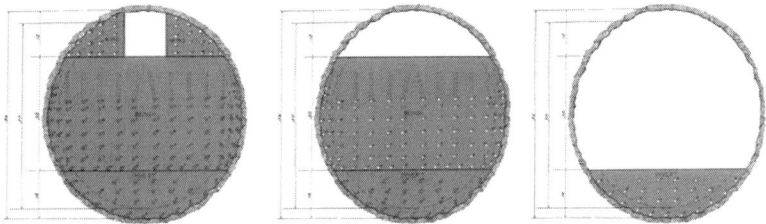

FIG. 3. Drilling stages of the diversion tunnels (Rogers, 2005).

The tunnels were lined with three feet of reinforced concrete, which was accomplished with the use of steel slip forms. After lining, the invert section was temporarily backfilled with a gravel bed to provide vehicular access until the tunnels were filled with water. Two of these tunnels are assigned to carry floodwaters, one on each side of the dam, connecting a spillway on that side with the downstream river.

The other two diversion tunnels, one on each side of the dam, were modified to hold a large penstock pipe.

The 100-foot-high cofferdams required 732,000 cubic yards of concrete in the upstream dike and 500,000 cubic yards in the downstream embankment. (See Figure 4)

FIG. 4. Cofferdams are being formed (USBR).

Each diversion tunnel was designed to convey 50,000 ft^3/s; a total of 200,000 ft^3/s during construction. A flash flood inundated the construction site on February 10, 1932, before any of the diversion tunnels were completed. The shutdown and clean up lasted five days. A milestone was reached on November 14, 1932, when river flow was diverted through Tunnel No. 4, 1.5 years ahead of schedule. This achievement was said to have guaranteed a profitable job.

The biggest flow the tunnels had to handle during construction occurred on June 16, 1933, when 73,000 ft^3/s was safely conveyed around the dam site, while the minimum flow of 1,000 ft^3/s was recorded on August 26, 1934.

The construction site was located in a remote area and the transportation of labor, materials and equipment was imperative to the completion of the job in a timely manner. Site Access was the most difficult obstacle to overcome. Building a railroad 34 miles from Las Vegas to the dam site was given high priority, as it was the only means by which to bring in large, heavy equipment along with masses of materials. Spider-like webs of steel wire cable were stretched from towers high over the dam to lower concrete, large pipe, and men down the sheer cliffs into the canyon.

Modified Trucks

Thousands of workers were mobilized and transported to the dam site around the clock. To carry these people to the site, large trucks were modified with decks of seats to hold 150 men at a time. (See Figure 5)

FIG. 5. Modified trucks to transport the workers (USBR).

Concrete

Gravel and sand for concrete were the only material items not supplied by government. The aggregate came from a site six miles upstream of Black Canyon on the Arizona side. Aggregate up to eight inches in diameter was used in the concrete. An aggregate washing, classification and storage plant was constructed in Hemenway Wash. (See Figure 6)

FIG. 6. Concrete aggregate plant at Hemenway Wash (USBR)

The first concrete was poured on June 6, 1933, 1.5 years ahead of schedule. Frank Crowe utilized the concept of high and low concrete mix batch plants working simultaneously, and doubling the efforts of supplying concrete where needed. These two plants contributed significantly in the two year early completion of the construction project. The large concrete plants had the capacity of mixing 16.5 tons of aggregate with cement and water in one minute. (See Figure 7) This operation delivered an average of 160,000 cubic yards a month to the dam site. A record was set with 10,462 cubic yards of concrete being poured on the peak day.

FIG. 7. Low Mix Plant (left) and High Mix Plant (right) (USBR).

Documentation shows that the Bureau of Reclamation had only used 5.8 million barrels of cement (4 bags = 1 barrel) in its 27 year history of construction before the construction of Hoover Dam. The 3.25 million cubic yards of concrete used in the construction of Hoover Dam went into 230 giant building blocks, each about sixty feet square at the upstream face of the dam and about 25 feet square at the downstream face.

Concrete was lowered from the unique cableway system created, into the dam and appurtenant works. Most of the mass concrete was placed using 8 cubic yard hopper buckets, delivered by the overhead cableways. Each bucket carried 16.5 tons of concrete. (See Figure 8)

FIG. 8. Overhead cableway transporting concrete (USBR).

First Use of Aggregate Chilling and Cooling of Mass Concrete

The purpose of a dam is to contain water, so a dam free of cracking is a must. Heat of hydration produced by the chemical combination of water and cement accumulates in massive concrete pours. In addition, the concrete tends to cool very slowly and unevenly. The resulting uneven contraction can easily lead to cracking in the concrete. In order to combat this problem at Hoover Dam, a series of one-inch-diameter pipes was embedded in the mixture, to allow for the circulation of air-cooled water first, and then water chilled to 42°F. The rated capacity of the cooling plant was 825 tons and in September 1934, the plant and cooling tower produced an average of 1,815 tons of refrigeration (BCP: Final Reports, 1941). This cooling system removed the heat generated by chemical hydration of concrete. Without artificial cooling, the concrete dam would have been 40°F hotter and taken 125 years to cure. In less than two months the cooling system adequately cooled the concrete blocks. (See Figure 9)

FIG. 9. Pipes are embedded in concrete mixture as part of cooling mechanism (USBR).

Concrete blocks were stacked on top of each other in vertical columns and cement grout was injected into the spaces between the blocks, creating a monolithic structure. In order to prevent water from seeping under the dam, the foundation was injected with cement grout under pressure into holes drilled into the rock.

The dam had required 3.25 million cubic yards of the project's 4.36 million cubic yards of mass concrete. This was placed between June 6, 1933 and May 29, 1935, 2-1/2 years before the contract required completion.

Labor

There were up to 5,200 workers employed during the construction of Hoover Dam. Men of all ages and walks of life came in droves to the construction site in Black Canyon hoping for any kind of paying work. Earlier in the construction project, the crews were not large, and most of the available jobs were dangerous, dirty, and low-paying. The men worked eight-hour days. The lowest wage was 50 cents an hour, and the highest was $1.25. The average for all of the workers at the dam was about 62.5 cents an hour. The average of 62.5 cents an hour works out to an annual income of $1,825, and the highest wage, $1.25, works out to $3,650 per year.

The High-Scalers and Hard Hats

Before construction could begin on the dam, loose rock that was eroded in the canyon walls had to be removed. Special men were required for the job; they were called "high-scalers." This was thought to be the most fascinating and highly paid work on the dam project. These men worked out in the fresh air and sun suspended hundreds of feet above the canyon floor, which was a stark contrast to the overheated spaces in the tunnels. Their job was to climb down the canyon walls on ropes and work with jackhammers and dynamite to remove the loose rock. (See Figure 10) It was hard and dangerous work, probably the most physically demanding work on the project. The high-scalers had to cautiously maneuver their way along the walls. The danger from falling rocks and dropped tools was intense and a common cause of injury and death during the building of the dam. Workers made makeshift hard hats for themselves by coating cloth hats with coal tar. These hats were extremely effective, as men who were hit by falling rocks did not receive skull fractures and, in some cases, escaped certain death. (See Figure 11) Later, Six Companies, Inc. contracted for commercially made hard hats and issued them to every man on the project. This is the first documented use of hard hats on a construction site.

FIG. 10. A "High-Scaler" in action (USBR).

FIG. 11. Improvised hard hats created by workers (left). First use of safety hats (USBR).

Death Toll

During July 1931, fourteen men died working in the overheated and under-ventilated tunnels. Many of the workers' wives and children died in the campsite area due to extreme heat and lack of sanitation. These events triggered a strike by the workers in August 1931, which led to changes and upgrades in safety precautions. An Emergency Room was built on the job site and Water Carriers were added in late summer of 1931. The official death toll of workers for the entire job was 96; the unofficial total was 112.

CONCLUSIONS

The construction of Hoover Dam was truly a formidable task. There were no precedents to follow in building very high dams and therefore precise planning was required on the part of the Six Companies, Inc. Frank Crowe supervised what must have been an extensive list of activities to comprehensively and effectively manage all available resources. There could be no mistakes as there was far too much at risk and the detractors were many.

At a height of 726.4 ft, a weight of 6.6 million tons, a total storage capacity of 32,471,000 acre-ft, power generating capacity at 2,080 megawatts, having seventeen generators, and being part of a system that provides water to over 25 million people in Southwest United States – the constructors of Hoover Dam pioneered strategies and innovations throughout the construction to successfully complete the project. Many of these were milestones in the evolution of construction and are still in use today. At 75 years of age, Hoover Dam remains a magnificent man-made wonder, both a great technical and a great social achievement. Its success has been emulated all over the world.

ACKNOWLEDGMENTS

The authors thank the Bureau of Reclamation, Technical Service Center for their assistance in the study and the Steering Committee for the opportunity to contribute to this historic celebration of Hoover Dam.

REFERENCES

Boulder Canyon Project: Final Reports, Part IV (1941). "Design and Construction", *Bulletin 2*, Boulder Dam. Denver:196.
Castro, F., Süer, H., Gonzalez, J., and Yates, A. (2009). "Construction project scheduling with time, cost, and material restrictions using fuzzy mathematical models and critical path method." *Journal of Construction Engineering and Management.*
Chapman, K. (Creator). (1998). "Bob the Builder." Retrieved on October 12, 2009, from http://www.bobthebuilder.com.
Chan, A.P., Scott, D., and Chan, P.L. (2004). "Factors affecting the success of a construction project." *Journal of Construction Engineering and Management*, 130, 153–155.
Dunar, A. and McBride, D. (1993). "Building Hoover Dam: an oral history of the great depression." New York: Twayne Publishers.
Haussler, T. and Rekentbaler, D. (1999). "The Hoover Dam Bypass." *Public Roads,* 63,30.
Hubbard, D.G. (1990). "Successful utility project management from lessons learned." *Project Management Journal, 21*, 19–23.
Jaselskis, E.J. and Ashley, D.B. (1991). "Optimal allocation of project management resources for achieving success." *Journal of Construction Engineering and Management,* Vol. 117: 321–40.
Mann, E. (2001). "Hoover Dam." New York: Mikaya Press.
McBride, D. (2003). "Desperate times: the building of the hoover dam." Retrieved October 14, 2009, from http://www.bbc.co.uk/history.
Rogers, J.D. (2005). "Hoover Dam: impacts on the engineering profession, America, and the world." *Assoc. of Engineering Geologists Annual Meeting Las Vegas, Nevada.*
Stevens, J. (1988). "Hoover dam: an american adventure." Univ. of Oklahoma Press.
U.S. Dept. of the Interior-Bureau of Reclamation: Lower Colorado Region-"Hoover Dam." http://www.usbr.gov.

Building Hoover Dam
(Men, Machines, and Methods)

Raymond Paul Giroux[1], M. ASCE

[1] ASCE Chairman Brooklyn Bridge 125th Anniversary; Lafayette, CA; paul.giroux@kiewit.com

ABSTRACT: Hoover Dam is rightly regarded as one of the greatest civil engineering achievements in history. While Hoover Dam is mired in some controversy for safety and labor practices, built during the great depression, it must be regarded as not only a triumph of engineering, but also a triumph of the human spirit. Large concrete dam construction was in its infancy in the early twentieth century. Hoover Dam's successful construction and enduring ability to harness the Colorado River are the result of the convergence of extraordinary men, machines, and methods. In the decades preceding Hoover Dam's construction, tremendous advancements were realized in every discipline of engineering. In the realm of heavy civil construction, these collective engineering advancements provided the construction technology that made Hoover Dam possible. This paper highlights how the right men, the right, machines, and the right methods all came together in the spring of 1931 to build a structure of unprecedented scope and challenges.

INTRODUCTION

On September 27, 2001, the ASCE dedicated a plaque at Hoover Dam recognizing the dam as a "Civil Engineering Monument of the Millennium." This honor has only been bestowed upon ten twentieth century civil engineering projects throughout the world. The plaque reads, "One of the finest examples of how civil engineering ingenuity shaped the development of society's quality of life in the 20th century." Hoover Dam is clearly worthy of such recognition as it not only set an example for future North American dam projects, but the entire world. Like many noteworthy structures, study and analysis of the history and construction of Hoover Dam continue to reveal relevant lessons.

Built in Black Canyon, Hoover Dam was first conceived and referred to as Boulder Canyon Dam, then Hoover Dam in September 1930, was completed and dedicated as Boulder Dam in 1935, before definitively having its name changed back to Hoover Dam in 1947. As a point of clarity, the dam will be generally referred to as Hoover Dam throughout this paper.

Lewis and Clark's Corps of Discovery expedition from 1804 to 1806 set in motion the westward expansion of the United States. Fueled by promise and opportunity, the west was settled by the end of the 19th Century, yet the west's mighty rivers were untamed. In order to harness the water resource potential of the West, the United States Federal Government formed the United States Reclamation Service (USRS) in 1902. The United States Reclamation Service would later change its name to the Bureau of Reclamation, an agency of the United States Department of the Interior in 1923. For clarity, the "Bureau" will generally be referred to as the United States Bureau of Reclamation (USBR) throughout this paper.

In the early twentieth century, the USBR proved itself to be a can-do agency successfully self-performing the construction of dozens of irrigation related projects throughout the western states including Arrowrock Dam on the Boise River in southern Idaho. Built between 1911 and 1915, the record setting 350-foot tall Arrowrock Dam was an opportunity for young ambitious engineers to gain large dam building experience. Taking on the challenge of Arrowrock, were 35 year old Frank Weymouth (USBR), 30 year old Jack Savage (then a design consultant), 26 year old Frank Crowe (USBR), and 23 year old Walker Young (USBR). Unknowingly, they were learning lessons and skills at Arrowrock that set them on a date with destiny, to dam the Colorado River twenty years later.

In the early twentieth century the USBR built dam after dam throughout the west, ushering in an era of managed, reliable water to support growing communities and agriculture. Yet, there was the mighty Colorado River, for ages, wild, untamed, and at times deadly. In 1905 the Southern California's Imperial Valley suffered a devastating flood due to high spring runoff in the Colorado River. This event set in motion a thirty year process to harness the Colorado River. The watershed of the Colorado River, known as the Colorado River Basin, covers land in seven states including Wyoming, Colorado, Utah, Nevada, California, Arizona, and New Mexico. With Arizona and New Mexico achieving statehood in 1912, all of the Colorado River Basin states could be brought together to work out a water allocation agreement.

It would take time to study the Colorado River and its channel to locate a suitable location for a dam. Finally, in the spring of 1921 Arthur Powell Davis, Reclamation Service director, announced to the world plans to build the "highest dam in the world". (Rocca, 2001)

The early USBR plans for damming the Colorado River were based upon building a dam in Boulder Canyon. Boulder Canyon is located about 20 miles upstream of Black Canyon, where Hoover dam was built.

On November 24, 1922 the Colorado River Compact was signed. During the next six years, several unsuccessful attempts to gain House and Senate approval for the "Boulder Canyon Project Act" to dam the Colorado River were made.

Unfortunately, it would be a tragic event, thousands of miles from Washington, D.C. that would finally help the Boulder Canyon Project get the political support it needed to pass. On March 12, 1928 the St. Francis Dam collapsed forty miles northwest of Los Angeles. The ensuing flood killed in excess of 450 people. So it was in December 1928, both the House and the Senate finally approved the bill for the Boulder Canyon Project, and on December 21, 1928, President Calvin Coolidge signed the bill into law. The July 21, 1930 issue of *Time Magazine* reported:

"Work on Boulder Dam, world's highest (727 ft.), was ready to start last week. Congress had appropriated $10,660,000 to get the $165,000,000 project under way. Secretary of the Interior Wilbur approved a construction order which was telegraphed to Las Vegas, Nev., where Walker R. Young, resident U. S. engineer, received it. Said Secretary Wilbur: 'With dollars, men and engineering brains we will build a great natural resource . . . make new geography . . . start a new era ... conquer the Great American Desert. To bring about this transformation requires a dam higher than any the engineer has hitherto conceived or attempted to build.' Secretary Wilbur warned against a great rush of workmen to the barren dam site where their services are not yet needed."

At the outset of building Hoover Dam, there was really no precedent for its size and scope of work. The USBR had built many dams prior to Hoover Dam, but nothing came close to matching its size, complexity, and technical challenges. The United States Army Corps of Engineers' (USACE) largest dam of the day was Wilson Dam on the Tennessee River in Alabama. Completed in 1927 it too was dwarfed by Hoover Dam. Table 1 provides some contemporary comparisons of other record breaking dams preceding Hoover Dam. Arrowrock Dam in Idaho would be the closest completed dam at 350 feet high, as the 417 foot-high Owyhee Dam in Oregon was still under construction.

Table 1. Comparison of Hoover Dam to Contemporary Record Breaking Dams

Dam	Owner	Dam Type	Years Built	Height, ft.	Volume Concrete, CY
Arrowrock	USBR	Gravity Arch	1911-1915	350	527,300
Wilson	USACE	Gravity	1918-1927	137	1,300,000
Owyhee	USBR	Gravity Arch	1928-1932	417	537,500
Hoover	**USBR**	**Gravity Arch**	**1931-1936**	**726.4**	**3,250,000**

The design of Hoover Dam was completed at the USBR's Denver Office. It took years of site investigations, studies, and design and involved the efforts of some 200 engineers and other workers and consulting firms. The scope of work for Hoover Dam provided in the USBR's plans included the following major divisions of permanent work: (*Compressed Air Magazine*, 1931-1935)

River Diversion Works This work consisted of installing upstream and downstream cofferdams, installing four 50-foot diameter tunnels, each approximately 4,000-feet long through rock, and the lining of those tunnels with 3-feet of concrete.

Concrete Arch Gravity Dam The 726-foot-high dam would require 3,250,000 cubic yards of concrete along with concrete cooling requirements.

Twin Spillways Each side of the river would require a concrete-lined open channel spillway structure be built, each consisting of a 50x50 -foot stoney gate, and a concrete ogee overflow crest 700-feet long.

Twin Outlet Works   Providing water for the powerhouse, these massive cylindrical structures were equipped with cylinder gates in the bottom of the intake towers to regulate flow to the powerhouse.

Powerhouse   A U-shaped power house of concrete and structural steel immediately below the dam.

The Challenge

The quantities of work involved in building the 726-foot-high Hoover Dam were staggering. Construction of the dam would require 3,250,000 cubic yards of concrete, with another 1,150,000 cubic yards in tunnel lining, spillways, powerhouse, intake towers, and outlet structures (Figure 1). In order to produce the concrete, every cubic yard would require on-site mining, processing, and batching to unprecedented standards for consistency and quality. This was more concrete than previously placed in all of the USBR dams in its 29 years of existence prior to 1931. Hoover Dam would require over six times more concrete than both the current world height record gravity arch dam, Arrowrock Dam, and the soon to be record holder Owyhee Dam. Hoover Dam would require two and a half times more concrete than was placed in the 1,300,000 cubic yard Wilson Dam in Alabama.

The quantities of construction materials for Hoover Dam are unfathomable. The basic work required 4,400,000 cubic yards of concrete, 45,000,000 pounds of reinforcement steel, nearly 22,000,000 pounds of gates and valves, over 44,000 tons of steel (14,800 feet) of penstock and outlet pipes, and over 5,000,000 barrels of cement. The average number of people employed on the project was 3,500 with over 5,200 employed during the peak of construction.

Additionally, prior to building the dam a massive excavation operation would be necessary which required approximately 5,500,000 cubic yards of material excavation for the foundation of the dam. Additionally, there was over 1,500,000 cubic yards of rock excavation in the tunnels.

Hoover Dam would also require vast amounts of temporary structures and facilities to support the work; including miles of new railroad lines, river trestles, cableway crane systems, the biggest aggregate processing plant in the world, the biggest concrete batch plant in the world, a massive electrical substation, water pumping and treatment plants, air compressor plants, access roads, bridges, and a man camp capable of housing over 4,000 men plus families.

While there was some precedent for the design details and the construction methods with which to build Hoover Dam, the scale and scope of the work were unprecedented.

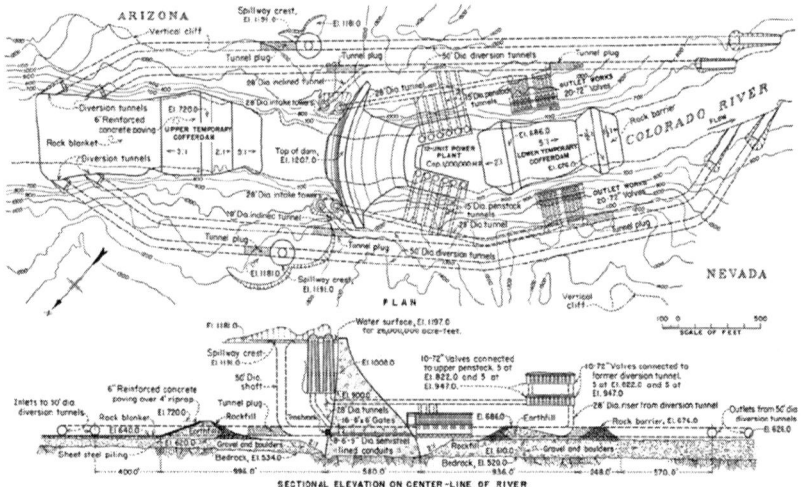

FIG. 1. The General Arrangement for Hoover Dam. Courtesy Bureau of Reclamation

Machines, Means, and Methods

Hoover Dam, the great concrete monolith which dams the Colorado River is widely regarded as one of the greatest civil engineering achievements in history. Insightfully, in his book *The Essential Engineer*, Henry Petroski describes the interdisciplinary nature of great engineering achievements. Petroski writes, "No great achievement is wholly the province of a single engineering discipline, but each can take pride in its contributions to the greater good."

In the decades preceding the start of the Hoover Dam tremendous advancements were made in not only civil engineering, but also mechanical, electrical, chemical, metallurgical, and other branches of engineering.

Collectively, multi-disciplinary engineering advancements paved the way for rapid progress in construction technology in the early twentieth century. Excavation equipment, blasting technology, concrete production and methods, and tools were constantly evolving driven to some degree by the World War I effort, but more significantly by the advent of the automobile and American's growing desire for more roads. Because of the speed of technological change in the early twentieth century, it is safe to say, Hoover Dam could not have been built more efficiently or to the standards achieved even a decade earlier in history.

Grading and Hauling Operations

During the United States led construction of the Panama Canal between 1904 and 1914, Chief Engineer John Frank Stevens (in Panama 1905-1907) developed very efficient material handling techniques. The Culebra Cut alone, is estimated have required the excavation and disposal of 100,000,000 cubic yards of soil. While efficient, Steven's methods relied heavily on railroad tracks, locomotives, and rail-based cranes to excavate and haul its excavated spoils.

Due to Hoover Dam's tight access, steep grades, and the quick paced nature of the work, more efficient excavation and hauling methods would be necessary to handle the large volumes of blasted rock and river sediments. The hauling methods used at Hoover Dam relied heavily on the dump truck which was first devised and patented in 1920. Several truck brands were used at Hoover Dam including International, White, GMC and Mack. Mack's popular "AC" model truck evolved through the twenties and the "B Series" was introduced in 1927 in response to demand for larger capacity and higher haulage speeds. The Mack trucks were the biggest in the world, "powered with 250-horsepower motors (sic) and equipped with special duralumin bodies, they are capable of waddling away with sixteen cubic yards of earth–just twice the capacity of the biggest truck hitherto." (*Fortune*, 1933).

As trucks improved, so did the tires for them with Goodyear and Dunlop both introduced pneumatic truck tires in 1919.

Caterpillar's first diesel powered "Diesel Sixty" tractor was produced in 1931, with a new efficient source of power for track-type tractors. Although not exclusively, both the B Series Mack trucks and the Diesel Sixty were extensively used on Hoover Dam.

Blasting Operations

By the time Hoover Dam began, the blasting industry was already using "controlled blasting" methods where blasts were designed to detonate sequentially using delay primers. Detonation of blasts with delay primers keeps the entire blast from detonating all together in one instant, which improves results, and limits detrimental vibrations. Undoubtedly, the 1912 U.S. Government ordered breakup of DuPont, led to increased innovation and competition for a fledgling explosives industry. In 1913 the Institute of Makers of Explosives (IME) was organized to gather and distribute accurate information regarding the industry, set standards to maintain high and uniform explosives manufacturing standards, and inform the public regarding the proper use of explosives.

The first portable seismograph came on the scene in 1917, and in the twenties the seismographs continued to evolve. In 1926 a blasting cap safety education program was launched was by IME. It is certain that the increased competition in the blasting industry and the founding of IME did much to improve the state of the blasting art prior to the start of work on Hoover Dam.

Compressors and Air Tools

Hoover Dam was a huge drill and shoot rock excavation project. It would require the latest in technology from the compressed air industry.

In 1912 Chicago Pneumatic perfected the "simplate valve", which replaced mechanical valves on compressors – a breakthrough that soon became adopted by all compressor manufacturers. Consistent improvements in the compressed air industry continued and by the time Hoover Dam was ready to begin the industry offered many different sizes of stationary and portable compressors in diesel, gasoline, and electrically powered models. At peak construction, Hoover Dam would employ the use of several air compressor plants producing over 16,000 cubic feet per minute. Also, in 1912 the revolutionary lightweight, hand-held (jackhammer) sinker drill was invented. Many types of compressed air tools were used on Hoover Dam including "air tugger" winches, and air powered concrete vibrators.

Excavation Operations

Crane shovel capacity and mobility were an important consideration for contractors at Hoover Dam, and a key element in developing an efficient excavation plan. The first full-rotation, crawler-mounted crane shovels were coming on the scene around 1920. Through the twenties crawler-mounted cranes gained in capacity and were manufactured in front shovel and dragline configurations. Power options for cranes also improved greatly in the twenties with crane manufacturers offering various engine options including steam, gasoline, and diesel. Manufacturers also offered electrically powered shovel cranes. Tunnel mucking (excavation) machines were also evolving in the twenties.

Concrete Operations

The 4,400,000 cubic yards of concrete required for the Hoover Dam project were unprecedented, requiring many innovations. In the early 1900s, Portland cement manufacturing techniques continued to evolve with improvements to rotary kilns, ball mills, and other aspects of manufacture. In 1928 the grate pre-heater kiln was introduced providing the first major improvement in thermal efficiency from the previous long, wet kilns. Additionally, in the thirties, roller mills were first applied to cement manufacture.
The American Concrete Institute (ACI) was established in 1913 as the successor to the National Association of Cement Users. It was during the twenties that the ACI really came of age with a broad range of technical committees for evaluation of the design and arrangement of the site mixing plants and machinery, handling and delivery of materials, concrete placing, concrete reinforcement, and formwork.
In the twenties central mixing plants appeared on the construction scene and established their economic viability, rapidly gaining in popularity and ushering in the ready-mixed concrete business. By the late twenties, old volumetric methods for batching concrete materials had given way to weighing methods of controlling material proportions.
In 1927, an ACI committee reported on the economic advantages of "field control of concrete." The committee made recommendations on a wide range in topics, including the supervision of all the processes of concrete manufacture; quality of constituent materials; design of the mixture; mixing, placing, and curing; and field testing.
By 1929 bulk cement transport and handling methods were also developed and

along with batch plant weighing methods eliminated the need for labor-intensive use of bagged cement. The rapid increase in concrete plants in the twenties and the growing use of ready-mixed concrete led to proposed specifications for the design and operation of central mixing plants. In 1930, ACI Committee 504 was believed to have introduced the first formal specifications for ready-mixed concrete.

Also, with the aid of concrete vibration techniques emerging in the late twenties, dry concrete [low slump, low water-cement ratio concrete] was being economically placed on large jobs and by 1930 the ACI established several new committees on the effect of vibration on properties of concrete. And, as the start of Hoover Dam loomed, studies were being conducted on mass concrete, its thermal characteristics, and volumetric changes on the USBR's Owyhee Dam. It was at Owyhee (built 1928-1932), that USBR engineer Jack Savage introduced artificial cooling methods for mass concrete, thus removing the detrimental effects from the heat of hydration in months instead of a hundred years otherwise required.

Driven primarily by the need for more and more roads, larger and more efficient aggregate processing technology evolved through the twenties which allowed for increased production and material quality. Conveyor belts made from layers of cotton and rubber were common in the twenties, yet conveyor belt technology also underwent tremendous changes allowing greater capacity and longer transport distances in the decade preceding Hoover Dam.

In the early twenties, concrete formwork was primarily built in place, piece-by-piece from rough-cut dimensional lumber. These methods were inconsistent, inefficient, and generally did not provide an adequate finished surface. Williams Form Engineering came on the scene in the mid-twenties with their innovative "She-bolt" form tie system. This system allowed for stronger, more efficient reusable formwork designs. The Six Companies, Inc. employed the "She-bolt" system at Hoover Dam with great success.

By any measure, the decades preceding the construction of Hoover Dam were a revolutionary time for all of the processes of concrete manufacture: cement manufacture, aggregate processing, mix designs, batching, transporting, placing, curing, and field testing.

Welding

In 1920, automatic welding was introduced by General Electric Company. In the late twenties demand for higher quality welding methods increased and welding codes began to appear which required higher-quality weld metal through the use of covered electrodes. In 1924 a process for welding using a shielding gas was developed. In 1929, Lincoln Electric Company began producing and marketing extruded electrode rods for sale to the public. By 1930, covered electrodes were widely used.

Material Handling

The massive quantities of materials required for the Hoover Dam would dictate efficient material handling methods. The material handling workhorses at Hoover

Dam were the cableway cranes. These cranes relied on the latest wire rope technology, winch systems, and communication systems.

In spite of all of the modern machines and methods the Six Companies brought to bear to build Hoover Dam, there was still need for traditional methods. Men would still perform much back-breaking work by hand. During early mobilization burros were used to haul an air compressor in sections up steep terrain. Traditional steam powered trains were also an essential part of the overall transportation plan.

THE MEN OF HOOVER DAM

Frank Crowe, General Superintendent for the Six Companies, Inc. commented at the start of work "This will be a job of machines." (Stevens, 1998) Yet, technology and modern machines would not be enough to build Hoover Dam. There would also be a convergence of unprecedented engineering skill, knowledge, innovation, determination, and hard work in Black Canyon.

It was not enough to have a dream to build the world's largest dam. This dream had to be founded on sound engineering principles. This dream had to be founded on practice – man's collective skill to utilize men and machines to transform raw materials for the benefit of society.

Generally, large heavy civil engineering projects rely on a combination of new and traditional means and methods. Deciding which means and methods to use for each aspect of the work is driven by a complex set of criteria including the site conditions, geology, scope of the work, quantity of work, quality requirements, available access and staging areas for the work, schedule demands, cost of labor intensive methods compared to equipment intensive methods, skill and availability of labor and supervision, equipment availability and dependability, suitability of innovative methods, proper analysis of risk vs. reward, and other criteria.

To be sure, there are many ways to approach building a project, but there is only one approach which is optimum and provides the lowest cost. The challenge during a project bid proposal is to analyze a complex set of plans and specifications, evaluate which means and methods will deliver each operation at the lowest cost, merge all of these direct costs with the proper supervision and general project indirect costs, or overhead, to obtain a total project cost.

So in building a dam or anything else for that matter, the success of a project hinges on the skill of the men, the people who design, estimate, and build the work. Like the Hoover Dam, its men, its people were remarkable. There would be many men in the story of Hoover Dam. Thousands would work on Hoover Dam, yet Messrs. Davis, Mead, Weymouth, Savage, Young, and Crowe warrant commentary.

Arthur Powell Davis, F. ASCE and ASCE President 1920

Arthur Davis is regarded by many as the chief unsung hero of Hoover Dam. He was born in 1861 in Decatur, Illinois and earned a Bachelor of Science degree in Civil Engineering from George Washington University (then called Columbian University) in 1888. Davis was the nephew of explorer John Wesley Powell who led the Powell

Geographic Expedition down the Colorado River in 1869. Collectively between Powell, and primarily Davis a vision emerged to tame the Colorado River.

Davis worked for the U.S. Geological Survey from 1884 to 1894. In 1898 he also spent time working on the Nicaragua and Panama Canal routes. In 1906, Davis became Chief Engineer of the United States Reclamation Service, a position he held until his appointment to Director on December 10, 1914. During his tenure as Director, the USBR developed formal studies and plans to tame the Colorado River. Davis is credited to be the first to recommend construction of multi-purpose dams with hydroelectric powerplants to generate revenues to offset construction costs. When the Reclamation Service changed to Bureau of Reclamation on June 18, 1923, Davis retired the following day. Davis was appointed as a consulting engineer on Hoover Dam in 1933 and died later that year in Oakland, California.

Elwood Mead, F. ASCE

Elwood Mead was Commissioner of the USBR. Mead was born on January 16, 1858 and grew up in Indiana. He attended Purdue University, earned a Bachelor of Science degree in Agricultural Engineering in 1882, followed by a post-Bachelors degree in Civil Engineering from Iowa Agricultural College (presently Iowa State) in 1884. Purdue subsequently awarded Mead its first-ever honorary doctorate in 1904. Mead headed the USBR from 1924 until his death, January 26, 1936. During his tenure at the USBR he was responsible for overseeing some of the most complex projects the Bureau of Reclamation had ever undertaken including Hoover, Grand Coulee and Owyhee Dams.

Frank E. Weymouth, Hon. M. ASCE

Born on June 2, 1874 in Massachusetts, Weymouth attended the University of Maine and graduated in 1896 with a Bachelor of Science degree in Civil Engineering. Weymouth later joined the USBR in 1902 and led the construction of Idaho's Minidoka Project in 1904. In 1908 Weymouth became supervising engineer for the Idaho District of the USBR. During his time in the Idaho District he successfully oversaw the construction of then, the largest concrete dam in the world, Arrowrock Dam on the Boise River in Idaho. From 1916 to 1920, Weymouth was the Chief of Construction for all of the USBR work in the western states. He led the preparation of the feasibility studies for the Boulder Canyon Dam. Known as the "Weymouth Report", this work was used as the basis upon which the final decision on dam approval was based. Later, Weymouth would leave the USBR to become Southern California's Metropolitan Water District's (MWD) first general manager and chief engineer from 1929 to 1941 and oversaw the design and construction of MWD's initial distribution system and Colorado River Aqueduct.

John "Jack" Lucian Savage, Hon. M. ASCE

Jack Savage was the USBR's Chief Designing Engineer. Called the first "Billion Dollar" American engineer (for the cost of the structures designed under his supervision, not his fees), Savage was born on December 25, 1879 in Cooksville, Wisconsin. He attended the University of Wisconsin, Madison and graduated in 1903 with a Bachelor of Science degree in Civil Engineering. Savage was interested in designing large structures and decided to take a position with the newly created U.S. Reclamation Service. Savage's first assignment was in Boise, Idaho as a designer on the Minidoka Project in Idaho's Snake River Valley.

In 1908, Boise area engineer A.J. Wiley secured a large private sector consulting commission to design the world's highest dam at the time-the 350-foot Arrowrock Dam, a concrete gravity-arch structure to be built on the Boise River. Needing a skilled assistant, Wiley lured twenty-eight-year-old Savage away from the government. (Weingardt, 2008)

Frank Weymouth, the construction engineer for the Service on the Arrowrock Dam project was very familiar with Savage's talents and was able to convince Savage to return to the United States Reclamation Service in June of 1916 as a design engineer for the Bureau's Denver Center. With a record of sustained success, Savage was named the Chief Designing Engineer for the USBR in 1924.

While working at the USBR, Savage was responsible for design oversight for many projects throughout his career including Hoover (Boulder) Dam, Parker Dam, Shasta Dam, the All-American Canal System, Grand Coulee Dam, and dozens more.

Undoubtedly Savage and his team would face unprecedented challenges in constructing these monumental structures. Many construction related problems are the result of incomplete designs or unconstructable designs. Regarding constructability, it was said of Savage, "He was able to foresee unusual problems which could arise during the design and construction of Bureau structures" and "he required solutions to these problems before completion of the designs; and he recognized the importance of engineering research and development for providing data needed to design adequate structures." (Wolman, 1978)

Like many great men Savage was humble, self-effacing, and totally dedicated to his work, and his profession, and to public service. He left to posterity monuments as uplifting and permanent as any created in the history of civilization. (Weingardt, 2008) Savage died on December 28, 1967.

Walker "Brig" R. Young, Resident Engineer, F. ASCE

Walker Young graduated from the University of Idaho in 1908 with a degree in mining engineering. Young joined the USBR in 1911 to work for Frank Weymouth on Arrowrock Dam near Boise, then the highest dam in the world. Young's engineering skills were tested early, being assigned as assistant engineer in charge of design and inspection on Arrowrock Dam.

Between 1921 and 1924 Young supervised a USBR team responsible for preparing feasibility reports for dams in Boulder Canyon and Black Canyon. With the study completed by 1924, a detailed report submitted to the Secretary of the Interior who concurred that Black Canyon was the better site for a dam.

Young had proven himself a competent engineer and a good manager, so the USBR assigned him to be the engineer in charge of the Boulder Canyon Project (Figure 2). Young was assisted by Ralph Lowry, Field Engineer, and John C. Page,

Office Engineer, Hon. M. ASCE. Page went on to become the Commissioner of the USBR in January 1937. Young rose to the rank of Chief Engineer of the USBR before his retirement in 1948 at the age of 63.

Francis "Frank" T. Crowe, General Superintendent, Hon. M. ASCE

Remarkably, Frank Weymouth was not only a brilliant engineer, he also a good judge of engineering talent. It was in 1904 that Frank Weymouth took time to go back to the University of Maine to recruit engineers for the USRS. While at the University of Maine, Weymouth lectured students on the challenges and goals of the USRS to transform the west by building ambitious irrigation and dam projects. It was during one of Weymouth's lectures that civil engineering student Frank Crowe became intensely interested in the work of the USRS and approached Weymouth for a summer job. Weymouth agreed and at summer's end Crowe knew building dams was what he wanted to do.

FIG. 2. Walker "Brig" Young (left) and Francis "Frank" Crowe. Courtesy Bureau of Reclamation

After graduating in 1905, Crowe returned to the west to start working full time for the USRS under Weymouth as an instrumentman. Crowe's first assignment was on the Yellowstone irrigation project. Growing impatient, Crowe left the USRS in 1906 to work with contractor James Munn on private sector opportunities, and soon Munn introduced Crowe to two local builders, Harry Morrison and Morris Knudsen. In the fall of 1908 Weymouth contacted Crowe about working on the Minidoka Project in Idaho. On July 5, 1910, the Jackson Lake Dam failed and Weymouth sent Crowe to investigate. Crowe seized the opportunity and by the fall was promoted to full engineer and led efforts to repair the dam. With no time to rest, Weymouth assigned Crowe to the record breaking Arrowrock Dam in Idaho, in late 1910. Finally Crowe would get his chance to build large dams; Arrowrock would be the largest concrete dam in the world, measuring 350 feet in height and requiring over 500,000 cubic yards of concrete.

It was at Arrowrock that Crowe honed his skills and refined many construction techniques such as the use of cableway cranes for material handling. Working with James Munn, Crowe developed and installed a 12 ton capacity cableway with moveable tail towers to allow complete coverage of the work area. The cableway was capable of handling 8-ton concrete placing buckets with a vertical hoist speed of 300 feet per minute and a horizontal travel speed of 1,200 feet per minute. In order to provide directions and signals for the cranes, a new Bell Telephone system was installed in addition to a backup bell system.

Arrowrock Dam was completed in 1915 and during the next ten years Crowe continued to build on his reputation in the USBR as an innovative and hard-charging construction man working on a succession of USBR projects such as the Pioneer Drainage Works, Tieton Dam, and the Yakima Project. With each project, Crowe not only proved his strength in the field, but also proved his worth in the office assembling large complex cost estimates and work schedules.

Crowe's ability to solve problems, improve the efficiency of men and machines, and make schedule was only overshadowed by Crowe's desire to build the biggest dams of the day. However, in 1925 the USBR was directed to contract dam and irrigation projects to private companies, and Crowe soon became stuck behind a desk instead of pushing work in the field. Crowe grew impatient and commented, "Do you know what I feel like sitting at this desk shuffling papers? I feel like a bull in a china shop." (Rocca, 2001)

Frustrated, Crowe resigned from the USBR on June 1, 1925 and went to work for Morrison-Knudsen (MK) of Boise, Idaho, who had a subcontract with Utah Construction Company (UCC) to supply the labor on the Guernsey Project in Mitchell, Nebraska. "Within days, everyone in and around Guernsey was talking about Frank T. Crowe. He would be the boss, and the boss was fair, but he wanted results. Nothing less would be tolerated." (Rocca, 2001) While at the USBR Crowe as always known for his ability to reach and sustain high levels of production. Now in the private sector, Crowe seems to further sharpen his ability for optimizing construction operations by attacking work on multiple headings, optimizing linear operations, and constantly pushing for day-by-day improvement of operations.

In September 1927 Crowe prepared an estimate for the Van Giesen Dam in Nevada and MK was the successful low bidder. Crowe, along with his carpenter superintendent Charlie Williams further honed their concrete construction methods and completed the job by May of 1928. Crowe then returned to Boise to serve as an advisor to Harry Morrison. They spent considerable time discussing the upcoming Hoover Dam Project and Crowe encouraged Harry Morrison and Morris Knudsen to seriously think about preparing to bid on the project. During 1929 and 1930 Crowe made many visits to Las Vegas to gain firsthand knowledge about Black Canyon and the surrounding area.

In July 1929 Crowe once again demonstrated his estimating and bidding abilities when MK / UCC was low bidder on the Deadwood Dam project. And once again Crowe would rely on Charlie Williams as his carpenter superintendent. Early in the job, Chief Designing Engineer for the USBR, Jack Savage visited Crowe to help evaluate aggregate sources for concrete. Later in the project as excavation neared completion, Savage again returned meet with Crowe to inspect the dam foundation. Crowe and Savage were sure to have much in common, both knowing Frank Weymouth and working in Idaho on USBR projects some 25 years earlier. And, once again Crowe employed his signature cableway system for material handling and received a steady stream of USBR visitors to evaluate his methods.

BUILDING HOOVER DAM

"Now this is just a dam but it's a damn big dam..." William Wattis: (Rocca, 2001)

By late fall of 1930 the Deadwood Dam project was nearing completion and "Harry Morrison [MK] and the Wattis Brothers [UCC], once again impressed with the abilities of the ambitious Crowe, came away from the Deadwood project feeling that their man could build anything anywhere." (Rocca, 2001)

It was Crowe who guided the MK/UCC alliance into the arena of dam building with success at Guernsey and Deadwood Dams. Crowe's performance on these projects gave confidence to the alliance to seriously consider bidding the Hoover Dam project. Crowe projected Hoover Dam to cost between $40,000,000 and $50,000,000. To be able to realistically go after the project, Morrison and Crowe determined they would need working capital of $5,000,000, yet the MK/UCC alliance could only come up with $1,500,000. So the Hoover Dam team would have to be expanded. It would take much effort during 1930 to find other like-minded construction men capable and ready to take on such a challenge.

With persistence, an agreement to form a consortium was finally established in 1930. The team members and their share of the venture were as follows: (*Compressed Air Magazine*, 1931-1935)

1. Warren Bechtel, San Francisco, California and Henry Kaiser, Oakland, California, 30%
2. Utah Construction Company, Ogden, Utah, 20%
3. MacDonald & Kahn Los Angeles, California, 20%
4. Morrison-Knudsen Company, Boise, Idaho 10%
5. J.F. Shea, Portland, Oregon 10%
6. Pacific Bridge Company, Portland, Oregon 10%.

In mid-February, 1931, the representatives of the Six Companies met in San Francisco to perform a bid review. Crowe had prepared many bids before, but nothing like Hoover Dam. The project had 119 different bid items, with most of the project risk in the bids for placing over 3,250,000 cubic yards of dam concrete and excavating over 1,500,000 cubic yards of rock in the tunnels. The meticulous engineer had detailed these difficult tasks and the others, calculating down to the man-hours required to complete each task." (Rocca, 2001)

Three of the six companies had prepared bids for review; MK (Crowe), Kahn, and UCC. Surprisingly, all of the cost estimates were relatively close at about $40,000,000, yet, "It was agreed that the three bids would be read, discussed, and then compromised." (Rocca, 2001) They calculated a peak cash outlay of $3,200,000. The venture agreed on the name of Six Companies Incorporated (Six Companies). Although ailing, it was agreed W.H. Wattis would serve as president.

Crowe spent the second half of February back in Boise. Here, Crowe worked diligently to construct a detailed scale model of the dam, complete with "movable parts for demonstrating crucial operations." (Rocca, 2001) He planned to use the model in his final briefing and discussions with the Six Companies officials, hoping to convince them his low bid was realistic. Crowe took his model back to San Francisco to show the directors. William Wattis' health continued to fail, and so the Six Companies leadership met in Wattis' hospital room for final discussions on the bid and agreed on a profit margin of 25%.

In the days leading up to bid day, Crowe would work long hours checking and rechecking hundreds of pages of calculations in order to finalize the bid. Then on the morning of March 4, 1931, Crowe submitted the bid for the Six Companies to the USBR. It was reported 100 sets of drawings were issued by the USBR to prospective bidders. Assuredly, it would have been a nerve racking time for the Six Companies. How many bidders would there be? Would they be low bidder; and if so, how much money would they leave on the table? They soon got their answer, only three responsive bids, and they left $5,010,000 on the table.

Six Companies:	$48,890,000
Arundel Corporation:	$53,900,000
Woods Brothers:	$58,600,000

In September 1933 *Fortune* magazine reported, "For their $48,890,995 the Six Companies must foot all construction bills - for dynamite, for trucks, for digging mud and dumping mud, for bosses' salaries, and for labor's wage. The Six Companies do not pay for construction raw material - for the 5,500,000 barrels of cement consumed, or the 55,000 tons of steel plates and castings, or the turbines and generators in the power plant, or any of the permanent operating machinery of the dam."

Crowe was immediately announced as the General Superintendent. In the Los Angeles Examiner it was reported; "Frank T. Crowe of Boise, Idaho is to be the General Goethals of Boulder Dam. As Goethals was the dynamite force that kept his organization driving ahead during tropic heat and jungles to complete the Panama Canal, so Crowe will be the mainspring of the machine that will build the world's greatest dam." (Rocca, 2001)

Crowe wasted no time getting started and had a temporary office established in Las Vegas within a week. Crowe announced that 3,000 men would be employed. It was time for Crowe to start building his team.

On April 20, 1931 the USBR gave the Six Companies official notice to proceed with the work. In order to have a successful job, Crowe knew he would have to complete the four tunnels and divert the Colorado River by October 1, 1933, or pay a $3,000 fine for every day it ran over the deadline. (Stevens, 1998) Immediately, Crowe set the pace for the job, ordering work be performed 3 shifts per day, 7 days a week.

Access roads from the top of the Nevada side to the bottom of the canyon, and bridges from the Nevada side to the Arizona side would take precious months to complete. There was no time to wait for land based access, so Crowe ordered an amphibious assault on the walls of Black Canyon in order to get the tunnel operations going.

Management

Crowe was known for his tireless efforts and attention to detailed planning. Most assuredly, Crowe's mind would be racing day and night during the first months of the Hoover Dam construction. He would have been consumed by team building, supervising, planning the work, and dealing with the USBR. It had to be overwhelming at times, yet some of his biggest problems early in the project were not from building the work.

Each of the principals of the Six Companies consortium were proven construction men in their own right, headstrong and demanding, and they were all eager to offer Crowe advice during the spring of 1931 as work got underway. For example, Morrison thought of a job in terms of major equipment like draglines and steam shovels. Kaiser emphasized maximizing the efficiency of men and machines. Kahn's focus was on money and organization charts. And, Charlie Shea always thought in terms of men.

Most likely, the principals would offer their advice and directives with only the best of intentions. Yet, one week a principal would show up and give Crowe specific instructions on how to handle some aspect of the work, and a week later another would show up and give different instructions. Edgar Kaiser, son of Henry Kaiser recalled in 1968: "Frank Crowe often used to say that his most serious problem in the building of Hoover [Dam] was the guest house." (Stevens, 1998) Apparently, a lot of time during early project meetings was spent on where to build the guest house and who would stay in it. This had to be frustrating to Crowe with so many urgent issues to resolve.

Of all of the board members, Crowe enjoyed Charlie Shea's visits and felt comfortable talking with him about job problems including the board member's mixed messages. Crowe explained how the mixed messages led to confusion, wasted effort, and undermined his authority on site. Convinced that he had to protect Crowe, Shea went to San Francisco to convince the board their meddling was hindering the work.

The Six Companies directors had finally realized that, as Felix Kahn put it, "a board of directors can establish policy, but it can't build a dam." (Stevens, 1998)

In order to establish clear lines of communication with Crowe, the board voted to create a four-man executive committee. The Six Companies Executive Committee was comprised of the following members: (*Compressed Air Magazine,* 1931-1935)

- Henry Kaiser, Chairman
- Charles Shea, Director of Construction
- Felix Kahn, Manager of Boulder City Company
- Stephen Bechtel, Purchasing, auditing, and warehousing.

In order to build work momentum, Crowe employed the time-honored strategy of divide and conquer, by assigning some field supervision to complete early critical field mobilization activities including housing, access roads, and utilities. Still other superintendents and engineers were assigned to work in the office planning future work. One of the most important of these early planning activities was to develop a plan of attack to divert the Colorado River during the winter of 1932-33.

Finally by early summer of 1931, buoyed by the promise of no more executive interference and by the availability of ample [electric] power to drive machinery and illuminate the dam site Crowe let it be known that he was ready to start "hi-balling" the job. (Stevens, 1998)

Mobilization and Initial Temporary Works

In order to build any structure, it is necessary to mobilize the project and construct sufficient temporary works to facilitate construction. Typically, these efforts are directly proportional to the magnitude of the permanent works being built. So not only was the size and scope of the Hoover Dam permanent works unprecedented, the mobilization effort and temporary works were also unprecedented.

In the mobilization phase, the Six Companies estimated they spent $2,000,000 (over 4% of the original contract value) before the first shovelful of "pay dirt" was turned over. (*Compressed Air Magazine,* 1931-1935)

Housing

One of the Six Companies' most important early work items to complete was the construction of fully operational worker's camp known as Boulder City. Yet, Boulder City would not be ready for housing workers until nearly the end of the summer of 1931. This was not soon enough as rapid mobilization of men and machines was essential to Crowe's plan to attack the most critical work, the diversion of the river by October 1933.

In order to keep from having to bus workers all of the way to and from Las Vegas Crowe said, "We must have a place to eat and sleep before we can put men out there." (*Compressed Air Magazine,* 1931-1935) Further, Boulder City would take months to become operational and accessible for the first 500 men necessary to get the job started. So in the spring of 1931, Crowe put his carpenter superintendent, Charlie Williams ("the wizard with wood") to work building the "River Camp" a temporary worker's camp at a location called Cape Horn. The River Camp would need to house about 500 men there during the start of the job. This facility included six 80-man dormitories, a mess hall, commissary, and a recreation room. Located upstream of the dam, River Camp provided for better access to start operations in the canyon.

> "We found in our investigations that the temperature was at least 10 degrees lower than at the dam site. We also found that the air currents, particularly from the lake area, from around Hemenway Wash, created almost a continual breeze, very slight on some occasions, very stiff and very severe in other instances. It wasn't a very attractive site for us, but better than anyplace else in that particular area." (Dunar and McBride, 1993)

By the June 1931, 700 men were at work, and by the end of summer the camps could accommodate 2,000 men. In September 1933 *Fortune* magazine published an informative story about Hoover Dam and life in Boulder City:

"The gang on the job varies with the various steps in the dam's progress. The maximum estimated, but never reached, was 4,000. Less than 3,000 are at work this summer. This horny-handed army enjoys a tidy comfort that seems luxury compared to army camps of 1918. They eat and sleep in Boulder City, built on a U.S. reservation. Nobody can build houses or sell so much as a radish without a U.S. permit. And 80 per cent of the workers must live on the reservation. A community of some 5,000 is the result, with 1,050 houses on as parched and barren a patch of wind-swept rocky desert as could be found if one were seeking an ideal spot in which not to live. But the married men have trim cabins; bachelors live in huge refrigerated dormitories, each man with a seven-by-ten room for himself."

"Bachelors eat in a mess hall serving excellent food, including iced tea and ice cream. The food is cooked and served on contract by Anderson Brothers, caterers who feed thousands of western laborers a day, including some workers in the rich vineyards of the movie locations. The quality of the food is guaranteed by a twenty-four-hour cancellation clause in Anderson Brothers' contract. The quality of the housing was dictated by the discovery that the better the workers' conditions, the faster they dug the dam. For their first-rate food and private rooms, plus transportation eight miles to the job, the construction stiffs pay $1.60 per day."

FIG. 3. Boulder City, Nevada. Courtesy Bureau of Reclamation

Railroads

To be sure, establishing access to the work was one of the early challenges of building Hoover Dam, yet methods and machines for pioneering roads and building railroads were well established, so the Six Companies elected to subcontract most of this work. The quantities of materials required to be shipped to and around the work site were enormous. To build Hoover Dam, raw materials like cement and steel, and equipment like shovel cranes, draglines, bulldozers, trucks, concrete plants, gravel plants, and air compressors all in mass quantities had to be shipped to the site. Further, on the work site, railroads would play a key role in the Six Companies plan. Railroads would be relied upon for transport of 8,600,000 tons of sand and gravel to produce the concrete for the project. All the sand and gravel would have to be transported from the Arizona pits over 7 miles the gravel plant in Nevada. The rail link to the gravel pits required the construction of a 1,140-foot-long river crossing trestle. All of the Six Companies tracks were constructed using Standard gauge, 90 pound rails and Oregon Fir ties.

Prior to the Six Companies mobilizing, the USBR had already contracted with Union Pacific to install a 22-mile branch from Las Vegas to Boulder City, and with Lewis Construction Company to install a 10-mile railroad spur from Boulder City to edge of the Nevada canyon wall. The Six Companies would eventually build 20 miles of railroad to tie into the USBR's rail lines. It was estimated the Hoover Dam rail system would haul 33,000,000 tons of live load, and 440,000,000 ton miles of dead load and live load combined. Locomotives were estimated to travel an aggregate distance of 700,000 miles, hauling 63,000 trains. (*Compressed Air Magazine,* 1931-1935) Once the sand and gravel were processed, approximately 3,000,000 tons had to be transported nearly 5 miles to the low-level mixing plant, and approximately 5,000,000 tons transported 10 miles to the high-level mixing plant.

Railroads would also be used for moving excavated material from tunnels, cofferdams, and dam to disposal areas. Most tunnel excavation was hauled by truck, yet vast quantities were also moved by rail. It is estimated the Six Companies moved almost 3,000,000 cubic yards of muck (excavated spoil material) by rail. Railroads also delivered the cement for the project, as well as other long-haul transported materials. When concreting operations are at their height, this railroad will carry a volume of traffic heavier than that on any main line in the country. (*Compressed Air Magazine,* 1931-1935)

Roads

Crowe would look for as many access points as possible to get men and materials in and out of Black Canyon. As work mobilized in the spring of 1931, the Six Companies had a 2-mile highway built to connect the end of the USBR road on the top of the Nevada side of the canyon with river level at a point near the site of where the diversion tunnel outlets would be built. This was about a half mile downstream of the dam site. Building this access road was tough work with most of the alignment requiring rock excavation.

Upstream of the dam site the Six Companies installed another access road to river level at Hemenway Wash, which provided a natural approach to the river at a point some two miles above the dam site.

Temporary Construction Utilities; Electricity

To facilitate the aggressive construction schedule, the USBR awarded a separate $1,500,000 contract to build a 222-mile power line from San Bernardino, California to a substation on the Nevada rim of the canyon. This transmission line delivered 80,000 volts and was turned on June 25, 1931. Later, a 6.8-mile 33,000-volt, timber pole line was added to feed Boulder City, and a 0.7-mile 2,300-volt line to furnish power to Pump House No. 1 for the Boulder City water supply.

Some 30 miles of power lines had to be run from the substation to various points in the canyon to serve electric shovels, electric-driven air compressors, pumps, tunneling machinery, etc., and to furnish current for lights within the tunnels and outside flood lights for night work.

Temporary Construction Utilities; Compressed Air

Frank Crowe was already a strong believer in the use of compressed-air tools having studied their capabilities at the University of Maine and successfully employing them on his previous dam work.

During the estimate Crowe planned for a total installed compressor capacity of 25,000 cubic feet per minute. Actual job requirements turned out to be less, about 16,000 cubic feet per minute, still a huge challenge for a temporary construction site. Crowe and his equipment superintendents made a conscience decision to select one manufacturer of air compressors for ease of maintenance and spare parts inventory.

During the early stages of the project there were up to 18 Ingersoll-Rand Type 20 and Type XL portable air compressors. The first stationary compressors were Ingersoll-Rand Type XRE which had 1,302 cubic foot displacement. These stationary units were initially powered by 200 horsepower diesel engines, and then in June, 1931 these were switched to electric power. At the lower tunnel portal compressor plant, four each Type XRE and two each Class PRE machines were utilized during the tunneling work with a total displacement 9,598 cubic feet per minute.

Temporary Construction Utilities; Water System

Water pumping systems were widely used at Hoover Dam and essential to its construction. Pumping and distribution had to be installed and maintained for Boulder City, the aggregate processing plant, concrete batching plants, and other applications.

Personnel Access

With three shifts every day and thousands of workmen to get to and from Boulder City to their work operations, Crowe and his team would have to develop efficient means to transport men. Getting men to and from Boulder City to Black Canyon was done using various types of converted flat-bed trucks equipped with bench seating. The biggest of these was known as "Big Bertha" with a capacity of 150 men (Figure 4). Apparently, the Big Berthas were slower than the smaller truck transports so men would try to avoid them if possible.

As work progressed in the canyon, photographs show a whole array of access methods including stairs, ladders, and catwalks. Early photographs even show some stairs descending from the top of the canyon all the way to the bottom, one hell of a climb.

Aerial cableways, including the small "Joe Magee" cableways, were also used to get men into and out of the canyon. Additionally, there was a "Monkey-Slide" elevator which was a platform which slid on skids propped off of the canyon wall and was hoisted and lowered using an air operated winch known as an "air tugger". As fast paced as the work was, Crowe would need to have a considerable sized crew devoted to maintaining access around the project.

FIG. 4. The Six Companies used several types of personnel transports to ferry men to and from Boulder City including the slow-moving 150-passenger "Big Bertha" bus. Courtesy Bureau of Reclamation

Tunnel Excavation

The official USBR bid tabulation called for 1,563,000 cubic yards of Tunnel Excavation. The Six Companies bid this work for $8.50 per cubic yard. The primary challenge of the tunnel excavation was in the four 56-foot-diameter, 4,000-foot-long diversion tunnels. The total length of the four diversion tunnels was 15,900 feet. The scope of the tunneling task was unprecedented, and considering their overall length, the tunnels were a world record for their day. (*Compressed Air Magazine*, 1931-1935)

The primary rock being mined in the tunnels was andesite tuff breccia, an igneous formation. Throughout tunneling the tunnels remained dry with no major faults, maintenance grouting was minimal, and the average overbreak was 7 inches.

Crowe's tunneling plan called for opening up as many work headings as possible. To accomplish this, Crowe's plan called for mining cross adits perpendicular from the canyon walls to intersect with the diversion tunnels near the center of the future dam. In the spring of 1931 there was no land access to the dam area at the base of Black Canyon, so Crowe devised an amphibious assault plan to gain access for the early adit work. Barges were assembled upstream at Cape Horn and were moved downstream of Cape Horn laden with portable air compressors, hoses, drifter drills, jackhammers, and even a blacksmith shop for sharpening drill steel. The barges were able to make land in an area close to the main core of the dam on a relatively level area of loose rock. Remarkably, in just a little more than two months after bid day, the first production blasting holes were drilled for the Arizona cross adit on May 12, 1931.

After gaining a beach head on the Arizona side and the Arizona cross adit was underway, Crowe had another crew attack the Nevada cross adit, gaining access by having a suspension bridge installed from Arizona side to the Nevada cross adit. The Nevada cross adit was much trickier, due to its sheer rock face dropping off into the river. Crews had to work entirely off of barges until enough tunnel muck could be dumped into the river to create a foothold.

Prior to permanent electric power being established in Black Canyon the Six Companies used gas and diesel generators to supply electricity for the tunneling operations. The initial removal of tunnel spoils was done by hand loading into 1 cubic yard mine dump cars. Crowe's team constantly explored methods to improve productivity, and hand mucking was soon replaced with "Conway" tunnel shovels.

When the cross adits reached the diversion tunnels, 12-foot by 12-foot top headings were started in both the upstream and downstream directions. This strategy created eight headings to allow more men to be put to work. By starting with the top headings, tunnel ventilation and access were established for later tunnel enlargement. Further, by advancing the top headings, engineers were able to investigate the quality of the in-situ rock before the main tunnel heading was advanced. The bulk of top headings were accessed from the adits. As the summer of 1931 progressed, Crowe's team would also attack the top headings from the portals, thus opening as many as 16 headings, but in practice the most in production at one time was 12 headings. Common practice was for crews to alternate top heading operations, drilling one and mucking another.

The back-breaking work in the tunnels was really about all a man could bear, yet as work progressed through the summer temperatures soared in Black Canyon to over 130 degrees Fahrenheit. Crowe and his foreman noticed men beginning to slow and wither. Rock surfaces soon absorbed the searing heat requiring heavy gloves to be worn by cursing rock mucking miners. The canyon effectively cutoff air circulation, stifling workers who stripped to the waist and gasped continuously for relief. When the wind did come, it felt more like an oven door opening than a refreshing breeze. (Rocca, 2001)

During June and July of 1931, there were 13 recorded deaths due to heat prostration. Yet, the heat was not the only safety concern. There were hazards from handling explosives, high-voltage power, truck traffic, and access the workers would face.

For tunnels No. 1 and No. 2 on the Nevada side, Crowe put Bernard "Woody" Williams, son of carpentry boss Charlie Williams in charge. For tunnels No. 3 and No. 4 on the Arizona Side, Crowe put Floyd Hunnington in charge. When Crowe met him he realized "that here was a hard-driving determined construction stiff who had "pulled himself up by his own boot straps" - much as he had." (Rocca, 2001)

On September 21, 1931, the downstream portal of tunnel No. 4 was the first tunnel to begin enlargement.

Almost as if the Colorado River sensed man's plan to change its natural course, a flash flood came without warning into Black Canyon on the night of September 26, 1931. Crews quickly worked to build dikes around the tunnel entrances, and fortunately the flood receded as quickly as it came.

The 56-foot-diameter tunnels were excavated in multiple headings from multiple tunnel portals to allow a maximum amount of resources to be employed at once. Each 56-foot-diameter diversion tunnel was split into three headings after the initial 12-foot by 12-foot pilot bore had been opened. The top heading was 12 feet in height, the main heading or "bench" was 30 feet in height, and the bottom heading was 14 feet in height. The 30-foot main bench was the most challenging with access required to drill some 90 horizontal blast holes spread over a 15,000 square foot vertical face.

Early in the tunneling operations, scaffolding was erected to gain access to drill the blast holes. Scaffold erection and removal was very time consuming and inefficient. Universally, enterprising builders and engineers love a challenge and an opportunity to develop innovative solutions. In the case of the tunnel excavation, tunnel superintendent Woody Williams devised an access system which allowed for the drill scaffolding to be mounted on a truck, thus allowing all access to be simply driven into position, and driven away after completion. This innovation became known as the William's Jumbo (Figure 5).

It was an ingenious, highly effective piece of equipment that proved to be one of the keys to beating the deadline for tunnel completion. At peak tunneling operations, five jumbos were in use, four of which alternated between headings, and one as a back-up.

Each jumbo carried 25 sets of 1-¼-inch hollow drill steel. The abrasive rock would quickly dull the drill steel, consuming about a third of a pound of drill steel per cubic yard of rock excavated. Drill steel was sharpened after every round of production at one of blacksmith shops established by the Six Companies. Like production tunneling, the blacksmith shops worked on three shifts keeping 60 men busy.

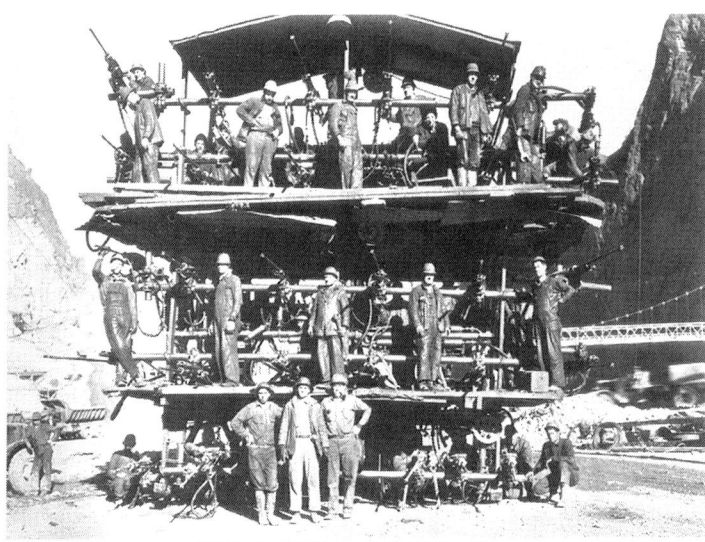

**FIG. 5. Williams' Drill Jumbo
Courtesy Bureau of Reclamation**

The main 30-foot-high production tunnel bench was split into two haves for drilling. The drill jumbo was positioned on the first half of the bench and a 6-inch air line and a 2-inch water line were hooked up to the jumbo for drilling. The average set up time to spot the jumbo and hook-up utilities was 20 minutes.

During the tunnel excavation phase of Hoover Dam, *Compressed Air Magazine* published a story and reported it was "possible to truthfully say that the rock drill, more than any other one class of tool or machine, has been the principal performer during the early stages of the work." Adding, "compressed air is the breath of life to the rock drill." (*Compressed Air Magazine,* 1931-1935)

After drilling, a truck loaded with blasting materials backed into the tunnel and the holes were loaded with 40 percent gelatin dynamite, tamped into place, and primers set. Typical density of dynamite loaded into the drill holes was 2.38 pounds per cubic yard. After the first half was loaded, the jumbo was repositioned to the second half and the process was repeated. Average drilling time on a bench was 4 hours for a 16-foot length. (*Compressed Air Magazine,* 1931-1935) The jumbo was then moved out of the tunnel to the adjacent tunnel bore to repeat the process. Each main bench crew was comprised of 50 men; 22 miners, 21 chuck tenders, 5 nippers (apprentices), 1 safety miner, 1 foreman or shifter. (*Compressed Air Magazine,* 1931-1935) In advance of the main bench excavation was the top heading drilling operation which employed two crews of 15 men.

Once ready for blasting, the tunnel was cleared and the shot initiated. Production tunnel excavation then started. A 100-ton Marion 490 electric-powered shovel was then moved into position to load tunnel muck directly into dump trucks. Unlike the jumbos, the shovels could not be quickly moved from tunnel to tunnel, so eight shovels were employed, one for each heading.

To optimize shovel excavation, Caterpillar tractors were also used to help feed muck to the Marion shovels and clean the floor of the bench. The tractors were equipped with "bulldozers" to push muck and "cowdozers" to pull muck. Initially, mucking of the top heading was done by hand, and later a Caterpillar 30 was used to push muck out to the main bench face for removal by the Marion shovel.

The Marion shovels were equipped with 3½-cubic-yard buckets which loaded the dump trucks. Various makes and models of dump trucks were used. Equipped with special rock bodies, trucks ranged in capacity from 7 to 14 cubic yards. Typically, there were four active tunnel excavation headings occurring at any given time, one Arizona upstream portal, one Nevada upstream portal, one Arizona downstream portal, and one Nevada downstream portal. Each of the four headings was serviced by 25 dump trucks. The average excavate and haul spread production was 110 cubic yards per hour and the average mucking time was 9 hours per round excavating 1,000 cubic yards per round. (*Compressed Air Magazine,* 1931-1935) The last phase of production tunnel excavation was the invert which used its own jumbo. A custom-built platform was used to provide access to trim and scale the tunnel before concrete operations.

Once in full swing, tunnel operations were conducted 24 hours a day in three 8-hour shifts, there were as many as 1,500 men engaged in tunneling activities, and the average number employed has been 1,200. (*Compressed Air Magazine,* 1931-1935)

Diversion tunnel No. 2 on the Nevada side and diversion tunnel No. 3 on the Arizona side would also be used as permanent tunnels to carry water in header pipes from the four upstream intake towers to the powerhouse. The intersecting tunnels between the intake towers and the header pipes were excavated 41-foot diameter on a slope. The header tunnels were drilled for excavation using a full face drilling jumbo mounted on rails. Normal advancement was one 12-foot round per shift. Electric shovels and trucks were used for the excavate and haul operation. The Nevada header was 1,322 feet long and required about 75,000 cubic yards of excavation. The Arizona header was 1,197 feet long and required 67,000 cubic yards of excavation.

Both Crowe and Shea were hands-on superintendents and visited the tunnel work almost daily to support superintendents Williams and Hunnington. Like their modern-day counterparts, Crowe and Shea would gladly offer advice on safety, production, quality, labor management, equipment, or any other topic which needed to be addressed. While Crowe and Shea were both known to be demanding, they were also very supportive of their men.

Speaking of Crowe and Shea, tunnel worker Saul "Red" Wixson said, "He was down on the job. He didn't want to listen to what was going on down there. He wanted to see it with his own eyes. I never saw Crowe get excited about anything. If something went wrong, he was there to get an eye on it, to explain what was wrong, fix it. He was there to help you, not fire you. He respected his men. And Charlie Shea'd come down - and he was usually with Crowe. "We'd always say, 'Here comes Crowe-Shea! Here comes Crowe-Shea!'" (Dunar and McBride, 1993)

Of course, it would take more than Crowe's bountiful energy to accomplish his goals; he needed support, and support he received. It is clear that Crowe's management policy of "tough, but fair" struck deeply in his staff of foremen, assistant superintendents, and office personnel. What he expected from them is what they had to expect from the workers under them. Regular staff meetings ensured that foremen knew exactly what the short and long-term work objectives were. (Rocca, 2001)

Crowe was known to also tap into the competitive nature of his men, setting a challenge to exceed another team's performance. Tunnel worker Marion Allen said, "The main thing was the challenge. You had to beat the other crew. It didn't make any difference what you did, you had to beat the other crew. You had to get more footage. 'We got two more feet than you did!' This was the whole conversation.'" (Dunar and McBride, 1993)

As the work in the tunnels advanced deeper and deeper there were also safety concerns about the fumes from the trucks. The noxious gas from the exhausts of the muck-hauling trucks and Caterpillar tractors mixed with the haze of dust and powder smoke that hovered in the tunnels and slowly poisoned the men who breathed it day in and day out. Nevada mining safety laws specifically forbade the use of gasoline engines underground, and on November 7, 1931, the state inspector of mines, A. J. Stinson, ordered Six Companies to stop using gasoline-powered trucks to haul muck out of the tunnels. (Stevens, 1998)

The injunction was filed, but failed, and the case was postponed for six months. During this time the tunneling continued. Eventually, the case was heard by a panel of federal judges in San Francisco. Not only did the Six Companies have concerns about the case, the USBR had much riding on the outcome of the hearing, so government lawyers assisted the Six Companies lawyers. Ultimately, a judge ruled the issue of fumes and carbon monoxide was a moot point because the diversion tunnels were nearly finished.

Regarding the tunnel ventilation, worker Harry Hall commented, "Later on they had air pumped into the tunnels, but there was a lot of exhaust fumes from the trucks so that the condition of the oxygen in there really was not too great. The most oxygen we had in the air was 16 percent, and I'm sure it was less than 16%. Miners used to take canaries and put them in a cage. They didn't realize that the exhaust fumes were as dangerous as they were." (Dunar and McBride, 1993)

Healthy air is 19.5 percent oxygen and considered a minimum permissible oxygen level to ensure no ill effects. If workers were in fact breathing 16 percent oxygen for an extended period of time, they would tire quickly and their coordination would have been impaired. At 12 to 15 percent oxygen, respiration and pulse would noticeably increase and men would have experienced impaired coordination, perception, and judgment. Depending on an employee's physical condition, some workers certainly could have been overcome by the air quality. Coupled with the heat during the summer of 1931 there is little doubt conditions would have been hazardous.

As the worked progressed tunnel ventilation was improved by the addition of high capacity blowers and help from natural air movement. It was reported the air in the tunnel could be cleared 5 minutes after a blast and the air quality was comparable to the carbon monoxide levels in New York's Holland Tunnel. (*Compressed Air Magazine,* 1931-1935)

Records do not indicate any fatalities in 1931 due to air quality related issues, yet men that were overcome by the fumes were sent to the Six Companies hospital. Falling ill from the fumes workers were then routinely diagnosed with pneumonia. If a man died from pneumonia, this was not considered a work related injury, so the employee and his family would not receive any workman's compensation payments.

"'The hospital had always been a topic of the workmen's conversation,' John Meursinge wrote. 'Up there a fellow always dies of pneumonia, never of anything else. That was a standard joke all over the job.'" (Stevens, 1998)

Tunnel worker Tex Nunley commented, "You had to be dead, absolutely dead, down on the job to get killed on the job. If they ever got you in the hospital, you didn't get killed on the job; you just died. Bad luck." (Dunar and McBride, 1993)

With Crowe and Shea pushing the work, day-by-day the tunnels neared holing through and finally on January 30, 1932, tunnel No. 3 on the Arizona side was the first tunnel to hole through. This was quickly followed by tunnel No. 2 holing through on February 3, 1932.

On February 9, 1932 the Colorado River once again unleashed its furry with a flash flood. Rapidly, the river rose 11 feet overtopping the river trestle, inundating the machine shops, and nearly breaching protective dikes before receding as quickly as it came. Undeterred, the Colorado River unleashed yet another and even more powerful flood on February 12th. Crowe feared it "would wash us right out of the canyon." The foaming crest rose with stunning and terrifying speed, seventeen feet in three hours, melted the levees on the Arizona bank like sugar, and poured into the tunnels. (Stevens, 1998)

In the aftermath of the flood, everything the flood waters touched was coated with a nasty smelling, slimy coating of yellow-brown silt. It would take a week to recover.

By the end of spring 1932 the main diversion tunnel excavation was completed. Since making the daring amphibious assault in May, 1931 Crowe's tunneling crews averaged over 150,000 cubic yards of tunnel excavation per month. With no time to rest, Crowe and his team would have to finalize their plans for the all important diversion tunnel concrete lining operation.

Aggregate and Concrete Production

From the start of work in the spring of 1931 through late winter of 1932, the men and machines that labored day and night in the bottom of Black Canyon were of singular purpose, to excavate the diversion tunnels. Yet, for miles around a flurry of preparation was taking place to allow for the eventual main event, the building of the 3,250,000 cubic yard dam with an additional 1,150,000 cubic yards of concrete in the concrete tunnel lining, spillways, intake structures, outlet structures and powerhouse.

In order to produce 4,400,000 cubic yards of concrete, three major challenges had to be overcome: 1) locating 8,600,000 tons of high quality sand and gravel, 2) designing and building the world's largest aggregate washing and screening plant, and 3) building the world's largest concrete batch plant.

FIG. 6. Six Companies, Inc. Aggregate Processing Plant
Courtesy Bureau of Reclamation

Fortunately, during the USBR's investigations they had discovered a 30-foot-deep bed of alluvial material deposited by the Colorado River on the Arizona side 8 miles upstream of the dam. There was plenty of raw material at the Arizona Pit to meet the needs of Hoover Dam. There is a certain irony that through the millennia, the untamed Colorado River deposited the very sand and gravel which would be used to produce the concrete to allow the river to be tamed.

Crowe's plan called for processing and producing concrete on the Nevada side, so 7.25 miles of railroad and a 1,140-foot-long trestle across the Colorado River were constructed to provide access from the pit to the processing plant. After stripping off a couple of feet of overburden, a 5-cubic-yard Marion Type 125 drag line excavator was used to mine the raw aggregates and load them into rail cars for transport to the processing plant. Pit material density was 3,800 pounds per cubic yard. Like the rest of the project, the Arizona Pit operations ran 3 shifts per day, seven days per week. The typical pit production was 250 car loads or 12,250 tons per day. (*Compressed Air Magazine*, 1931-1935)

Once loaded, the aggregate train delivered the raw material from the pit to the aggregate processing plant (Figure 6). Built during 1931, the aggregate processing plant was ready to supply aggregates for concrete production by early 1932. The plant had a capacity of up to 600 tons per hour.

The USBR's concrete mix design allowed for aggregate up to 9 inches in size, but required specific amounts of various gradations for each cubic yard of concrete produced. The Six Companies processing plant was equipped to screen, crush, cleanse, and segregate the aggregate. To accomplish this, the plant was equipped with a scalper station, four classification towers, a sand washer, a sand conveyor, and four live storage piles for gravel and sand. Storage capacity of the plant was 1,700 tons of cobbles, 1,500 tons of three gravel sizes, and 22,000 tons of sand. (*Compressed Air Magazine,* 1931-1935)

Aggregate processing plants require a lot of water to process the raw material. To supply the plant, water was pumped two miles from the Colorado River with a lift of 415 feet to an 800,000-gallon concrete reservoir. During production, the plant used 3,000 to 4,000 gallons of water per minute with 20% of the flow coming from water pumped from the river, and 80% of the flow coming from reclaimed plant water. The plant and rail transport operations from the pit were manned by fewer than 100 men.

The 4,400,000 cubic yards of concrete for the dam, tunnels, powerhouse and other structures on the Hoover Dam had to be produced to the most rigid concrete specifications ever applied to a large-scale project. Specifications required a minimum of 2,500 lb/in^2 compressive strength for the dam, and 3,500 lb/in^2 for slabs, beams, and other reinforced members.

FIG. 7. Himix Concrete Plant
Courtesy Bureau of Reclamation

Crowe's plan for concrete production was based upon a two plant strategy. First, a low-level (Lomix) concrete plant at elevation 720 ft was erected 4,000 feet upstream of the dam to provide all the concrete for the diversion tunnel linings, the powerhouse foundation, and two-thirds of the dam concrete. Crowe's Lomix concrete plant was strategically located down in the canyon to allow low elevation delivery of concrete to the dam site and thereby minimize crane hoisting time in the early high-volume areas of the dam. The elevation of 720 ft. was planned to allow for the concrete feeder rail lines to cross over the top of the upstream cofferdam, which was also at elevation 720 ft.

A second, high-level concrete plant (Himix) was erected at elevation 1235 ft on the

Nevada rim of the canyon adjacent to the dam to provide the balance of dam, powerhouse, and intake structure concrete (Figure 7). Crowe's plan was for all of the concrete in the dam up to elevation 720 ft to be batched in the Lomix concrete plant. From elevation 720 ft to 975 ft concrete would be produced in both plants. Finally, dam concrete above elevation 975 ft to the crest at 1232 ft would be batched exclusively in the Himix plant.

Construction of the Lomix concrete plant started in November of 1931 and was ready to produce concrete by early March of 1932. With diversion tunnels holing through in January, February, and March of 1932, completion of the Lomix concrete plant was on schedule to allow the next phase of critical work in Crowe's plan, placing the concrete lining in the diversion tunnels.

Aggregates for the Lomix concrete plant were transported 4.7 miles from the aggregate processing plant in 30-cubic-yard side dump cars and dumped into a track hopper set up for sand, 3 sizes of gravel, and cobbles. Material was then conveyed to the mixing plant. (*Compressed Air Magazine*, 1931-1935)

In order to produce the 4,400,000 cubic yards of concrete for the Hoover Dam project, the Six Companies had to rely on several cement suppliers. In total over 5,500,000 barrels of cement were required for the project. Each of the various cements being used had different structural properties and different colors. To overcome this challenge, the Six Companies built large blending silos in the location of the Himix concrete plant. All cement was delivered to the high-level plant and blended with other cements to produce an overall consistent cement for the project.

The Lomix concrete plant was equipped with as many as six 4-cubic-yard tilting mixers. Each mixer was capable of mixing 4 cubic yards in 3.5 minutes. Crowe sized the plant carefully to produce as much as 170,000 cubic yards in one month to meet his aggressive schedule. The plant batched concrete based on weight of materials for each batch of concrete. To ensure quality standards, the plant was equipped with state-of-the-art measuring and recording capabilities. Further, the plant was equipped with recording equipment to record the weight of ingredients and mixing time for every batch of concrete. While one 4-cubic yard batch was being mixed, aggregates for the following batch were being weighed and discharged into the adjacent mixer hoper.

Aerial Cableway Crane System

Crowe successfully used aerial cableway cranes on his previous dam projects. Each time Crowe used this method he learned and refined his understanding of cableway systems. By the time Hoover Dam came on the scene, Crowe was an expert in cableway crane systems, and their use was a key element in his overall attack plan to efficiently build Hoover Dam. On previous Crowe projects one cableway had usually been enough to get the work done. At Hoover Dam Crowe's plan would require nearly a dozen cableway cranes of various sizes and configurations.

FIG. 8. 20-ton Cableway Cranes
Courtesy Bureau of Reclamation

The main workhorses for Crowe were his five 20-ton cableway cranes strategically located over Black Canyon (Figure 8). The 20-ton cableway cranes had a traverse speed of 1,200 feet per minute, and a hoist speed of 300 feet per minute. The primary purpose of the cableway cranes was placing concrete, but they were also used for moving personnel into and out of the canyon, as well as any other material handling. Each main cableway crane was powered by a 500 horsepower, 2,200 volt electric motor to drive a three-drum Lidgerwood hoist. A separate travel winch was powered by 100 horsepower motor. All of the winches were equipped with air breaking systems. The cableways were vital to Crowe's plan and were maintained by a crew of 50 riggers.

In addition to the five 20-ton cranes there were also a half dozen small cableway cranes installed across the canyon primarily established for providing personnel access. The small cableway cranes were dubbed "Joe Magees". "One of the Joe Magees, the No. 11 cableway, was known as the 'Widow Maker,' as it was responsible for one of the most spectacular accidents during Hoover Dam's construction." (Dunar and McBride, 1993)

River Diversion Plan

In order to force the Colorado River out of its ancient river bed, the USBR plans prescribed the installation of massive rock-filled upstream and downstream cofferdams and four diversion tunnels. The Colorado River was thought to have peak flows in May and June capable of 200,000 cubic feet per second (CFS). Historically, November and December of each year provided a short window of low flows, generally below 7,500 CFS, so the fall would be the lowest risk period for river diversion.

The Six Companies Contract required the Colorado River to be diverted into the diversion tunnels by October 1, 1933. With tunnel No. 3 holing through on March 16, 1932, Crowe would have 18.5 months to complete the challenging concrete tunnel lining and tunnel gate work. Each diversion tunnel would require on average about 100,000 cubic yards of a 3-foot-thick concrete lining to rigid specifications in the 4,000-foot-long tunnels. It was a daunting challenge to be sure, but Crowe's plan wasn't to divert the Colorado during the 1933 low flow window, but instead the 1932 low flow window. This would mean completing a diversion tunnel in 6.5 months. It was a huge risk to be sure. If Crowe came up short, it would add a year to his schedule and millions of dollars to the cost.

Tunnel Concrete

The official USBR bid tabulation called for 312,000 cubic yards of tunnel lining concrete. The Six Companies bid this work for $11.00 per cubic yard. (*Compressed Air Magazine,* 1931-1935)

The basic scope of the tunnel lining work was to line the 56-foot rough diameter excavated diversion tunnels with a 3-foot-thick lining of concrete to make a smooth 50-foot-diameter water passage. Within the four main diversion tunnels there were nearly 400,000 cubic yards of concrete (including overbreak and waste) required to be placed, or about 25 cubic yards of concrete per linear foot of tunnel. The tunnel lining operation was broken into three main components, the invert, sidewalls, and arch.

With the holing through of the diversion tunnels and completion of the Lomix concrete plant in early 1932, the tunnel concrete lining operations began on March 16, 1932 in tunnel No. 3. Tunnel No. 2 concrete lining started on March 29, 1932, tunnel No. 4 started on April 7, 1932, and tunnel No. 1 started on August 3, 1932. There was no time to lose if Crowe and his team were going to meet the 1932 low river flow window.

Similar to Crowe's tunnel excavation plan which relied on establishing as many work headings as feasible, Crowe's tunnel lining plan broke the operation into six primary operations; access, rock preparation, invert concrete, side wall concrete, arch concrete, and pressure grouting. In order to provide access prior to production, railroad rails were set on each side of the tunnel to allow an electrically powered 10-ton traveling gantry crane to be used for the invert work to set forms, place concrete, and provide general material handling. Prior to concrete placement, crews thoroughly cleaned the rock to a near spotless condition to ensure a good bond with the concrete.

To form the inverts, crews used a combination of reusable steel forms and built-in-place wood forms. Once ready for concrete, the gantry crane hoisted concrete off of converted tunnel muck trucks which hauled two 2-cubic-yard bottom dump concrete buckets from the Lomix plant into the tunnel and the gantry crane for placement. For the side wall concrete, Crowe's plan called for using reusable steel forms that traveled from set to set forming 40 feet of tunnel per set. The sidewall form traveler was designed to roll from set to set on rails and was positioned using screw jacks. Once the forms were secured and the end bulkhead forms placed, concrete was placed using gantry cranes mounted on the formwork traveler. Once the placed concrete had gained sufficient strength, the forms were stripped and the form was advanced 40 feet to the next set.

Arch concrete followed the sidewall concrete and was formed using a separate formwork traveler. The arch traveler was capable of forming 40-foot sections for each set, and like the side wall traveler moved on rails, and was positioned using screwjacks. Concrete for the arch was placed using a pneumatic placing method. "The almost unbelievable sight of 4 cubic yards of freshly mixed concrete shooting upward into the form awed most everyone." (Rocca, 2001)

After completion of the tunnel concrete lining operation, a pressure grouting operation was performed to seal any voids around the perimeter of the tunnel lining. Low pressure grout was pumped above the arch concrete to fill any voids or shrinkage after concreting. A high pressure grouting operation at 300 to 500 lb/in^2 followed at 8 locations, 20 to 30 feet into the rock on a radial pattern.

At peak employment, tunnel lining operations employed about 300 men per shift, or 900 men total. Operations included, invert and wall cleaning, grouting, placing concrete, trucking concrete, mixing plant, erecting wood forms, fabricating forms, moving invert, side, and arch forms, and other miscellaneous operations.

On average, there were about 100 40-foot-long sections of tunnel to be formed and placed in each of the four diversion tunnels. Each 40-foot length of tunnel required roughly 1,000 cubic yards of concrete. In March 1932, during the first month of tunnel concreting, crews only placed about 1,300 cubic yards of concrete. At this pace Crowe knew they would never make the fall 1932 low flow window. With a lot on the line, Crowe spent considerable time in the tunnels during the lining operation. (Rocca, 2001)

He encouraged each shift of workers to challenge each other to ever-greater production in concrete placements. Much pride and enthusiasm went into these challenges and foremen constantly looked for ways to streamline steps without shortcutting quality and safety. (Rocca, 2001) Hard work and persistence paid off and during June and July crews were able to reach an average daily placement of just over 2,000 cubic yards and a record placement of 2,750 cubic yards on June 14, 1932.

In addition to the typical tunnel cross section, custom concrete forming had to be performed for the spillway tunnel tie-ins in tunnels No.1 and No. 4.

At the upstream end of the tunnels 50-foot by 35-foot "Stoney Gates" and concrete guide structures were built at the portals for tunnels No. 2 and No. 3. At the portals for tunnels No. 1 and No. 4, 50-foot by 50-foot steel bulkhead gates and concrete guide structures were installed. Each bulkhead weighed over 3,000,000 pounds. (*Compressed Air Magazine,* 1931-1935)

River Diversion

In early September, 1932 with the low river flow window approaching, muck was dumped from the Nevada spillway downstream from the inlet portals of diversion tunnels No. 1 and No.2 on the Nevada side. With the Colorado River crowded to the Arizona side, a 5-cubic-yard dragline was mobilized to begin river bottom excavation in order to expose bed rock for the new dam. The dragline was supported by a fleet of trucks. By the end of October, 1932 fill material began to be deposited for the upstream cofferdam.

Diversion tunnel No. 4 was the first to be completed. And so, on November 13,

1932, a blast was set off in front of tunnel No. 4 to remove a small dike which allowed the waters of the Colorado River to leave their natural course for the first time. Commenting on this historic milestone, President Herbert Hoover said,

> "The waters of this great river, instead of being wasted in the sea, will now be brought into use by man. Civilization advances with the practical application of knowledge in such structures as the one being built here in the pathway of one of the great rivers of the Continent. The spread of its values in human happiness is beyond computation."

With water flow established through the diversion tunnels, an access trestle was built across the river downstream of the upstream diversion tunnel inlets. After the November 13th river diversion, trucks end-dumped rock fill off of the trestle. Within 30 hours the river was fully flowing through the tunnels. (*Compressed Air Magazine,* 1931-1935)

In addition to the diversion tunnels, the other essential feature of the USBR's river diversion plan was the installation of upstream and downstream cofferdams to allow the foundation of the dam to by constructed nearly 200 feet below the Colorado River's high water elevation of elevation 707 feet. Based on the USBR's studies they designed the cofferdams to accommodate a high flow event of 200,000 cubic feet per minute and selected a top elevation of 720 feet for the upstream cofferdam.

While the cofferdams could not be completed until the river was diverted through the tunnels, they most certainly had to be completed before the following year's spring run-off or risk not only flooding out the job, but really obliterating anything in the river's path.

The upstream cofferdam was massive in scale; 98 feet high, 480 feet long, and 750 feet thick at the base (Figure 9).

FIG. 9. The Upstream cofferdam shown in December 1932
Courtesy Bureau of Reclamation

It required 500,000 cubic yards of earth and gravel, 151,000 cubic yards of rock, and 3,500 cubic yards of concrete. The upstream face was graded on a 3 to1 (horizontal to vertical) slope and the downstream face was graded on a 4 to 1 slope. To make the cofferdam water tight, three reinforced-concrete percolation stops were installed at the crest on each abutment, the upstream toe or base of the cofferdam was sealed into the river bottom using interlocking steel sheet piles up to 55 feet in length, the upstream face was sealed with a 6-inch-thick reinforced-concrete blanket, and a rubber seal was installed between cofferdam and canyon wall from base to crest on both abutments.

The downstream cofferdam was much smaller, yet constructed in a similar manner to the upstream cofferdam. It required 100,000 cubic yards of rock fill and was 54 feet high, 380 feet long, 210 feet thick at the base, and 50 feet wide at the crest. Both the upstream and downstream faces were sloped on 1.5 to 1.

Dam Excavation

In order to provide a solid foundation for the dam, its base and abutments would have to tie into solid rock. For the abutments of the dam, the USBR plans called for excavating loose and fractured rock from the canyon walls and creating large V-shaped keyways for the dam to lock into. At the base of the dam, below the river's waters laid naturally deposited sands and gravels which had to be excavated to reach bedrock. The excavation for the foundation of the dam necessitated the removal of substantially 7,000,000 cubic yards of material.

Preparing the canyon walls for the dam started in the summer of 1931. With over 700 feet of elevation difference between the top of the dam and the base, it was not at all practicable to erect scaffolding to gain access to perform the canyon wall excavation. Crowe's plan for access utilized specialized workmen known as "high scalers" (Figure 10). Dangling from long manila ropes anchored to the top of canyon walls, the high scalers descended the

FIG. 10. High Scalers at work.
Courtesy Bureau of Reclamation

walls.

The high scalers sat in bosun's chairs tied to the end of their ropes to free their hands for using pry bars, jackhammers, and other tools. Starting at the top of the canyon at elevation 1232 feet, the high scalers drilled into the canyon walls, loaded the holes with dynamite, and removed loose rock after the shot. Like an aerial circus, the high scalers were exciting to watch, swinging from point to point, picking away at the walls like a swarm of hornets.

Exciting? Yes, but it was also very dangerous for the men brave enough to perform this work. There was the height, the heat, and plenty of ways a rope could be cut or broken. Further, loose debris often rained down on the high scalers. Looking for some protection, it is commonly accepted the high scalers of Hoover Dam improvised the first construction hard hats by dipping cloth hats into hot tar and letting them set to a hard shell.

The contractors were sufficiently impressed to order thousands of factory manufactured "hard-boiled hats" and to suggest strongly that men working in exposed areas wear them, making the Boulder Canyon Project one of the first hard-hat jobs in American construction history. (Stevens, 1998)

The high scalers' wage, $5.60 per day, was 40 percent higher than that given men who did similar work in the tunnels or on the canyon floor, but few of the laborers toiling in relative safety on the riverbed begrudged the extra $1.60 received by the men dangling high above. As one laborer succinctly put it: "A fellow got (sic) to risk his life to make that money." (Stevens, 1998)

In Crowe's plan there was not time to wait for the diversion tunnels to complete before starting the dam abutment excavation and preparation. So it was in December 1932, high scaling operations began and continued through the spring of 1933, being substantially complete before the start of dam concrete operations.

The bulk of the dam foundation excavation was removal of naturally deposited sediments in the river bed. So, after the Colorado River was diverted in November 1933, Crowe had his team start the process of foundation excavation. Utilizing a large fleet of trucks and excavators crews worked around the clock to excavate the riverbed and prepare the foundation of the dam.

> "Those trucks never turned their ignition off unless they were broke (sic) down. When they were hauling that debris out of the river bottom, a man coming down assigned to a certain truck might take any truck until he found his own. They would change shifts right on the run." (Dunar and McBride, 1993)

Averaging 22,000 cubic yards per day, dam foundation excavation work continued through the winter of 1932-1933. As was the case during the drilling of the diversion tunnels, there was an intense push for speed by the superintendents and foremen, and fierce competition among the three shifts to see who could move the most foundation material. The winner of this hard-fought contest was the swing-shift crew, which on January 24, 1933, disposed of a staggering 1,841 truckloads, nearly four a minute, during its eight-hour stint. (Stevens, 1998)

Walker Young, Resident Engineer of the USBR would also have to work hard to ensure his staff of government inspectors were keeping watch over the contractor's work. *Fortune* magazine wrote:

"Clearly, such speed requires a traffic cop. He is on duty, a government inspector, who reports the contents of every batch of cement [concrete], the blast of every dynamite barrage, the loads on the cableways, and the depth of every hole. He even goes down the canyon wall on ropes to outline the rock to be moved, then later to report the tons of rock chipped off by high scalers. Over 150 men are paid government money to stick their noses into the contractor's business. Not until they are satisfied that the work conforms in the minutest detail to rigid U.S. specifications do the Six Companies get Washington's check for the preceding month's payroll and expenses."

The last blast in the foundation excavation was fired on May 31, 1933 and crews excavated the last remaining muck. At the base of the dam crews encountered a water seepage problem from a spring flowing through fissures in the rock. After sealing the seepage, final clean-up was performed to prepare for the dam's first concrete. Conducting this all important inspection was Jack Savage, the Bureau of Reclamation's Chief Designing Engineer. It is more than likely that Frank Crowe and Walker Young would have been there with Savage to gain his formal approval and mark the end of the preparatory phase.

Dam Concrete Placing

The official USBR bid tabulation called for 3,250,000 cubic yards of Dam Concrete. The Six Companies bid this work for $2.70 per cubic yard. (*Compressed Air Magazine*, 1931-1935)

Assuredly, the Six Companies directors would have been apprehensive about their bid price of $2.70 per cubic yard, a mere quarter of their bid price for the tunnel concrete. If Crowe made a mistake in his bid, or his plan for the work, the Six Companies would have to live with it 3,250,000 times for every cubic yard of dam concrete.

With the completion of dam excavation the end of May, 1933, Crowe could finally move into the primary purpose of the project, to place the concrete for the dam. It had taken over two years from notice to proceed to reach this point. Crowe's plan for building the dam would be primarily based upon methods previously used, yet at an unprecedented scale. The dam would be placed systematically, block by block in 5-foot-high lifts of hundreds of interlocking concrete blocks which ranged in size from about 25 foot square to about 60 foot square.

There were many operations required for every concrete block. First crews set heavy-duty wooden forms ranging in height from 6 to 11 feet high that were capable of withstanding the liquid head (form pressure due to wet concrete) from concrete during placement. To speed formwork, Crowe took advantage of the newly invented "she-bolt" form tie system which allowed forms to be erected and stripped in large panels. To ensure a watertight structure, the USBR's design required labor intensive keyways and waterstops to be placed between adjoining blocks and concrete cooling pipes installed on top of the previous lift.

Everything would have to work as planned: the men, the aggregate processing plant, the concrete batch plants, the aerial cableway cranes, the railroad system, the temporary electric, air, water, and lighting systems, the concrete formwork, personnel access schemes, and countless other details. Further, all of the overhead support functions such as engineering, survey, payroll, purchasing, and staffing would have to figure out how to keep pace with Crowe's superintendents.

It was time to find out if the planning and preparation were enough, and so on the morning of June 6, 1933 the first concrete was placed in Hoover Dam. Crews quickly learned the ropes. Crowe's cableways proved to be fast and efficient, easily delivering 8-cubic-yard concrete buckets from the rail concrete shuttle trains at elevation 720 ft to the dam. June of 1933 would have been a period of learning, and the tedious task of getting the dam concrete built up off the bedrock. By the end of June 1933 crews had only placed 25,000 cubic yards of concrete in the dam. At this pace, it would take over 10 years to finish the dam. With so many new operations occurring and men learning their duties, the slow pace of June 1933 must have been frustrating for Crowe.

And then, in July of 1933 word arrived that Warren Bechtel, President of the Six Companies died while on a trip to Russia. E.O. Wattis was soon elected to replace Bechtel as president. With challenges both in and out of Black Canyon, Crowe would have clearly been in a pressure cooker.

In order to meet Crowe's schedule, employment would peak at over 5,000 men working in Black Canyon on the three shifts. There was a lot to plan and coordinate each and every day. The mass of humanity and machines called for a strict vigilance on the part of Crowe's management team. Frequent performance and safety meetings occurred, and often, Crowe led the discussions in all three shift meetings. (Rocca, 2001) Crowe described the challenge: "We had 5,000 men in a 4,000 foot canyon. The problem, which was a problem in materials flow, was to set up the right sequence of jobs so [the workers] wouldn't kill each other off." (Stevens, 1998)

Through the summer of 1933, cableway crews perfected the handling, loading, and spotting of the specially designed 8-cubic-yard buckets (Figure 11). "The 'Old Man' had designed to dump the full eight yards as quickly as possible or as workers remembered later, 'like crap from a goose.'" (Rocca, 2001) Once the concrete was placed, it was allowed to take an initial set and the laitance was removed using a blast of high-pressure water and air. Forms were generally left in place for 24 hours after a pour. Once the concrete gained sufficient structural strength, the she-bolts were removed allowing the forms to be raised to the next lift by using A-frames and hand winches.

The entire process of building a dam is complex and labor intensive. To improve production, Crowe was known to constantly seek out alternative methods from the crews to get the work done more efficiently. "He enjoyed nothing more than talking with the 'construction stiffs,' finding out where they had come from." (Rocca, 2001)

When we look at Hoover Dam today from the outside, we only see a small fraction of the formed surfaces which were necessary to build the dam. In addition to the exterior concrete formwork, the USBR's dam design called for the construction of a complex array of interior access tunnels, inspection galleries, elevator shafts, and ventilation shafts. These shafts and tunnels would intersect the blocks at varying locations and elevations as the dam gained in elevation.

Just as Crowe had used in other parts of the job, he offered bonuses to increase production. There were inter- and intra-shift competitions to see who could place the most concrete in an eight-hour shift.

FIG. 11. Cableway cranes were efficient and able to safely position 8 cubic yard concrete buckets for placing the dam concrete. Courtesy Bureau of Reclamation

After the slow start of 25,000 cubic yards of concrete in June 1932, month-by-month the Hoover Dam team improved. By August the figure had jumped to 149,000 cubic yards. Two months later it had passed 200,000, and in March, 1934, more than 262,000 cubic yards, 1,100 buckets a day, one every 78 seconds-gushed out of the mixing plants and into Black Canyon. (Stevens, 1998)

As Crowe pushed harder and harder, Walker Young of the USBR would have to rally his army of inspectors to keep pace.

While Crowe emphasized quality work, he also pushed hard for production. At the workman's level, the push for production most likely caused some cutting of corners from time to time. Walker Young's men had to be vigilant to ensure the quality of the work, like concrete batch plant inspector, Bob Parker recounted:

> "One night I had trouble down there. One of the mixer concrete batcher men in concrete plant was goosing the concrete batches with water. That would make the concrete a lot thinner. They poured a lot faster; you didn't have to tamp it as much. When I caught this batcher doing that, I climbed his frame about it. He said I couldn't prove it. I said, 'If I catch you doing that one more time, we're shutting this plant down.' So he told me he was going to call Frank Crowe. I said, 'Better yet, I'll call Walker Young and Frank Crowe both, and get them down here.' 'You'll have to answer to Frank Crowe because I know Frank Crowe and Walker Young are just like peas in a pod. They are here to work together and get this job done.'" (Dunar and McBride, 1993)

Crowe called Young, "The great delayer." Young called Crowe: "Hurry up Crowe" (Figure 12).

FIG. 12. Brig Young (r), a.k.a., "The great delayer", and Frank Crowe (l), a.k.a., "Hurry up Crowe" pose to commemorate the placing of the two millionth cubic yard of concrete on June 2, 1934, one year after starting dam concrete. Courtesy Bureau of Reclamation

"Of course", says Frank Crowe (General Superintendent, Six Companies), speaking of Walker Young (U.S. Construction Engineer in charge), 'we like to cry at each other and raise hell. He says my foremen are no good, but he don't mean anything.' 'Yes', agrees Young, with such circumspection as befits a great government engineer, 'sometimes we fight with each other for the fun of it.'" *Fortune* magazine 1933

Crowe had always been a hands-on, field-oriented superintendent. As Hoover Dam peaked he believed he could still get around to see all of the operations on a regular basis. But Hoover Dam proved to be too big even for Crowe and he would learn to rely on regular weekly meetings to keep his job superintendents and assistant superintendents informed about concerns and upcoming changes.

During the second half of 1933, news of Crowe's dam building machine became widespread. There was a steady stream of tourists and journalists making their way to Black Canyon to witness history. Upon visiting the dam site in 1933, journalist Duncan Aikman wrote:

"There is a stark and uncompromising efficiency which one feels at every point of contact with the dam work, Everywhere men move fast, throw all their power of muscle and machinery into what they are doing, waste no time in workmanly sociabilities on the job. There is no gayety about the scene, no sense of men colorfully enjoying their work. Instead, a kind of surly determination broods over their labor. ... By now the sheer, nervous drive of the task and the ruthless hiring and firing policy of the Six Companies contracting organization, which leaves only the supremely efficient on the job, have achieved a coordination that is practically flawless." (Stevens, 1998)

News of Hoover Dam's construction spread around the world and many were eager to learn its lessons (Figure 13). And so they came; in 1933 Crowe met with engineers and statesmen from Scotland, Japan, Canada, England, Switzerland, China, Mexico, India, Austria, Germany, Congo, South Africa, Italy, and other countries." (Rocca, 2001)

Not only did news about Hoover Dam spread around the world, but also its folklore. Perhaps the most popular folklore of Hoover Dam suggests men are still buried within the mass of Hoover Dam. Records do not indicate any men were ever permanently buried in the concrete.

FIG. 13. News of the Six Companies' progress on Hoover Dam spread around the world. This photo taken August 31, 1933 shows the progress roughly three months into the dam concrete operations. Courtesy Bureau of Reclamation

However, on November 8, 1933 there was a form collapse which caused a fatality. This accident was the closest anyone ever came to being permanently entombed in Hoover Dam. Formwork failures occurred occasionally and for the most part only caused inconvenience to the men to clean up the wet concrete and make repairs. However, on the November 8th graveyard shift, a concrete block placement adjacent to the central 8-foot-wide slot was being placed when the form suddenly collapsed. The collapse allowed more than a hundred tons of fresh concrete to rain down into the slot. As the concrete mass fell, it gained speed, wrecking havoc and destroying pipes and platforms in its path. Tragically one man was swept away to his death in the flow of concrete. It took men 16 hours to recover the body, with Crowe personally supervising the recovery.

Innovations were many at Hoover Dam, but the refinement of concrete cooling is noteworthy. When the Hoover Dam concrete cooling specifications were issued; under Jack Savage's direction, the USBR was investigating concrete cooling at Owyhee Dam (1928-1932) in Oregon which was being built by General Construction

Co. of Seattle, Washington (As of 2001, a wholly-owned subsidiary of Kiewit Corporation).

As concrete cures it generates heat from hydration of the cement, normally about 40 degrees temperature rise after placement. As the concrete eventually cools to ambient temperature concrete will shrink, albeit to a small degree. This shrinkage can set up internal stresses and cause cracking. Cracking and shrinkage are two very undesirable things for a dam. The Hoover Dam was so massive, the USBR calculated it could take 125 years for the heat to dissipate. The idea of artificially cooling the concrete was to speed up the shrinkage process, thereby allowing crews to pressure grout the joints between the blocks to tighten up the structure. To cool the concrete crews installed 800,000 lineal feet of 1-inch-diameter steel cooling pipes in the dam through which chilled water would be circulated.

To remove from the mass the additional heat imparted to it by setting would therefore require the extraction of approximately 38,600 BTUs for each cubic yard of the 3,250,000 cubic yards in the dam.

As the dam was built pour-by-pour, the 1-inch cooling pipes were placed in each block connected to adjacent blocks of concrete forming continuous loops to carry the chilled water. Access to the cooling pipes was gained from an 8-foot-wide central slot left at the center of the dam in "I-Block". From the slot the pipes could be plumbed, and upon completion filled with cementitious grout. Upon completion of the pressure grouting, the central slot gap was filled with concrete in 50-foot-high lifts.

To monitor the dam during construction, an unprecedented arsenal of hundreds of instruments were installed in the dam including: hydrostatic uplift monitors, electrical resistance thermometers, recording thermometers, rock temperature resistance thermometers, electrical joint meters, the strain meters, mercury-surface optical tilt-meters, strong-motion accelerometers, and triangulation surveys targets on the downstream face of the dam.

On February 3, 1934 word arrived that E.O. Wattis, President of Six Companies had died, and Harry Morrison was elected as his replacement. While this would have been sad news for the Six Companies, the men in the canyon were a well-oiled machine now. They would push on.

FIG. 14. Dam Progress shown on April 1, 1934. Work on the powerhouse can also be seen taking place.
Courtesy Bureau of Reclamation

Crowe's men kept pushing harder and harder through early 1934. And then on March 20, 1934, a team of 458 men set a record concrete placement rate of 10,417 cubic yards in one 24-hour day (Figure 14).

On December 5, 1934 photographers snapped away as the three-millionth cubic yard of concrete was deposited in the dam. Newspapers reported that the Bureau of Reclamation's bid specifications had called for concrete pouring to start on December 4, 1934, and that instead the structure's mass concrete was 92 percent complete.

Sixty-three days later, with almost no fanfare, cableway 6 lowered a concrete bucket to a shallow form on column K and the block was topped out, the first to be brought to the dam's final height of 726.4 feet. Five more columns were nearly finished, needing only a two-foot pour to complete the roadway that would run along the dam's crest. Rapid progress was also being made in the construction of the powerhouse and canyon-wall outlet works, where scaffolding was in place and the valve house floors were being poured. (Stevens, 1998)

The final bucket of concrete was placed in the main dam on May 29, 1935, less than three years after starting the dam concrete. Crowe and his team had averaged about 94,000 cubic yards a month.

Pressure grouting of the dam foundation was required to consolidate the rock and to fill existing cracks. Holes for introducing the grout were drilled in the bottom of the canyon and in the abutments to depths of 50 to 150 feet. Pressures from 100 to approximately 600 lb/in^2 were used in applying the grout.

FIG. 15. Four reinforced-concrete intake towers were also required to be built to supply water to the turbine generators and the outlet structures. Courtesy Bureau of Reclamation

Intake Towers To supply water to the turbine generators and the outlet structures, four reinforced-concrete intake towers were also required to be built, two on the Nevada side and two on the Arizona side (Figure 15). These structures had a complex geometry and tough access, so they would have been very challenging to build. The upstream towers were built over the top of the 37-foot-diameter sloped concrete lined tunnels which intersect the 50-foot-diameter diversion tunnel No. 2 on the Nevada side and diversion tunnel No. 3 on the Arizona side. These tunnels were excavated 41-foot-diameter.

The downstream intake towers were built over the top of 37-foot-diameter inclined shafts which intersected a header tunnel at elevation 820 on each side of the river. The Nevada header was 1,322 feet long and required about 75,000 cubic yards of excavation. The Arizona header was 1,197 feet long and required 67,000 cubic yards of excavation. The header tunnels were drilled for excavation using a full face drilling jumbo mounted on rails. Normal advancement was one 12-foot round per shift. Electric shovels and trucks were used for excavate and haul operations. (*Compressed Air Magazine,* 1931-1935)

The intake towers were constructed in lifts and serviced by stiff-leg derrick cranes on each side of the canyon. From their perch, the derrick cranes could easily hoist eight-cubic-yard concrete buckets off of the flatcars which were pushed by Dinkey locomotives that serviced the Himix concrete plant.

Steel Header Pipes and Penstocks

In order to supply water to the powerhouse turbines the USBR design called for the installation of a complex plumbing system comprised of steel header pipes and penstocks. Like the dam concrete, the scope of the steel pipe production was unprecedented. In total there were 4,700 feet of 30-foot-diameter headers, 1,900 feet of 25-foot-diameter headers, 5,600 feet of 13-foot-diameter penstocks, and 2,300 feet of 8.5-foot-diameter outlet conduits. In total there were two and eight-tenths miles of steel headers and penstocks.

The USBR issued a separate contract for this work to Babcock & Wilcox (B&W) of Barberton, Ohio. B&W's scope was to fabricate and install the pipe for $10,908,000 with a contract duration of 1,975 days. Even though the USBR had established excellent railroad infrastructure to Hoover Dam, the size and scope of the work made it unfeasible to manufacture the pipe work at existing plants and ship them to the site. To overcome this challenge B&W built a new 500-foot long by 85-foot wide plant about a mile and a half from the dam to manufacture the pipe.

The plant was set up to manufacture the pipe using state-of-the-art techniques. Flat steel plate was delivered by rail to the plant, unloaded, rolled, and welded into pipe sections. After making the sections, the ends were prepped for girth welds using large lathes. All of the welding was completed using a new technology, the fusion-welding process to assure high-quality welds. For quality control, engineers used the latest in X-ray technology from General Electric to inspect every inch of completed seams and joints. It was the world's record X-ray job and required 159,000 separate X-ray pictures.

As the pipe sections were manufactured they were temporarily braced with "spider bracing" on the inside diameter to ensure roundness tolerances were maintained during transport to the job site.

Beginning in August of 1933, B&W began installing penstocks. Utilizing specially designed 200-ton capacity, 16-wheel transport trailers, B&W transported the completed header and penstock sections from their plant to the Government's 150-ton cableway at the edge of Black Canyon on the Nevada side. The trailers were moved using two caterpillar tractors, one pulling, with the other attached to the rear used for braking.

Once below the 150-ton cableway, steel pipe sections were rigged for lifting using a specialized lifting beam, hoisted, and lowered to either side of the canyon to the cableway landing areas at elevation 670 ft. From the landing area, pipes were set on special transport trailers and rolled into the tunnels for placement. Rigging of these large heavy pieces within the tunnels would have been extremely challenging requiring the proper equipment, know-how, and patience.

For this portion of the work B&W subcontracted with Eichleay Engineering Corp. of Pittsburgh to supply specialized tools and equipment and to move the pipe sections to within one foot of their final location. Full rotation of the header and penstock sections in the tunnels was impossible, yet it was necessary to manipulate sections to get them installed. To solve this problem, a custom-made two-part transporter was used equipped with a lower car and an upper car. The transport's lower car was used to move sections in and out of the adits and the upper car was used to move back and forth through the tunnels. The upper car sat on a section of track that was mounted atop the lower car. The pipe sections were placed atop the upper car and the entire unit was moved through the adit to the tunnel. Once in place, the upper transport car was rolled from atop the lower transport car onto a set of tracks that ran through the tunnel and the section was moved to its final location.

Setting the horizontal sections of header and penstock pipes was challenging enough, yet the sloped sections which tied into the intake towers were especially tough work. The tunnels on the Arizona side had slopes of about 52 degrees while the Nevada tunnels had slopes of about 37 degrees. Some of the sections had to be hoisted nearly 240 feet from the lower tunnels to the base of the intake towers. Using specialized rigging and hoists, the sections were hoisted up the inclined portions using rubber wheeled cars that rode on the tunnel lining. Some of these pipe sections were so difficult, they required about 12 hours to be lifted into place from the lower tunnels. After all of the sections of inclined header pipes were installed the entire section was encased in concrete.

Once a section had been moved to its final location it was attached to the previously installed sections using a specially designed assembly rig consisting of three "spider" sections mounted on a 20-foot-long central hub. The spider sections consisted of 16 steel H-beams extending outward from the hub like spokes. To aid in aligning section-to-section hydraulic jacks were mounted at the end of each spoke.

B&W worked four years on the manufacture and installation of the penstocks completing the work by August, 1937.

Spillway Structures

As part of the USBR's plan to ensure the long-term safety of Hoover Dam two spillway structures were built (Figure 16). Their purpose is to prevent overtopping of the dam during high flow periods when the upstream pool is at high levels. The Six Companies built two spillway structures, one on the Nevada side and one on the Arizona side. Each spillway has a maximum discharge capacity of 200,000 cubic feet per second. When in use, water from the Nevada Spillway can flow down the 50-foot-diameter inclined shaft to intersect with diversion tunnel No 1, and water from the Arizona Spillway can flow down the 50-foot-diameter inclined shaft to intersect with diversion tunnel No. 4.

Built upstream of the dam, high on the canyon walls, the two spillways required over 600,000 cubic yards of rock excavation.

FIG. 16. Complex geometry added to the challenge of building the spillway structures. The Nevada Spillway is shown under construction. Courtesy Bureau of Reclamation

The reinforced-concrete spillways were built with an ogee-shaped crest and have complex geometry to provide efficient hydraulic flow. Each spillway consists of a concrete-lined open channel, about 650 feet long, 150 feet wide, and 170 feet deep. A total of 127,500 cubic yards of concrete were placed in the spillways with the walls being lined with 18 inches of concrete, and the floors 24 inches. The intersection of the spillway's inclined shafts with the diversion tunnels also required complex built-in-place concrete formwork to create hydraulically efficient flow surfaces.

The spillways were serviced by cableway cranes No. 5 and No. 6 and supplemented with crawler cranes. Concrete was supplied from the Himix concrete plant.

Outlet Works

Reinforced-concrete outlet works were also constructed on both sides of the canyon, about 800 feet downstream from the dam at elevation 820. They are 175 feet above the downstream river elevation. Built on benches excavated into the rock, these structures each housed six, 84-inch needle valves which are protected by 96-

inch emergency gates. The needle valves are fed off of the 30-foot-diameter penstock by 102-inch-diameter steel outlet pipes. They are designed to control the release of water from the upper headers when power operations are suspended or to dewater the penstocks for inspection.

Excavation for the valve houses began in November 1932 and placement of concrete in the structures began in January 1935. All work on the canyon wall outlet works was completed in early 1936.

Powerhouse

In addition to the dam, the Six Companies also built the powerhouse, a U-shaped structure 650-foot-long capable of housing seventeen turbine-generators (Figure 17). The powerhouse rises 229 feet above its lowest foundation elevation, or 154 feet above the tailrace low-water elevation. Concrete work began on the powerhouse in November 1932.

Dwarfed by the dam, the powerhouse tends to be also overshadowed in published accounts of Hoover Dam's construction. Yet, drawings indicate a complex structure requiring multiple levels, galleries, and water passages to be built. The powerhouse was serviced by cableway crane No. 9 and mobile crawler cranes. Concrete was produced at the Himix plant. After the Six Companies had completed the first-stage concrete operations, USBR crews installed the turbine-generators and all other mechanical-electrical systems for the powerhouse. After the installation of embedded turbine-generator components, the Six Companies placed the second-stage concrete.

FIG. 17. The Powerhouse presented its own challenges for concrete operations as well as Turbine-Generator installation.
Courtesy Bureau of Reclamation

Project Close-out

January 31, 1935, was the Colorado's last day of freedom; the next morning, a steel bulkhead gate weighing more than one thousand tons was lowered, closing the portal of tunnel No.4 and forcing the river into tunnel No. 1 through the honeycombed plug. By manipulating the four valves, water sufficient to meet the irrigation needs of Imperial Valley, two hundred miles downstream, was allowed to pass through while the remainder began to pool behind the dam, forming the murky nucleus of a lake that eventually would cover 210 square miles. (Stevens, 1998)

On the morning of February 29, 1936, Ralph Lowry, who five months earlier had succeeded Walker Young as construction engineer, met Frank Crowe on the crest of the dam. While several reporters stood by and a lone newsreel camera recorded the scene for posterity, Crowe shook Lowry's hand and announced: "As representative of Six Companies, builders of [this] dam, I am very happy to turn over the job for your acceptance." The next day in Washington, Secretary Ickes formally accepted the dam and powerhouse on behalf of the government, terminating the contract and ending construction exactly two years, one month, and twenty-eight days ahead of schedule. (Stevens, 1998)

On May 1, 1936, the temporary gates in the Nevada tunnel plug were closed, and the bulkhead gate at the opening of the Nevada tunnel was closed on May 6. From that point on, all normal releases have been made through the gates of the four intake towers.

Dam Dedication

The dedication of Hoover Dam was conducted September 30, 1935. President Franklin D. Roosevelt delivered the dedication speech, in-part:

> "When we behold them it is fitting that we pay tribute to the genius of their designers. We recognize also the energy, resourcefulness, and zeal of the builders .... But especially we express our gratitude to the thousands of workers who gave brain and brawn to work of construction."
>
> "This is an engineering victory of the first order—another great achievement of American resourcefulness, skill and determination. This is why I congratulate you who have created Boulder [Hoover] Dam and on behalf of the Nation say to you 'Well done.'"

CONCLUSIONS

As a society it is vital to evaluate history, to learn of our forefathers' journeys, their successes, and their failures. Writing of the importance of bridge history, Henry Petroski wrote, "…knowing the story of any bridge and its builders invariably reveals a rich and rewarding chapter in the history of a place, its people and their dreams." (Petroski, 1996)

Like our bridges, the story of our civil engineering infrastructure: our buildings, roads, ports, canals, levees, tunnels, water and wastewater systems, and our dams, their designers, and their builders all invariably reveal an important chapter of each place, its people and their dreams. These chapters are not just about boring logs, calculations, stresses, and strains. Like other great civil engineering achievements, Hoover Dam's chapter speaks to the story of remarkable men and women, their character, courage, and leadership.

It so happens that Hoover Dam's chapter isn't just about the dreams of the people in the southwest United States to harness the Colorado River. Hoover Dam was, and really still is an important chapter about our world in the twentieth century, a chapter about man's ability transform dreams into concrete reality with engineering, creativity, hard work, determination, and sacrifice.

Officially, there were 112 fatalities during the construction of Hoover Dam. At Hoover Dam, the inscription at the base of the flagpole reads:

"It is fitting that the flag of our country should fly here in honor of those men who, inspired by a vision of lonely lands made fruitful, conceived this great work and of those others whose genius and labor made that vision a reality."

There were many remarkable people involved with the design and construction of Hoover Dam, yet Frank Crowe is the most intriguing, a kid from Maine who grew up to build the biggest dam in the world, the guy Frank Weymouth once called, "best construction man I have ever known." (Rocca, 2001) The write-up on Crowe in the September 1933 issue of *Fortune* magazine provides an interesting glimpse of Frank Crowe, the man:

"He has one hobby-the development of men; specifically, the men who follow him by hundreds to work on his dams. His principal exhibit is Bernard (Woody) Williams, who first worked for him at thirteen, and now, at thirty, is in complete charge when Crowe leaves Black Canyon for Boulder City. For Williams and his foremen he has only one working rule: "To hell with excuses–get results!" He is tall, talks loudly, and laughs hard. He is noted for his humor.

Frank Crowe's last vacation was his honeymoon twenty years ago. He avoids cities except for required directors' meetings and an occasional football game. He plays the stock market a bit, buys Buicks exclusively for work on the job, and can be seen matching quarters with $4-a-day 'muckers' while waiting for a big dynamite explosion. He twists around in a chair a lot while he talks, preferring the outdoors, and makes an absolute rule that no letter shall go out of his office over one page long. He believes any idea can be expressed in that space and that anything longer is a waste of words. He had one dominant desire in life--to work on dams--and has gratified that desire almost steadily since Arrowrock."

In later years, fellow engineers would look back on Crowe's successful career and point to his ability to reach and sustain high levels of production throughout each construction assignment, despite the high risk to men and machines. Much of this success can be attributed to his constant concern to maintain or improve each job responsibility and shift quotas. (Rocca, 2001)

Modern day builders would be wise to take note of Frank Crowe's lessons. He knew the details of his projects and was a master of recognizing and managing risk. He was passionate about his work, a hard pusher to be sure, but loyal and fair to those that gave their best. He had a straight forward and honest management style. He constantly kept his finger on the pulse of the work and showed his respect for the everyday "construction stiffs" by walking the job every day and getting his boots dirty and simply saying thank you for a good job. Many men loyal to Crowe followed him for the remainder of his career: Parker Dam (1936-1938), and later Shasta Dam (1938-1945). Working his entire adult life, some say Frank Crowe died in his boots on February 26, 1946 in Redding, California. At Crowe's funeral Hoover Dam colleague, A.E. Cahlan eulogized Crowe, "Perhaps the greatest builder of dams in the world, he was one of the humblest human beings." (Rocca 2001)

In many ways Hoover Dam pushed the limits of engineering and construction technology. Like other great engineering achievements, Hoover Dam assuredly had its detractors who would label it an impossible achievement. Science-fiction author, inventor, and futurist Arthur C. Clarke's is well known for his three "laws of prediction," which declare: (Petroski, 2010)

- When a distinguished but elderly scientist states that something is possible, he most certainly is right. When he states that something is impossible, he is probably wrong.
- The only way of discovering the limits of the possible is to venture a little way past them into the impossible.
- Any sufficiently advanced technology is indistinguishable from magic.

Frank Crowe, Brig Young, and all of those who worked on Hoover Dam collectively discovered the limits of the possible and perhaps even ventured into the impossible. They also advanced technology, and most certainly to the layman of the 1930s their acts were indistinguishable from magic. And now, seventy-five years after its completion, modern-day engineers and builders remain humbled and still find something magical about the story of building Hoover Dam.

ACKNOWLEDGMENTS

I would to thank Dick Wiltshire for his invaluable help. Thank you to Brit Storey of the United States Bureau of Reclamation. A special thanks to Henry Petroski for his inspiration and encouragement.

REFERENCES

Dunar, A. and McBride, D. (1993). "Building Hoover Dam, An Oral History." University of Nevada Press. Las Vegas, NV
Fortune (September 1933)
Petroski, H. (1996). "Engineers of Dreams." Alfred A. Knopf. New York
Petroski, H. (2010). "The Essential Engineer." Alfred A. Knopf. New York

Rocca, A. (2001) "America's Master Dam Builder." University Press of America. New York.
Stevens, J. (1988). "Hoover Dam An American Adventure." University of Oklahoma Press. Norman, OK.
Time (July 21, 1930)
Various Authors and Titles. (1931-1935). *Compressed Air Magazine.* (A compilation of 23 articles published between 1931-1935, reprinted by Nevada Publications) Las Vegas, NV
Weingardt, Richard C. (July 2008). "John Lucian Savage." *Civil Engineering.* Reston, VA.
Wilde, R. (2003) "ACI – A Century of Progress."
Wolman, A. Lyles, W.H. (1978). "John Lucian Savage." National Academy of Sciences.

Author Index

Page number refers to the first page of paper

Bailey, Jim, 48
Bartojay, Katie, 74
Brown, Tim, 288
Burgi, Philip H., 249

Dolen, Timothy P., 58
Dunn, Philip, Jr., 307

Fiedler, William R., 267

Giroux, Raymond Paul, 360
Gold, David, 288

Hein, Michael, 346

Jackson, Donald C., 1
Joy, Westin, 74

Parrish, Charles R., 340
Powell-Melhado, Tamiko, 346

Rogers, J. David, 85, 124, 163, 189, 216
Rogers, Jerry R., 40
Ruth, Linda Cain, 346

Sadatsafavi, Hessam, 329
Schuetz, Mary Beth, 288
Storey, Brit Allan, 25

Toothman, Adam, 288

Walewski, John, 329
Willis, Alfred, 318

Subject Index

Page number refers to the first page of paper

Aggregates, 124

Bench marks, 267
Bureau of Reclamation, 48

California, 1
Case studies, 346
Colorado River, 340
Concrete construction, 58, 163
Construction, 124, 329, 346, 340
Construction equipment, 360

Dam failures, 1
Dam safety, 1
Design, 85

Economic factors, 25
Engineers, 307
Excavation, 124

Fabrication, 163

Geology, 124

History, 25, 48, 58, 74, 85, 124, 163, 189, 216, 249, 329, 346, 360
Human factors, 360

Hydraulic structures, 267
Hydraulics, 249

Infrastructure, 40

Material properties, 74
Monuments, 318

Nevada, 25, 40, 48, 58, 74, 85, 124, 163, 189, 216, 267, 288, 307, 318, 329, 346, 360

Political factors, 1, 25
Power plants, 288

Safety, 163, 360
Seismic analysis, 288
Spillways, 267

Tourism, 216

United States Army, 48
Urban development, 40

Wyoming, 340

413